PLANT- AND MARINE-BASED PHYTOCHEMICALS FOR HUMAN HEALTH

Attributes, Potential, and Use

PLANT- AND MARINE-BASED PHYTOCHEMICALS FOR HUMAN HEALTH

Attributes, Potential, and Use

Edited by

Megh R. Goyal, PhD
Durgesh Nandini Chauhan, MPharm

AAP APPLE
ACADEMIC
PRESS

Apple Academic Press Inc.
3333 Mistwell Crescent
Oakville, ON L6L 0A2
Canada

Apple Academic Press Inc.
9 Spinnaker Way
Waretown, NJ 08758
USA

First issued in paperback 2021

Exclusive worldwide distribution by CRC Press, a member of Taylor & Francis Group
No claim to original U.S. Government works

ISBN 13: 978-1-77463-152-2 (pbk)
ISBN 13: 978-1-77188-670-3 (hbk)

Library and Archives Canada Cataloguing in Publication

Plant- and marine-based phytochemicals for human health : attributes, potential, and use / edited by Megh R. Goyal, PhD, Durgesh Nandini Chauhan, MPharm.

(Innovations in plant science for better health : from soil to fork book series)
Includes bibliographical references and index.
Issued in print and electronic formats.
ISBN 978-1-77188-670-3 (hardcover).--ISBN 978-1-351-25198-3 (PDF)

1. Medicinal plants. 2. Phytochemicals. I. Goyal, Megh Raj, editor
II. Chauhan, Durgesh Nandini, editor III. Series: Innovations in plant science for better health

| QK99.A1P63 2018 | 581.6'34 | C2018-905075-6 | C2018-905076-4 |

CIP data on file with US Library of Congress

Apple Academic Press also publishes its books in a variety of electronic formats. Some content that appears in print may not be available in electronic format. For information about Apple Academic Press products, visit our website at **www.appleacademicpress.com** and the CRC Press website at **www.crcpress.com**

INNOVATIONS IN PLANT SCIENCE FOR BETTER HEALTH: FROM SOIL TO FORK BOOK SERIES

Series Editor-in-Chief:

Dr. Hafiz Suleria
Honorary Fellow at the Diamantina Institute,
Faculty of Medicine, The University of Queensland (UQ), Australia
email: hafiz.suleria@uqconnect.edu.au

The objective of this new book series is to offer academia, engineers, technologists, and users from different disciplines information to gain knowledge on the breadth and depth of this multifaceted field. The volumes will explore the fields of phytochemistry, along with its potential and extraction techniques. The volumes will discuss the therapeutic perspectives of biochemical compounds in plants and animal and marine sources in an interdisciplinary manner because the field requires knowledge of many areas, including agricultural, food, and chemical engineering; manufacturing technology along with applications from diverse fields like chemistry; herbal drug technology; microbiology; animal husbandry; and food science; etc. There is an urgent need to explore and investigate the innovations, current shortcomings, and future challenges in this growing area of research.

We welcome chapters on the following specialty areas (but not limited to):

- Food and function
- Nutritional composition of different foods materials
- Functional and nutraceutical perspectives of foods
- Extraction of bioactive molecules
- Phytomedicinal properties of different plants
- Traditional plants used as functional foods
- Phytopharmacology of plants used in metabolic disorders

- Importance of spices and medicinal and functional foods
- Natural products chemistry
- Food processing waste/byproducts: management and utilization
- Herbals as potential bioavailability enhancer and herbal cosmetics
- Phytopharmaceuticals for the delivery of bioactives
- Alternative and complementary medicines
- Ethnopharmacology and ethnomedicine
- Marine phytochemistry
- Marine microbial chemistry
- Other related areas include nuts, seed spices, wild flora, etc.

BOOKS IN THE INNOVATIONS IN PLANT SCIENCE FOR BETTER HEALTH: FROM SOIL TO FORK BOOK SERIES

Bioactive Compounds of Medicinal Plants: Properties and Potential for Human Health
Editors: Megh R. Goyal, PhD, and Ademola O. Ayeleso, PhD

Plant- and Marine-Based Phytochemicals for Human Health: Attributes, Potential, and Use
Editors: Megh R. Goyal, PhD, and Durgesh Nandini Chauhan, MPharm

Human Health Benefits of Plant Bioactive Compounds: Potentials and Prospects
Editors: Megh R. Goyal, PhD, and Hafiz Ansar Rasul Suleria, PhD

CONTENTS

ABOUT THE EDITORS

Megh R. Goyal, PhD, PE
Retired Professor in Agricultural and Biomedical Engineering, University of Puerto Rico–Mayaguez Campus; Senior Acquisitions Editor, Biomedical Engineering and Agricultural Science, Apple Academic Press, Inc.
Email: goyalmegh@gmail.com

Megh R. Goyal, PhD, PE, is a Retired Professor of Agricultural and Biomedical Engineering in the General Engineering Department at the College of Engineering, University of Puerto Rico–Mayaguez Campus; and Senior Acquisitions Editor and Senior Technical Editor-in-Chief of Agriculture and Biomedical Engineering for Apple Academic Press, Inc. He has worked as a Soil Conservation Inspector and as a Research Assistant at Haryana Agricultural University, India, and Ohio State University, USA.

During his professional career of 45 years, Dr. Goyal has received many prestigious awards and honors. He was the first agricultural engineer to receive the professional license in Agricultural Engineering in 1986 from the College of Engineers and Surveyors of Puerto Rico. In 2005, he was proclaimed as the "Father of Irrigation Engineering in Puerto Rico for the twentieth century" by the American Society of Agricultural and Biological Engineers (ASABE), Puerto Rico Section, for his pioneering work on micro irrigation, evapotranspiration, agroclimatology, and soil and water engineering. The Water Technology Centre of Tamil Nadu Agricultural University in Coimbatore, India, recognized Dr. Goyal as one of the experts "who rendered meritorious service for the development of the micro irrigation sector in India" by bestowing him with the Award of Outstanding Contribution in Micro Irrigation. This award was presented to Dr. Goyal during the inaugural session of the National Congress on "New Challenges and Advances in Sustainable Micro Irrigation" on March 1, 2017, held at Tamil Nadu Agricultural University. Dr. Goyal is slated to receive the Netafim Award for Advancements in Microirrigation: 2018 from the American Society of Agricultural Engineers at the ASABE International Meeting in August 2018.

A prolific author and editor, he has written more than 200 journal articles and textbooks and has edited over 55 books. He is the editor of several book series, including Innovations in Agricultural & Biological Engineering, Innovations and Challenges in Micro Irrigation, and Research Advances in Sustainable Micro Irrigation.

He received his BSc degree in Engineering from Punjab Agricultural University, Ludhiana, India; his MSc and PhD degrees from Ohio State University, Columbus, USA; and his Master of Divinity degree from Puerto Rico Evangelical Seminary, Hato Rey, Puerto Rico, USA.

Durgesh Nandini Chauhan, MPharm, has several years of academic experience at institutes in India in pharmaceutical sciences. She taught subjects such as pharmaceutics, pharmacognosy, traditional concepts of medicinal plants, drug delivery, phytochemistry, cosmetic technology, pharmaceutical engineering, pharmaceutical packaging, quality assurance, dosage form designing, and anatomy and physiology.

She is a member of the Association of Pharmaceutical Teachers of India, SILAE: *Società Italo-Latinoamericana di Etnomedicina* (The Scientific Network on Ethnomedicine, Italy), and others. Her previous research work included "Penetration Enhancement Studies on Organogel of Oxytetracycline HCL." She has attended several workshops, conferences, and symposiums, including the AICTE-Sponsored Staff Development Program on "Effects of Teaching and Learning Skills in Pharmacy: tools for Improvement of Young Pharmacy Teachers." She has written more than 10 articles published in national and international journals, 11 book chapters, and authored in one book (*Optimization and Evaluation of an Organogel*). She is also active as a reviewer for several international scientific journals and an active participant at national and international conferences, including Bhartiya Vigyan Sammelan and the International Convention of the Society of Pharmacognosy. Presently she joined the Ishita Research Organization, Raipur, India as a freelance writer and guide for students in pharmacy, Ayurvedic, and science in their research projects.

Mrs. Chauhan earned her BPharm degree in pharmacy from the Rajiv Gandhi Proudyogiki Vishwavidyalaya, Bhopal, India, and her MPharm in pharmaceutics from Uttar Pradesh Technical University (currently Dr. A.P.J. Abdul Kalam Technical University), Lucknow, India.

CONTRIBUTORS

Munawar Abbas, MSc
Research Associate, Food Science & Technology, Institute of Home & Food Sciences, Government College University, Faisalabad (GCUF), Allama Iqbal Road, Faisalabad 38000, Pakistan, Mobile: +92-332-0677170, E-mail: foodian2007@gmail.com

Ashika Advankar, BSc
MS II-Year Student, Department of Natural Product, National Institute of Pharmaceutical Education and Research (NIPER) – Ahmedabad, Opposite Air force StationPalaj, Gandhinagar, Gujarat 382355, India, Mobile: +91-9029029698, E-mail:ashika.advankar@gmail.com

Himani Agrawal, MSc
Ph.D. candidate, Academy of Scientific & Innovative Research (AcSIR), Council of Scientific and Industrial Research, Institute of Himalayan and Bioresource Technology, Palampur 176061, India, Mobile: +91-8894110769, E-mail: himanicsirihbt@gmail.com

S. Aishwarya, PhD
Assistant Professor, Department of Bioinformatics, Stella Maris College, Chennai, Tamil Nadu 600062, India, Mobile: +91-8903428058, E-mail: s.aishwaryabiotech@gmail.com

Naif Abdullah Al-Dhabi, PhD
Professor, Department of Botany and Microbiology, Addiriyah Chair for Environmental Studies, College 16 of Science, King Saud University, P.O. Box 2455, Riyadh 11451, Kingdom of Saudi Arabia, Mobile: +966-535996677, E-mail: naifaldhabi2014@gmail.com

Sahar Y. Al-Okbi, PhD
Professor of Biochemistry and Nutrition, Nutrition and Food Sciences Department, National Research Centre, El Buhouth Street, Dokki 12622, Cairo, Egypt, Mobile: 00201003785152, E-mail: S_Y_alokbi@hotmail.com

Neelam Athawale, BSc
MS Pharm II-Year Student, Department of Natural Product, National Institute of Pharmaceutical Education and Research (NIPER) – Ahmedabad, Opposite Air force Station Palaj, Gandhinagar, Gujarat 382355, India, Mobile: +91-9763256561, E-mail: neelamathawale@gmail.com

Kalpana Kumari Barhwal, PhD
Assistant Professor, All India Institute of Medical Sciences, Sijua, Patrapada, Bhubaneswar 751019, India, Mobile: +91-9438884026, E-mail: drkalpana2009@hotmail.com

A. Bhardwaj, PhD
Ph.D. Candidate, Senior Research Scholar, Defense Institute of Physiology and Allied Sciences (DIPAS), Lucknow Road, Timarpur, Delhi 110054, India, Mobile: 91-9971544026, E-mail: anujabhardwaj75@gmail.com

Shriya Bhatt, MSc
Project Fellow, Council of Scientific and Industrial Research, Institute of Himalayan and Bioresource Technology, Palampur 176061, India, Mobile: +91-9459786397, E-mail: shriyabhatt070@gmail.com

Suryanarayan Biswal, PhD
Ph.D. Fellow, Defence Institute of High Altitude Research (DIHAR), Ministry of Defence- Govt.
of India, c/o 56 APO, Leh-Ladakh 901205, Jammu and Kashmir, India, Mobile: +91-7589124065,
E-mail: surya08bio@gmail.com

Masood Sadiq Butt, PhD
Dean, National Institute of Food Science and Technology, University of Agriculture, Faisalabad,
Pakistan, Mobile: +92-3006622685, E-mail: drmsbutt@yahoo.com

Vikas Dadwal, MSc
Ph.D. candidate, Academy of Scientific & Innovative Research (AcSIR), Council of Scientific and
Industrial Research, Institute of Himalayan and Bioresource Technology, Palampur 176061, India,
Mobile: +91-8988457494, E-mail: vikasihbt@yahoo.com

V. Duraipandiyan, MPhil, PhD, BLIS
Assistant Professor, Department of Botany and Microbiology, Addiriyah Chair for Environmental
Studies, College 16 of Science, King Saud University, P.O. Box 2455, Riyadh 11451, Kingdom of
Saudi Arabia; Division of Ethnopharmacology, Entomology Research Institute, Loyola College,
Chennai, Tamil Nadu 600034, India, Mobile: +91-9444921032,E-mail: avdpandiyan@gmail.com;
avdpandiyan@yahoo.co.in; avdpandiyan@hotmail.com;

Sayed A. El-Toumy, PhD
Professor and Head, Chemistry of Tanning Materials, National Research Centre, Bohouth St., Dokki,
Cairo 12622, Egypt, Mobile: +01-000923320, E-mail: sayedeltomy@yahoo.com

Ashwini Ghagare, BSc
MS II-Year Student, Department of Natural Product, National Institute of Pharmaceutical Education
and Research (NIPER) – Ahmedabad, Opposite Air force StationPalaj, Gandhinagar,
Gujarat 382355, India, Mobile: +91-9860863846, E-mail: ashwinighagare@rediffmail.com

Anand Gugale, BSc
MS Pharm II-Year Student, Department of Natural Product, National Institute of Pharmaceutical
Education and Research (NIPER) – Ahmedabad, Opposite Air force StationPalaj, Gandhinagar,
Gujarat 382355, India, Mobile: +91-8275201515, E-mail:anandgugale1804@gmail.com

Mahesh Gupta, PhD
Scientist, Council of Scientific and Industrial Research, Institute of Himalayan and Bioresourse
Technology, Palampur 176061, India, Mobile: +91-9736440442, E-mail: mgupta@ihbt.res.in

Sunil Kumar Hota, PhD
Scientist – D and Head, Experimental Biology Division, Defence Institute of High Altitude Research
(DIHAR), Ministry of Defence - Govt. of India, c/o 56 APO, Leh-Ladakh 901205, Jammu and
Kashmir, India, Mobile: +91-9463998315, E-mail: drsunilhota@yahoo.co.in

Muhammad Jawad Iqbal, PhD
Lecturer, National Institute of Food Science and Technology, University of Agriculture, Faisalabad,
Pakistan, Mobile: +92-3346503007, E-mail: mjkamboh@gmail.com

Robin Joshi, PhD
Senior Technical Officer II, Council of Scientific and Industrial Research, Institute of Himalayan and
Bioresourse Technology, Palampur 176061, India, Mobile: +91-9816658482, E-mail: robinsjoshi@
gmail.com

Bhagyashree Kamble, PhD
Assistant Professor, Department of Natural Product, National Institute of Pharmaceutical Education
and Research (NIPER) – Ahmedabad, Opposite Air force Station Palaj, Gandhinagar,
Gujarat 382355, India, Mobile: +91-8469816232, E-mail: k.bhagyashree@gmail.com

Priyanka Kanukuntla, BSc
MS II-Year Student, Department of Natural Product, National Institute of Pharmaceutical Education
and Research (NIPER) – Ahmedabad, Opposite Air force Station Palaj, Gandhinagar,
Gujarat 382355, India, Mobile: +91-7674832175, E-mail:kanukuntlapriya@gmail.com

Shankar Katekhaye, PhD
Assistant Professor, School of Biological Sciences and Biotechnology, Institute of Advanced
Research, Koba Institutional Area, Gandhinagar, Gujarat 382007, India, Mobile: +91-9924461079,
E-mail: shankar.katekhaye@gmail.com

Abhishek Kulkarni, BSc
MS II-Year Student, Department of Natural Product, National Institute of Pharmaceutical Education
and Research (NIPER) – Ahmedabad, Opposite Air force Station Palaj, Gandhinagar 382007,
 Gujarat 382007, India, Mobile: +91-8007973539, E-mail:abhiks198@gmail.com

Kushal Kumar, PhD
Ph.D. Fellow, Defence Institute of High Altitude Research (DIHAR), Ministry of Defence - Govt.
of India, c/o 56 APO, Leh-Ladakh 901205, Jammu and Kashmir, India, Mobile:+91-9882229025,
E-mail: kushal1kumar@gmail.com

A. Anita Margret, PhD
Assistant Professor, Department of Biotechnology and Bioinformatics, Bishop Heber College,
Tiruchirappalli, Tamil Nadu 620017, India, Mobile: +91-9159626250, E-mail: anitamargret@gmail.
com

K. Misra, PhD
Scientist – F, Additional Director and Head, Department of Biochemical Sciences (DBCS), Defense
Institute of Physiology and Allied Sciences (DIPAS), Lucknow Road, Timarpur, Delhi - 110054,
India. Tel. (office): +91-11-23883303; Fax: 91-11-23914790; Mobile: +91-9871372350;
E-mail: kmisra99@yahoo.com

Izabela Nowak, PhD
Full Professor, Faculty of Chemistry, Laboratory of Applied Chemistry, Adam Mickiewicz
University in Poznań, ul. Umultowska 89b, 61-614 Poznań, Poland, Mobile: +480-618291580;
E-mail: nowakiza@amu.edu.pl

Irina Pereira, MSc
Graduate Student, Faculty of Pharmacy at University of Coimbra, Azinhaga de Santa Comba, Pólo
da Saúde (III), 3000-548 Coimbra; and Department of Biology and Environment, School of Life and
Environmental Sciences, (ECVA, UTAD), University of Trás-os-Montes and Alto Douro, P.O. Box
1013; 5001-801 Vila Real, Portugal, Mobile: +351-239488400; E-mail: irina.pereira@live.com.pt

T. William Raja, PhD
Research scholar, Division of Ethnopharmacology, Entomology Research Institute, Chennai, Tamil
Nadu 600034, India, Mobile: +91-9566858185, E-mail: williameri2020@gmail.com

Farhan Saeed, PhD
Assistant Professor, Department of Food Science, Institute of Home & Food Sciences, Government
College University – Faisalabad (GCUF), Allama Iqbal Road, Faisalabad 38000, Pakistan,
Mobile: +92-3338040311, E-mail: F.saeed@gcuf.edu.pk; far1552@yahoo.com

Josline Y. Salib, PhD
Professor, Chemistry of Tanning Materials, National Research Centre, Dokki, Cairo, Egypt.
Mobile: +20-1223340221, e-mail: joslineysalib@gmail.com (corresponding author); Chemistry of
Tanning Materials and Leather Technology Department National Research Centre, El-Bohous St.,
Dokki, Cairo 12622, Egypt, Fax: (20)(2)3370931, Tel.: +20-2-3371615, E-mail: joslines@hotmail.com

Ana C. Santos, PhD

Graduate Student, Faculty of Pharmacy of University of Coimbra, Azinhaga de Santa Comba, Pólo da Saúde (III), 3000-548 Coimbra; and Institute for Innovation and Health Research, Group Genetics of Cognitive Dysfunction, Institute for Molecular and Cell Biology, Rua do Campo Alegre, 823, 4150-180 Porto, Portugal, Mobile: +351-239488400; E-mail: ana.cl.santos@gmail.com

Ignacimuthu Savarimuthu, MSc, MPhil, PhD, DSc

Director, Division of Ethnopharmacology, Entomology Research Institute, Chennai, Tamil Nadu 600034, India, Mobile: +91-9840337667, E-mail: eriloyola@hotmail.com

Amélia M. Silva, PhD

Assistant Professor, Department of Biology and Environment, School of Life and Environmental Sciences, (ECVA, UTAD), University of Trás-os-Montes and Alto Douro, P.O. Box 1013; 5001-801 Vila Real, Portugal; and Centre for the Research and Technology of Agro-Environmental and Biological Sciences, University of Trás-os-Montes and Alto Douro (CITAB-UTAD), Vila-Real, Portugal, Mobile: +351-259350106; E-mail: amsilva@utad.pt

Eliana B. Souto, PhD

Assistant Professor (Habilitation), Faculty of Pharmacy of University of Coimbra, Azinhaga de Santa Comba, Pólo da Saúde (III), 3000-548 Coimbra; and REQUIMTE/LAQV, Group of Pharmaceutical Technology, Faculty of Pharmacy, University of Coimbra, Portugal, Mobile: +351-239488400; E-mail: ebsouto@ff.uc.pt; ebsouto@ebsouto.pt

Hafiz Ansar Rasul Suleria, PhD

RHD (Research Higher Degree) Fellow in Food & nutrition, The University of Queensland, School of Medicine, Translational Research Institute, Brisbane, QLD 4072, Australia; Mobile: +61-470439670; Tel.: +61-732142207; E-mail: hafiz.suleria@uqconnect.edu.au; h.suleria@hotmail.com

J. Theboral, PhD

Assistant Professor, Department of Biotechnology and Bioinformatics, Bishop Heber College, Tiruchirappalli, Tamil Nadu620017, India, Mobile: +91-9894543963, E-mail: jtheboral@gmail.com

Francisco J. Veiga, PhD

Professor, Faculty of Pharmacy of University of Coimbra, Azinhaga de Santa Comba, Pólo da Saúde (III), 3000-548 Coimbra; and REQUIMTE/LAQV, Group of Pharmaceutical Technology, Faculty of Pharmacy, University of Coimbra, Portugal, Mobile: +351-239488400, E-mail: fveiga@ff.uc.pt

Aleksandra Zielińska, PhD

Graduate Student, Faculty of Chemistry, Laboratory of Applied Chemistry, Adam Mickiewicz University in Poznań, ul. Umultowska 89b, 61-614 Poznań, Poland; Faculty of Pharmacy of University of Coimbra, Azinhaga de Santa Comba, Pólo da Saúde (III), 3000-548 Coimbra, Portugal, Mobile: +480-618291580, E-mail: ola.zielinska@amu.edu.pl

ABBREVIATIONS

5-LO	5-lipoxygenase
6-4PPs	6-4 photoproducts
AA	archidonic acid
AB	absolute bioavailability
ABTS	2,'-azino-bis-3-ethylbenzothiazoline-6-sulphonic acid
ACE	angiotensin-converting enzyme
Acetyl-CoA	acetyl coenzyme A
AcSIR	Academy of Scientific & Innovative Research
ADI	acceptable daily intake
ADME	Absorption, distribution, metabolism, excretion
AF	Activation function
Akt	serine/threonine-protein kinases
Ala	alanine
ALT	alanine transaminase
AmB	amphotericin B
AMPK	activated protein kinase-dependent pathway
AP-1	activation protein-1
API	Active pharmacetucials constituents
Arg	arginine
Asn	asparagine
ASSOCHEM	Associated Chamber of Commerce and Industry of India
ATP	adenosine triphosphate
AUC	area under the curve
Bax	Bcl-2-associated X protein
Bcl	B-cell lymphoma
Bcl-2	B-cell lymphoma 2
BCRP2	breakpoint cluster region pseudogene 2
BLA	blood lactic acid
BMI	body mass index
BUN	blood urea nitrogen
Ca	calcium
CAL	citronellal

CAM	chick chorioallantoic membrane
CAM	complementary and alternative medicine
CaSki	human caucasian cervical epidermoid carcinoma
CAT	catalase
CCN	cloud condensation nuclei
CDKIs	cyclin-dependent kinases inhibitors
CE-DAD	capillary electrophoresis diode array detector
C-H-R	cold-hypoxia-restraint
C_{max}	peak concentration
CME	crude methanolic extracts
CO_2	carbon dioxide
COPD	chronic obstructive pulmonary disease
COX	cyclooxygenase
COX-2	cyclooxygenase 2
CPDs	cyclobutane pyrimidine dimers
CTS	carpal tunnel syndrome
CVD	cardiovascular disease
CYP-1A1	cytochrome P450 1A1
CYP-1B2	cytochrome P450 1B2
CYP-3A4	cytochrome P450 3A4
CYP450	cytochrome P450
Cys	cystine
Cyt c	cytochrome c
DAD	diode array detector
DBD	DNA-binding domain
DHA	docosahexaenoic acid
DHC	dihydrocitronellal
DMADP	dimethylallyl diphosphate
DMBA	methylbenz[α]anthracene
DMSO	dimethyl sulfoxide
DNA	deoxyribonucleic acid
DOX	doxorubicin
DPP	differential pulse polarography
DPP	dipeptidyl peptidase
DPPH	2,2-diphenyl-1-picrylhydrazyl
DSC	differential scanning calorimetry
DTA	differential thermal analysis
DW	dry weight

EAE	encephalomyelitis
EBC_{50}	half maximal effective concentration for the biomass
EC_{50}	half maximal effective concentration
ECM	extracellular matrix
EGCG	epigallocatechin-3-gallate
EL	limit value
ELSD	evaporative light scattering detector
EMEA	European Medicinal Agency
EORP	extract of G. lucidum polysaccharides
EPA	eicosapentaenoic acid
ERE	estrogen-response-elements
ERK1/2	extracellular signal-regulated kinase1/2
ERs	estrogen receptor
$E\mu C_{50}$	half maximal effective concentration for the growth rate
FAK	focal adhesion kinase
FAs	fatty acids
FDA	US Food and Drug Administration
FPS	fingerprinting spectra
FRAP	ferric-reducing antioxidant power
G6PDH	glucose-6-phosphatedehydrogenase
GA-A	ganoderic acid A
GACP	good agricultural and collection practices
GAE	gallic acid equivalent
GA-H	ganoderic acid H
Gas	ganoderic acids
GA-T	ganoderic acid T
GBE	Ginkgo biloba extract
GC	gas chromatography
GC-FID	gas chromatography with flame ionization detection
GGT	gamma glutamyl transpeptidase
GIT	gastrointestinal tract
GLA	γ-Linolenic acid
Gln	glutamine
GLP	glucagon-like peptide
GLP-1	glucagon-like peptide-1
Gly	glycine
GPP	geranyl diphosphate

GPX	glutathione peroxidase
GS	ganoderma total sterol
GSH	reduced glutathione
GST	glutathione S-transferase
H/R	hypoxia/reoxygenation
HBV	hepatitis B virus
HCT	human colorectal
HCV	hepatitis C virus
HDL	high-density lipoprotein
HDL-ch	high-density lipoprotein cholesterol
Hep 3B	human hepatoma
HepG2	human hepatocellular liver carcinoma
HI	hazard index
His	histidine
HL-60	human promyelocytic leukemia
HPLC	high-performance liquid chromatography
HPLC–DAD-MS	high-performance liquid chromatography–diode array detector–mass spectrometry
HPTLC	high-performance thin-layer chromatography
HPV	human papillomavirus
HRT	hormone replacement therapy
HS-SPME-GC-MS	headspace solid-phase microextraction combined with gas chromatography–mass spectrometry
HT29	human colon adenocarcinoma grade II
HUVECs	human umbilical vein endothelial cells
i.v.	intravenous
I/R	ischemia–reperfusion
IBV	infectious bronchitis virus
IC50	half maximal inhibitory concentration
IFN-γ	interferon-γ
IFRA	International Fragrance Association
IgE	immunoglobulin E
IKK	IκB kinase complex
IL	interleukins
IL-6	interleukin 6
IL-10	interleukin 10
Ile	isoleucine
IPP or IDP	isopentenyl diphosphate

IR	infrared spectroscopy
ISSR	inter simple sequence repeat
JNK	c-JunN-terminal kinase
LAB	lucidenic acid B
LBD	ligand-binding domain
LC_{50}	median lethal concentration
LDH	lactate dehydrogenase
LDL	low-density lipoprotein
LDL-ch	low-density lipoprotein cholesterol
Leu	leucine
LIN-NPs	linalool-loaded nanoparticles
LLC	Lewis lung carcinoma
LLNA	local lymph node assay
LL-NLCs	Linalool-loaded nanostructured lipid carriers
LO	lipoxygenasee
LOAEL	lowest-observed-adverse effect level
LPS	lipopolysaccharide
LTs	leukotrienes
Lys	lysine
LZ-8	Ling Zhi-8
MAE	microwave-assisted extraction
MALDI-TOF	matrix-assisted laser desorption-time of flight
MAPK	mitogen-activated protein kinase
MBC	minimum bacterial concentration
MCAO	middle cerebral artery occlusion
MCF-7	Michigan Cancer Foundation-7
MDA	malondialdehyde
MDS's	Marine-derived substances
MEP	2-C-methyl-D-erythritol 4-phosphate pathway (non-mevalonate pathway)
Met	methionine
MFC	minimum fungicidal concentration
MHA	Mueller Hinton Agar
MHC	major histocompatibility complex
MHRA	medicines and health care product regulatory agency
MIC	minimum inhibitory concentration
MLC	minimum lethal concentration
MMP2/9	matrix metalloproteinase 2/9

MMPs	matrix metalloproteinases
Mn-SOD	manganese superoxide dismutase
MOE	margin of exposure
MRP	multidrug resistance-associated protien
MRP21	mitochondrial 37S ribosomal protein
MRSA	methicillin-resistant Staphylococcus aureus
MS	mass spectrometry
MTT	3-(4,5-dimethylthiazol-2-yl)-2,5-diphenyltetrazolium bromide
MVA	mevalonate pathway
N protein	nucleocapsid protein
NAFLD	non-alcoholic fatty liver diseases
NASH	non-alcoholic steatohepatitis
NF-kappa B	nuclear factor kappa B
NFO	non-functional overreaching
NLC	nanostructured lipid carriers
NMDA	N-methyl-D-aspartate receptor
NMR	nuclear magnetic resonance
NMU	N-nitroso-N-methylurea
NO	nitric oxide
NOAEL	no-observed-adverse effect level
NOEC	no-observed-adverse effect concentration
NOS	nitric oxide synthases
NSAID	non-steroidal anti-inflammatory drug
NTR	neuroprophin receptor
OECD	Organization for Economic Co-operation and Development
OGD	oxygen and glucose deprivation
OTS	overtraining syndrome
PO_2	partial pressure of oxygen
PABA	4-aminobenzoic acid
PARP	poly (ADP-ribose) polymerase
PC-3	prostate cancer-3
PDA	pancreatic ductal adenocarcinoma
PEBP	phosphatidylethanolamine-binding protein
P-gp	P-glycoprotein
Phe	phenylalanine
PI3K	phosphatidylinositol-3-kinases

PLA	phosphlipase A
PLNA	popliteal lymph node assay
PMA	phorbol-12-myristate-13-acetate
PPAR	peroxisome proliferator activating receptors
PrAlPO-5	praseodymium incorporated aluminophosphate molecular sieves
Pro	proline
PS	phosphatidylserine
PUFAs	polyunsaturated fatty acids
RA	rheumatoid arthritis
RfD	reference dose
RNA	ribonucleic acid
ROAT	repeated open application test
ROS	reactive oxygen species
RT-PCR	reverse transcription polymerase chain reaction
SC	stratum corneum
SCGE	single-cell gel electrophoresis assay
SD	Sprague–Dawley
SDA	Sabouraud dextrose agar
SDG	secoisolariciresinol diglucoside
SDH	succinate dehydrogenase
SDS-PAGE	sodium dodecyl sulfate-polyacrylamide gel electrophoresis
SECO	secoisolariciresinol
SED	systemic exposure dose
SeGLP-2B-1	Se-enriched G. lucidum
SEM	scanning electron microscopy
Ser	serine
SERMs	selective estrogen receptor modulators
SFE	supercritical fluid extraction
SiHa	human cervical cancer
SIRT	silent mating type information regulation
SLN	solid lipid nanoparticles
SOD	superoxide dismutase
SP	sulfated polysaccharides
Span-80	sorbitane monooleate
STZ	streptozotocin
T.Ch	total cholesterol

T-BOOH	t-butyl hydroperoxide
TCM	traditional Chinese medicine
TEM	transmission electron microscopy
TGF-β1	transforming growth factor beta 1
TGs	triacylglycerols
Thr	threonine
TiO$_2$	titanium dioxide
TLC	thin-layer chromatography
TLR-4	toll-like receptor-4
TM	traditional medicine
TNBS	5,6-trinitrobenzene sulfonic acid
TNFα	tumor necrosis factor alpha
Tps	terpene synthase
TQ	thymoquinone
T$_{rec}$	rectal temperature
Trp	tryptophan
Tween 20	polyoxyethylene sorbitan monooleate 20
Tween 80	polyoxyethylene sorbitan monooleate 80
Tyr	tyrosine
UDP	uridine diphosphate glucose
uPA	urokinase-plasminogen activator
uPAR	urokinase-plasminogen activator receptor
UV	ultraviolet
UVB	ultraviolet-B
Val	valine
VEGF	vascular endothelial growth factor
VLDL-ch	very low-density lipoproteincholesterol
VOCs	volatile organic compounds
VPH	virus del papiloma humano
WF	water fraction
WHO	World Health Organization
WST-1	water-soluble tetrazolium salts
XRD	X-ray diffraction
ZIC-HILIC	zwitterionic hydrophilic interaction chromatographic

SYMBOLS

α	alpha
β	beta
γ	gamma
κ	kappa
%	percent

PREFACE 1

"25 grams of soy protein a day,
As part of a diet low in saturated fat and cholesterol,
May reduce the risk of heart disease." [USFDA 21CFR101.82]

—Ramabhau Patil, PhD

If wealth is lost, nothing is lost,
If health is lost, something is lost and
If character is lost, everything is lost.

—P. P. Joy, Kerala

https://www.researchgate.net/profile/Pp_Joy

Quality and nutritional value of foods are highly dependent on environment, agricultural practices, production conditions, and consumer preferences, which all may result in different effects on human health. One of the main challenges of the food science and technology is to optimize food production to have a minimum environmental footprint, to lower production costs, and to improved quality and nutritional value.

Therefore, we introduce this book volume on *Bioactive Compounds of Medicinal Plants: Properties and Potential for Human Health* under book series *Innovations in Agricultural and Biological Engineering.* This book covers mainly the current scenario of the research and case studies under conditions of the African continent.

This book volume sheds light on the potential of medicinal plants on the African continent for human health for different technological aspects, and it contributes to the ocean of knowledge on food science and technology. We hope that this compendium will be useful for students, faculty, and researchers as well as for those working in the food, nutraceuticals, and herbal industries.

The contributions by the cooperating authors to this book volume have been most valuable. Their names are mentioned in each chapter and in the list of contributors. We appreciate their patience with our editorial skills. This book would not have been written without the valuable cooperation of these investigators, many of whom are renowned scientists who have worked in the field of food engineering and food science throughout their professional careers.

I am glad to introduce Mrs. Durgesh Nandini Chauhan, MPharm, who has several years of academic (teaching) experience at institutes in India in pharmaceutical sciences. She is currently with the Ishita Research Organization, Raipur, India, and has taught subjects such as pharmaceutics, pharmacognosy, traditional concepts of medicinal plants, drug delivery, phytochemistry, cosmetic technology, pharmaceutical engineering, pharmaceutical packaging, quality assurance, dosage form designing, and anatomy and physiology, and we appreciate her bringing her knowledge to this volume.

We would like to thank editorial staff, Sandy Jones Sickels, Vice President, and Ashish Kumar, Publisher and President at Apple Academic Press, Inc., for making every effort to publish the book when the food resources are a major issue worldwide. Special thanks are due to the AAP production staff also.

We request that the readers offer your constructive suggestions that may help to improve the next edition.

We express our admiration to our families and colleagues for their understanding and collaboration during the preparation of this book volume.

As an educator, there is a piece of advice to one and all in the world: *"Permit that our almighty God, our Creator, provider of all and excellent Teacher, feed our life with Healthy Food Products and His Grace...; and Get married to your profession..."*

—**Megh R. Goyal**, PhD, PE
Senior Editor-in-Chief

PREFACE 2

With the increasing use of the allopathic system of medicine, alternative medicine gradually lost its popularity among people due to the fast therapeutic actions of synthetic drugs. Almost a century has passed, and we have witnessed the limitations of allopathic system of medicine. Lately, phytomedicine has gained new momentum, and it is evident from the fact that certain herbal remedies are more effective as compared to synthetic drugs. Phytomedicines have played a miraculous role in the development of new effective therapeutic agents. The plant-derived drugs have led to the discovery of some interesting clinically useful molecules. Tu Youyou discovered the anti-malarial compound artemisin and was awarded the Nobel Prize in Physiology or Medicine.

This book addresses the importance of phytochemicals derived from marine and plants for use in therapeutics. It divided in four parts:

Part I: Bioactive Compounds in Medicinal Plants: Status and Potential covers topics on Herbal Drug Development: Challenges and Opportunities, Antimicrobial Properties of Traditional Medicinal Plants: Status and Potential, and Marine Bioactive Compounds: Innovative Trends in Food and Medicine.

Part II: Plant-Based Pharmaceuticals in Human Health: Review covers topics on Monoterpenes-Based Pharmaceuticals: A Review of Applications in Human Health and Drug-Delivery Systems; Role of Nutraceuticals in Prevention of Non-alcoholic Fatty Liver: A Review; and Black Cumin: A Review on Phytochemistry, Antioxidant Potential, Extraction Techniques and Therapeutic Perspectives.

Part III: Therapeutic Attributes of Mushroom, Cereal Grains and Legumes covers topics on Therapeutic Medicinal Mushroom *(Ganoderma lucidum)*: A Review of Bioactive Compounds and Their Applications; and Nutritional Attributes of Traditional Cereal Grains and Legumes as Functional Food: A Review.

Part IV: Innovative Use of Medicinal Plants covers topics on Therapeutic Implications of *Rhodiola* sp. for High-Altitude Maladies: A Review, Herbal Plants as Potential Bioavailability Enhancers; and

Formulated Natural Selective Estrogen Receptor Modulators (SERM): A Key to Restoring Women's Health.

This book is unique because of its focus on the use of phytochemicals of plant and marine origin for human health, and it is important because it provides evidence on the potentials of phytochemicals in curing certain diseases. This book will provide knowledge to readers about the importance of phytochemicals for various therapeutic effect and drug delivery.

This work could not have been completed in a timely manner without the cooperation of the contributors, their expertise, and time to the production of this volume. Individually the authors are the leaders in their field and collectively they embody an international collection of knowledge and experience in the phytochemical of plant and animal origin, to whom I am very grateful.

I am grateful to Dr. Megh R. Goyal for patiently guiding me in the preparation of final manuscripts, chapters. and making this project successful. I deeply appreciate my long-term mentor, Prof. V. K. Dixit, who taught me something new in every conversation. Finally, I would like to thank my husband, Dr. Nagendra S. Chauhan, my daughters Harshita and Ishita, for their love, understanding, support, and encouragement while this book was written. We hope that you enjoy reading our book.

—**Durgesh Nandini Chauhan**
Coeditor

PART I
Bioactive Compounds in Medicinal Plants: Status and Potential

CHAPTER 1

HERBAL DRUG DEVELOPMENT: CHALLENGES AND OPPORTUNITIES

BHAGYASHREE KAMBLE, NEELAM ATHAWALE,
ANAND GUGALE, ASHIKA ADVANKAR, ASHWINI GHAGARE,
SHANKAR KATEKHAYE, ABHISHEK KULKARNI, and
PRIYANKA KANUKUNTLA

CONTENTS

ABSTRACT

This chapter focuses on the challenges and opportunities, which should be briefly discussed while considering herbal drug development. Quality control and the standardization of herbal extracts[77] are important to protect the integrity of the herbal extracts for pharmaceutical quality. Several qualitative tests such as morphology, macroscopy, and microscopy have been used routinely for monitoring the quality of raw materials along with processed extracts during herbal drug development. Chemical evaluation of marker compounds and total fingerprint analysis are considered quantitative and semiquantitative measures, respectively, for monitoring the quality of several herbal drugs. Marker compounds suggest identification and quantitative measurement of known marker compounds in the raw material, processed extracts, and finished products using chromatographic techniques such as high-performance thin-layer chromatography (HPTLC), high-performance liquid chromatograph (HPLC), and gas chromatography (GC).

1.1 INTRODUCTION

"Herb" is a plant or part of a plant that is valued for its medicinal, aromatic, or savory qualities.[24] In the 21st century, medicinal herbs are gaining importance in the main stream of healthcare as more people are seeking relatively safe remedies and approaches for the treatment, and preventive measures. Plants have been a source of medicinal agents since ancient time and continue to be an abundant source of novel therapeutic agents. It was estimated that approximately 5–15% of the total 250,000 species of higher plants have been systematically investigated and yet the potential of novel bioactive compounds has barely been tapped.[17] There are countries such as India, China, Korea, Japan, and Africa that are fortunate enough to have various herbal remedies.[55] There is huge demand for herbal products in the market due to their therapeutic potential.[20] Herbal medicinal agents aid to cure a variety of diseases such as diabetes, cardiovascular disease, liver diseases, arthritis, and so forth.[13] It can also be used as an adaptogen, memory enhancer, and as dietary supplement.

Herbal drug development process includes various aspects such as plant identification, authentication, extraction, isolation, standardization, safety and efficacy evaluation, metabolic profiling, and quality control.[10,22]

This entire process has many challenges. There are very few companies that are involved in drug development from herbal plants due to various issues listed below:

- Although these herbs are generally regarded as safe, unexpected effects of many popular herbal medicines have been reported.
- Challenges in applying modern technologies for quality control of herbal products because of complex nature.
- Challenges such as social values, scientific validity.
- Difficulty in collection and authentication due to various geographical locations.
- High cost and complexity involved in the isolation of markers.
- Lack of data for safety evaluation on product and process development.
- Lack of knowledge about the stability of markers and plant material, that is, after plant collection enzymatic processes may destroy active constituents.

Although there are challenges for herbal drug development, some opportunities in the form of quality control and standardization tools (conventional and modern), newer approaches, such as establishing hyphenated techniques, are motivating features to invest in it.

This chapter focuses on the challenges and opportunities, which should be briefly discussed while considering herbal drug development.

1.2 CHALLENGES IN HERBAL DRUG STANDARDIZATION AND DEVELOPMENT

1.2.1 STANDARDIZATION ISSUES

Herbal medicines can help to fulfill primary healthcare needs of people.[48] Standardization of herbal medicines means confirmation of identity and determination of quality. "Standardization is a system to ensure that every packet of medicine is being sold has the correct amount labeled on it and will induce its therapeutic effect".[13] The subject of herbal drug standardization is very complex and often remains unaddressed. As herb consists of many constituents, therefore, preparation of standardization parameters itself is a challenge.[24]

Although it is believed that the herbal formulations are safe for consumption being originated from natural source, yet some biologically active constituents are known to be harmful for health and produce toxic and undesirable effects. In most countries (particularly unregulated markets), herbal preparations are sold without proper evaluation, safety, and toxicological studies that may compromise the quality of the product.[79] The herbal medicinal products are not completely free from contamination (toxic metals, toxins), adulteration and substituted herbal ingredients, or improperly processed products.[44] Presence of non-declared drugs and pesticides may cause toxicity.[7] Standardization is also necessary to avoid microbial contamination.[30] Many herbal preparations are sold as over the counter drugs under the name of dietary health supplements. They may be sold without prescription and the people might not be able to recognize the potential hazards in an inferior product.[13]

1.2.2   COLLECTION AND IDENTIFICATION OF RAW MATERIAL

World Health Organization (WHO) recently released guidelines on good collection practices for medicinal plant collection to avoid fluctuations in quality due to the heterogeneity of material obtained from wild source.[92] Reports by *The Associated Chambers of Commerce and Industry of India* divulge that 700 plant species are commonly used in India. Only 20% have been cultivated on a commercial scale and other 90% are collected from the wild resource. To obtain the quality herbal product, care should be taken while authentication, season and area of collection, purification process, and rationalizing the combination in case of polyherbal drugs. Overexploitation or unscientific collection may result in the destruction of species. This leads to genetic stocks and medicinal diversity, therefore sustainable natural product collection is necessary.[76] Difficulties in identification of plants and differences in genes, environmental and harvesting conditions create problems in developing uniformity in evaluation and standardization of data.[5]

1.2.3   PHARMACOLOGY OF HERBAL DRUGS: SAFETY

Safety is the main issue, which must be considered while using any drug. Although most of the herbal drugs are generally regarded as safe, yet many pharmacological studies have raised questions about their safety due to various toxicities reported in animals as well as in human beings.[61] Pharmacology of herbal medicines mainly deals with target identification and finding out mechanisms by which herbal drug is acting and also includes studies on biological activity and side effects. It is important to know whether given herbal drug is safe for its use or not (Fig. 1.1).

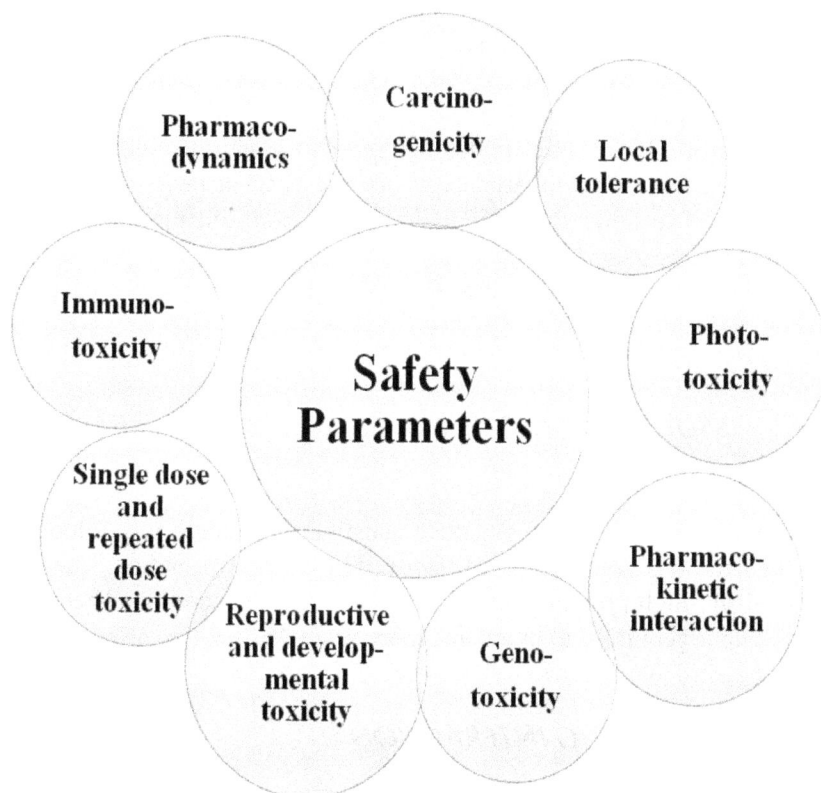

FIGURE 1.1 Safety parameters for herbal drugs.

1.2.4 PHARMACOKINETIC AND BIOAVAILABILITY OF NATURAL PRODUCTS: ISSUES

A general belief, that herbal drugs are safe and free from side effects, has gained much popularity in recent years with tremendous growth and usage of phytopharmaceuticals.[10] For a long time, natural product scientists have been studying pharmacodynamics but less attention has been paid on the pharmacokinetics of plant extracts and their phyto-constituents.[18] Unlike pharmaceuticals, the pharmacokinetics of herbal products, mixture of known and unknown components, is always challenging due to the complexity, unavailability, or inadequacy of standards and methods.[33] Pharmacokinetics is of prime importance prior to clinical trials of herbal products to make herbal remedies evidence-based drugs. The pharmacokinetic study includes: absorption, distribution, metabolism, and excretion (ADME). Therapeutic outcome depends on the rate and extent at which drug reaches at the site of action and its bioavailability.[27,37] A better understanding of the pharmacokinetics and bioavailability of phytopharmaceuticals can help in designing dosage regimen; and to understand the action of the body on the drug, which have numerous useful applications both in toxicology and biopharmaceutics.[58] Table 1.1 presents few examples of research studies on pharmacokinetics of herbs.

1.2.5 ISSUES OF POOR BIOAVAILABILITY OF HERBAL DRUGS

The poor bioavailability of bioactive constituent is due to low solubility, low penetration, gastrointestinal tract (GIT) biotransformation,[35] hepatic metabolism, high protein binding, and renal or bile excretion. Therefore, there is an urgent need to overcome these issues.

1.2.6 HERB–DRUG INTERACTION

The efficiency of herbal drug depends on many factors such as pharmacokinetic and pharmacodynamic properties of drugs. Coadministration of various drugs causes unexpected interactions.[25]

1.2.6.1 PHARMACODYNAMIC INTERACTION

It involves the interaction of conventional drugs with herbs for the same targets such as enzymes or receptors leading to additive, synergistic,

TABLE 1.1 Pharmacokinetic Study of Natural Products.

Name of the plant extract/ phytoconstituent	Pharmacokinetic parameter	Biomarker	Reference
Aqueous extract of *Paeoniae radix*	Paeoniflorin showed higher binding activity to organs and lower blood distribution, the low bioavailability was also possible because of low paeoniflorin concentration in plasma (about 3%)	Paeoniflorin	[11]
Ethanolic extract of fruit of *Piper sarmentosum*	Pellitorine, sarmentine; showed good oral bioavailability, tissue distribution and both get excreted in urine as metabolites, whereas sarmentosine excreted unchanged in feces and was not absorbed from the intestine.	Pellitorine, sarmentine, sarmentosie	[40]
Ethanolic extract of *Panaxginseng* berry	Ginsenoside Re showed very short half-life and low oral bioavailability	Ginsenoside Re, Rg1, and Rh1	[43, 67]
Herbal preparation of *Swertia chirata*	Pure markers were determined up to 48 h after IV administration of % mg/kg dose. The plasma concentration-time profile of mangiferine and amarogenine was traceable upto 24 and 0.75 h, respectively	Mangiferine and amarogentin	[46, 81]
Pinostrobin (PI) (5-hydroxyl-7-methoxy-Flavanone)	After intragastric administration, the half-life (t1/2) was 6.26 ± 0.31 h. The area under the curve (AUC) curve of Pinostorbin after intragastric administration was 3817.80 ± 352.89 ng·min/ml	Pinostorbine	[38]
Rheum undulatum L.	The plasma rhein C_{max} reached between 1 and 2 h after administration and the concentration decreased thereafter	Rhein	[51]
Standardization of leaf extract of *Ginkgo biloba*	The mean plasma concentration time profile of ginkgolides A, B, C, and bilobalide have been reported	Ginkgolides A, B, C, and bilobalide	[53, 76, 90]

TABLE 1.2 Examples of Pharmacodynamic Interactions.

Herb	Drug	Effect	Mechanism	Risk/benefit	Study model
Allium sativum	Chloropropamide	↑in antidiabetic effect	Additive	Beneficial	Case report
Ginkgo biloba	Haloperidol	↑efficiency	Synergistic	Beneficial	Clinical trial
Hypericum-perforatum	Warfarine	↓in INR values	Antagonistic	Risk	Case report
Piper Methysticm	Levodopa	Dopamine antagonistic	Antagonistic	Risk	Case report

↑ indicate increase in activity, and ↓ indicate decrease in activity; INR: international normalized ratio.

antagonistic effect. It has been considered that herbs contain thousands of biocompounds with diverse chemical nature and thereof, they have different affinities toward these primary or secondary therapeutic targets. The probability of possible interaction would be beneficial or unwanted. Table 1.2 gives specific examples.

1.2.6.2 PHARMACOKINETIC INTERACTIONS

Concomitant administration of herbs with conventional drugs affects ADME leading to toxic or subtherapeutic effect of drugs. Herbs affect the ADME of conventional drug upon coadministration. Herbs modulate transporter proteins such as p-glycoprotein (an efflux protein) and/or organic anionic (OATP) as well as cationic transporter proteins require (OCTP) for transport of conventional drugs present in gut, liver, kidney, brain, thus affecting pharmacokinetics of coadministered drugs. Examples of pharmacokinetic interactions are shown in Table 1.3.

The herbal drugs also have modulatory effect on several metabolizing enzymes[74] including CYP450 (e.g., CYP 3A4, 2C9, 2C19, 2D6 and 2B, etc.) which act as receptor and required for the metabolism of conventional drugs, leading to pharmacokinetic interaction when coadministered with conventional drugs.[21] Some plant extract have inhibitory effect on CYP enzyme, as shown in Table 1.4.

TABLE 1.3 Examples of Pharmacokinetic Interactions.

Herb	Drugs	Effect	Mechanisms
Ginkgo Biloba	Theophylline	↑Elimination	Ginkgo biloba extract pretreatment increased CYP1A4 metabolic pathway
Glycyrrhiza Glabra[84–86]	Methotrexate	↓Elimination	Multidrug resistance-associated protein inhibitor of substrate for OATP
Hypericam perforatum	Cyclosporine	↑Metabolism	Induction of p-gp and CYP3A4
H. perforatum	Digoxin	Affect absorption and distribution	Induction of p-gp
Rheum palatum	Digoxin	↓Absorption	Cause diarrhea

TABLE 1.4   Inhibitory Effects of Herbs on Pharmacokinetic of Drugs.

Herb	Drug	Effect on pharmacokinetic of drug	Reference
Echinacea	Darunavir	No effect	[60]
Garlic	Alprazolam	No effect	[31]
Ginkgo biloba	Alprazolam	AUC of alprazolam decreased	[56]
Kava	Digoxin	No effect	[32]
Milk thistle	Metronidazole	Increased clearance of metronidazole and decreased t½, Cmax, and AUC	[68]
Panax quinquefolius	Indinavir	No effect	[2]
Panax quinquefolius	Zidovudine	No effect	[52]

1.2.7   LIMITATIONS ON RESEARCH STUDIES FOR HERB–DRUG INTERACTIONS

Most of the available information on the interaction between the herbal product and the prescribed drug is obtained from case reports, although clinical studies now also beginning to appear in the literature. The published case reports are often incomplete as they do not allow us to conclude that the causal relationship exists. Even documented case reports have to be interpreted with great caution, as causality is not usually established beyond reasonable doubt. The majority of interactions may represent potential risk to the patient taking conventional medicine. The majority of interactions identified to date involve medicines that require regular monitoring of blood level (e.g., warfarin, cyclosporine). Although some herb–drug interaction may be clinically insignificant, (e.g., interaction between guar gum and penicillin V), yet others may have serious consequences (e.g., interaction between St. John's wort and cyclosporine)[63] knowing that millions of patients take herbal and conventional medicines concomitantly, often without the knowledge of their physicians and more research in this area seems to be urgent. It is interesting as well as important to study herb—drug interaction likelihood of which is more than drug–drug interaction.[26]

1.2.8 REGULATORY ASPECTS FOR DEVELOPMENT OF HERBAL DRUGS

Widespread and growing use of herbal drugs has created global health challenges in terms of quality, safety, and efficacy. Scientific validation and technological standardization of herbal medicines[8] are needed for the further advancement of traditional medicines (TMs). The safety problems emerging with herbal medicinal products are due to a largely unregulated growing market where there is lack of effective quality control.[75] Lack of strict guidelines is the main issue on the assessment of safety and efficacy, quality control, safety monitoring, and knowledge of TM/complementary and alternative medicine. In the United States, herbal medicines are regulated under dietary supplements. Similarly in the Europe, the European Medicines Evaluation Agency (EMEA) defines *herbal drugs as the whole, fragmented or cut, plants, parts of plants, algae, fungi, lichen in an unprocessed state usually in dried form or a fresh*. Under some regulatory system, they are categorized as food or food supplement whereas some countries such as Sri Lanka, Bhutan do not have regulatory system for herbal drug development.[72] Since, the regulation and legislation of herbal medicines has been enacted in only few countries, most countries do not have any proper regulation of botanicals, and the quality of herbal products sold in market is generally not guaranteed.[87] In view of this, there is an urgent need for the development of national policies for standardization, the establishment of regulatory mechanisms to control the safety and quality of herbal medicinal products for wider use of TMs, particularly in developed countries.

1.2.9 COMMON CHALLENGES

Absence of various reference standards for identification

- Active constituents are diverse and unknown quality of various batches difficult to control and maintain.
- Although there is availability of modern analytical technique, yet it is difficult to establish quality control parameter.
- Constraints on clinical trial and availability of people
- Difficult to evaluate drug interaction
- Inadequate pharmacological, toxicological, and clinical documentation
- Lack of facility

- Lack of research and development on product and process development[83]
- Lack of research on development of herbal drugs
- Lack of trained personnel and equipment
- No consistency in product because of batch to batch variation. It might occur due to lack of knowledge regarding collection process.
- Safety and efficacy assessment
- The availability and consistency in quality of raw material frequently faced problems.
- To regulate manufacturing practices and quality standards there is no effective machinery.
- Variability of the constituents of herbs due to genetic, cultural, and environmental factors make the use of herbal preparation more challenging.

1.3 OPPORTUNITIES IN HERBAL DRUG DEVELOPMENT

1.3.1 QUALITY CONTROL TOOLS FOR STANDARDIZATION

WHO under *"General Guidelines for methodologies on Research and Evaluation of TM"* issued a statement concerning TM. It says *"Despite its existence and continued use over many centuries, its popularity and extensive use during the last decade, TM has not been officially recognized in most countries"*.[95] The reasons for the lack of research data are not only due to healthcare policies, but also due to lack of adequate or accepted research methodology for evaluating TMs. Quality control (QC) and the standardization of herbal extracts are very important to protect the integrity of herbal extracts for pharmaceutical applications. It also forms part of a prerequisite for the reproducibility of the effect of active ingredients from one batch to another. Several international regulatory agencies (such as WHO, USFDA, and EMEA) recommend such studies at various stages of herbal drug development.[64,66] WHO guidelines on Good Agricultural and Collection Practices and Good Manufacturing Practices regulation can help to prevent contamination and adulteration.[97] In general, the methods for quality control of herbal medicines involve macroscopic, microscopic, physicochemical (qualitative and quantitative), toxicological, and analytical inspection. For example, for simple comparing, clustering, and principal component analysis, widely used techniques are high-performance liquid chromatography (HPLC), gas chromatography (GC), and finger-print spectra.[9,34,39]

1.3.2 QUALITY CONTROL PARAMETERS FOR HERBAL DRUGS (FIG. 1.2)

1.3.2.1 MORPHOLOGY AND ORGANOLEPTIC EVALUATION

Morphology and organoleptic evaluation is necessary for differentiation, authentication, standardization, and identification of drug. Parameters such as color, taste, odor, texture, fracture, size are studied in this evaluation.[28,49]

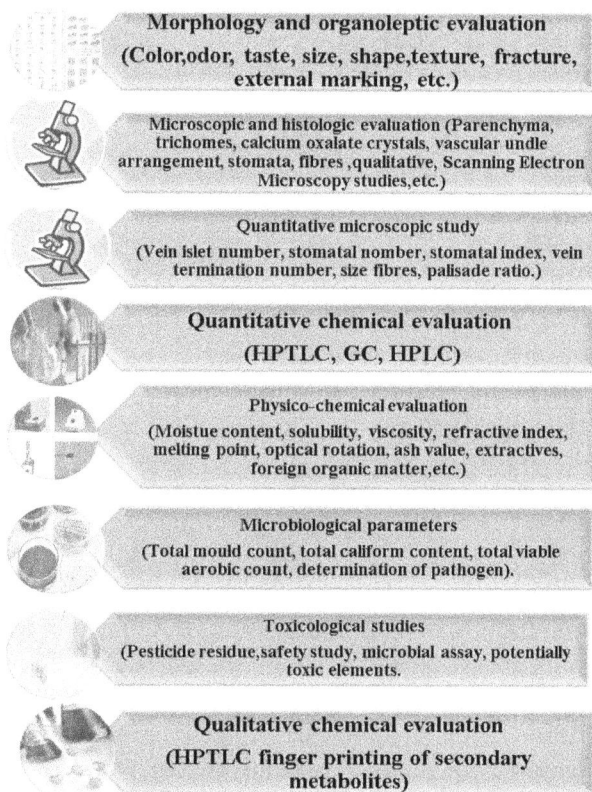

FIGURE 1.2 **(See color insert.)** Parameters for standardization and quality evaluation of herbal drugs.

1.3.2.2   MICROSCOPIC AND HISTOLOGICAL EVALUATION

It helps for more detailed examination of drugs by their own histological characteristics and to check for their quality and purity. It is
simple method of evaluation due to easy availability of techniques and
cost-effectiveness.[76,98]

1.3.2.3   QUANTITATIVE MICROSCOPIC STUDY

It involves quantitative microscopy and linear measurement. Various
parameters that are studied in this are as follows:

- **Palisade ratio** is defined as an average number of palisade cells
  beneath each epidermal cell. It can be determined with powdered
  drugs.[28,47]
- **Stomatal number** refers to the number of stomata per millimetre
  square of epidermis of the leaf.[28,47]
- **Stomatal index** is the percentage which the number of stomata
  forms to the total number of epidermal cells. Each stoma is counted
  as one cell.[28,47]
- **Vein islet number** refers to the number of vein islets per millimetre
  square of the leaf surface midway between the midrib and margin.[28,47]
- **Vein termination number** is defined as the number of vein islets
  termination per millimetre square of the leaf surface midway
  between midrib and margin.[3,28,47]

1.3.2.4   PHYSICOCHEMICAL EVALUATION

- **Ash value** is criterion to judge the identity and purity of crude drug.
  It is the residue remained after complete incineration which simply
  represents inorganic salt naturally present in drug or intentionally
  added to it as a form of adulterant.
- **Extractive value** refers to extract obtained from crude drug is an
  inactive form of chemical constituents present in it.[65]
- **Melting point** refers to the parameter that is useful to judge the
  purity of crude drug. The pure drug has very sharp and constant
  melting point.

- **Moisture content** is quantity of water present in given raw material. It is generally determined by oven drying or air drying. During storage initial moisture content aid to determine quality of herbal raw material.[70]
- **Refractive index (RI)** refers to when the beam of light changes its path while passing from one medium to another medium with different density.
- **Solubility** refers to the presence of adulterant in crude drug indicated by solubility.
- **Viscosity** is a state of being thick, sticky, and semifluid in consistency due to internal friction. It is a good tool for physical evaluation of liquid and semisolid samples.

1.3.2.5 QUALITATIVE CHEMICAL EVALUATION

The qualitative chemical evaluation includes the identification and characterization of crude drug in relation to phytochemicals constituents by using high-performance thin-layer chromatography (HPTLC) fingerprinting as a tool.

1.3.2.6 MICROBIOLOGICAL PARAMETER

Microbiological studies are carried out to determine viable aerobic count as well as total mold and coliform content in herbal product.

1.3.2.7 TOXICOLOGICAL STUDY

This study helps to estimate pesticide residue, toxic elements, and toxins by safety studies in animal.

1.3.2.8 QUANTITATIVE CHEMICAL EVALUATION

It involves determination of various classes of compounds by using various analytical techniques such as HPTLC, GC, and HPLC.

1.3.3   METHODS OF STANDARDIZATION

There are various methods of standardization (Fig. 1.3) that are catego-rized as conventional methods and modern methods.

1.3.3.1   CONVENTIONAL METHODS OF STANDARDIZATION

There are seven conventional methods (Fig. 1.3) that are described here which play important role in standardization of herbal drugs.

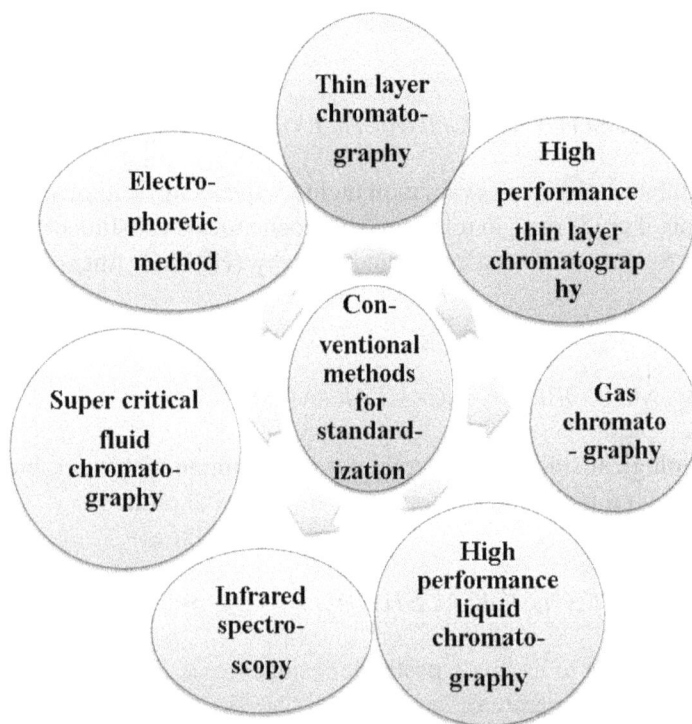

FIGURE 1.3   Conventional methods for standardization.

1.3.3.1.1   Thin-Layer Chromatography (TLC)

It is a separation technique used for qualitative and quantitative estimation of compounds and complex mixtures. Thin-layer chromatography (TLC) provide very valuable information to establish identity and purity of plant material.[24] Methods developed using TLC can be employed for multiple sample analysis and has huge possibilities in analyzing herbal medicine. It is very convenient method to detect the impurity and adulteration.

1.3.3.1.2 High-Performance Thin-Layer Chromatography (HPTLC)

HPTLC enables us to handle the most complicated separation. HPTLC method has been reported for quantitative estimation of many markers in various plants such as swertiamarin in different marketed polyherbal formulations.[13] Simultaneous estimation of diosgenin and levodopa has been done by HPTLC. The proposed HPTLC method has provided faster and effective quantitative control for analysis of diosgenin and levodopa.[69] Advantages of HPTLC include: extremely flexible design, multiple detections can takes place at a time, cost effective and time saving, process monitoring is very easy, results can be easily compared.

1.3.3.1.3 High-Performance Liquid Chromatography (HPLC)

HPLC is fully automated with high rate of precision, resolution, selectivity, and sensitivity.[4] Major advantages are: HPLC can be hyphenated with different detector such as ultraviolet (UV) for UV-absorbing compounds, diode array detector (DAD) for herbal fingerprinting, evaporative light-scattering detector, and chemiluminescence detector for non-UV absorbance, RI detector for terpenes and carbohydrates, nuclear magnetic resonance (NMR) for metabolic profiling, mass spectrometry for identification of separated compound.

1.3.3.1.4 Gas Chromatography (GC)

GC is a well-established analytical technique commonly used for characterization, isolation, and quantization of volatile components.[4] GC analysis of volatile oil has number of applications such as obtaining a fingerprint and detection of impurity.[61]

1.3.3.1.5 Fourier-Transform Infrared Spectroscopy

Fourier-transform infrared spectroscopy with statistical analysis is applied to identify and discriminate herbal medicines. Near and mid-IR techniques can be used for immediate determination of active components.

1.3.3.1.6 Supercritical Fluid Chromatography (SFC)

Supercritical fluid chromatography (SFC) allows separation and determination of group of compound which cannot be handled by either GC or liquid chromatography (LC).[29,59,91] SFC can be applied for materials such as natural product, drugs, food, and pesticide.

1.3.3.1.7 Electrophoretic Method

Capillary electrophoresis (CE) was introduced in early 1980s. It is a powerful separation tool. CE has been identified as an important alternative or complementary tool in field of herbal drug analysis. CE is economical technique and has many advantages such as small injection volume, high efficiency, and short analysis time, which can be useful in rapid and efficient determination of active components in complex system. Several studies have been reported on flavonoids and alkaloids by this method.[12,54]

1.3.3.2 MODERN METHODS FOR STANDARDIZATION OF HERBAL DRUGS (FIG. 1.4)

1.3.3.2.1 DNA Fingerprinting

This technique is useful to identify phytochemicals, which are indistinguishable from adulterant or substituted one. Literature indicates that DNA fingerprint region remains same irrespective of the plant part, whereas chemical content will vary with plant part in use. DNA markers are helpful to identify cells, individuals, species, and so forth, which are responsible to produce normal functioning of proteins to replace defective one. These markers are helpful in treatment of various diseases and helpful to distinguish them.[69]

- DNA-based marker helps to distinguish different species such as *Taxus wallichiana, Azadirchta indica, Juniperu scommunis, Andrographis paniculata* collected from different geographical origin.[13]
- Inter simple sequence repeats is polymerase chain reaction-based unique application and inexpensive popular technique of DNA fingerprinting which include DNA tapping, detection of clonal variation, phytogenetic analysis, detection of genomic instability, and assessment of hybridization.[69]
- DNA sequencing used for identification of various species due to transverse, insertion, or deletion.[92]

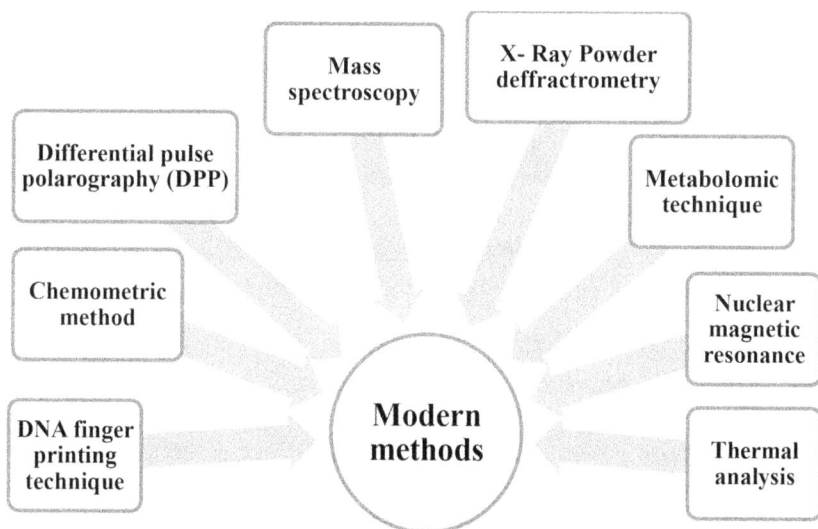

FIGURE 1.4   (See color insert.) Modern methods for standardization of herbal drugs.

1.3.3.2.2   Chemometric Method

Chemometrics is a statistical approach to analyze instrumental data. It gives more faster and precise assessment of components present in product or physical or sensory properties. These components under assessment can be carbohydrate, fat, fiber, or dairy product. Applications include: classification of sample into several categories and prediction of property of interested compound.

1.3.3.2.3 Metabolomic Technique

Metabolomic study involves qualitative and quantitative estimation of whole metabolite found in the cellular organism system and many compounds such as extracts of tissue, body fluids such as serum, urine in case of humans. It is very important tool for quality determination of botanicals; and for characterization of complex phenotype affected by both genetic and environmental factors.[61] There are four major steps of analysis of unknown compounds present in herbal medicine:

- **Metabolomic determination**: qualitative and quantitative data for estimation of unknown compound or metabolic pathway.
- **Metabolomic examination**: qualitative and quantitative estimation of whole metabolite.
- **Metabolomic fingerprinting**: process of sample classification by the rapid global investigation.
- **Targeted investigation of the compound**: in this quantification and determination of definite metabolite takes place.

1.3.3.2.4 Differential Pulse Polarography

Differential pulse polarography is used to study trace amount of chemicals with very small detection limits and some heavy metals such as cadmium, lead, zinc, copper, and iron. Content of total hypericin has been determined in *Hypericum perforatum*[88] using this method.[6,57]

1.3.3.2.5 X-Ray Powder Diffractrometry

This technique is used to identify minerals, crystalline material, and metal that are present in herbal formulation.

1.3.3.2.6   Thermal Analysis

The chemical and physical changes in product are studied by thermo-gravimetric analysis (TGA), differential thermal analysis, and differential scanning calorimetry. It helps to study pre-formulation or drug-excipient compatibility. TGA is operated under controlled conditions to analyze alcohol content in various herbal formulations such as Asava and Arishta and metals in Bhasmas.[82]

1.3.3.2.7   Nuclear Magnetic Resonance (NMR)

NMR is a powerful and time-saving method for separation and structural elucidation of unknown compound and quality control of herbal medicine.[36]

1.3.3.2.8   Mass Spectrometry

Mass spectrometry is an effective tool for quality control of herbal medicines and it helps to detect both active and nonactive ingredients with their qualitative and quantitative information.[94] It is useful to study aspects, such as high-throughput metabolomics, explosives detection, natural products discovery;[16] and plant metabolomics.[19]

1.3.4   HYPHENATED TECHNIQUE

TLC, HPTLC, HPLC, gas chromatography–mass spectroscopy (GC–MS), and spectrophotometer, and so forth are still not sensitive or reproducible methods enough for detecting trace-level parent natural products or their metabolites in biological fluids. Many alkaloids, flavonoids, or sesquiterpenoids are very unstable and decompose fast in the human body.[23,62,96] It is required to determine their metabolic fates including structure characterization and quantitative analysis of metabolites. Many terpenoid saponins have a narrow therapeutic index with serious side effects,[15] which makes it essential to accurately measure them in blood samples from a safety point of view. Since the introduction of new approaches of hyphenated chromatography and spectrometry such as HPLC–DAD, GC–MS,

CE–DAD, HPLC–MS, and HPLC–NMR provides the additional spectral information, which is a major contributing tool for quality control of herbal medicines.[55] LC coupled to MS technique is a feasible technique in the identification and quantification of natural products (flavonoids, alkaloids, saponins, and sesquiterpenoids) in biological fluids.

1.3.5 OPPORTUNITIES RELATED TO REGULATORY GUIDELINES

Recently, several European countries adopted the *European Directive on Traditional Herbal Medicinal Products* (formally The Directive 2004/24/EC amending) to assure the quality and regulation of herbal products. In the United Kingdom, this is implemented by the *Medicines and Health care products Regulatory Agency* formed in April 2003 by the merger of the *Medicines Control Agency* and *Medical Device Agency*.[45] Indian government has taken different initiatives apart from taking legislative measures, by strengthening laboratories for testing TMs, which is expected to promote TMs in national and international markets. They are also providing attention to intellectual property issues to stop biopiracy of indigenous traditional knowledge and/or natural resources that are used in TM products.[89]

1.3.6 NOVEL APPROACHES TO OVERCOME BIOAVAILABILITY ISSUES OF HERBAL MEDICINES

To overcome this poor solubility or erratic bioavailability issues, various approaches have been developed such as pharmaceutical salt, prodrugs, pH adjustment, complexation, emulsions, micellization/surfactant system.

Novel drug delivery[42] approaches such as development of novel dosage forms (polymeric nanoparticles and nanocapsules, liposomes, solid lipid nanoparticles, phytosomes and nanoemulsions, and so forth) have number of advantages on herbal drugs, including: enhancement of solubility and bioavailability, protection from toxicity, enhancement of pharmacological activity, and improvement of stability.[41,73] There are many reports, where researchers have developed various novel formulations in order to overcome bioavailability issues. Few of these have been enlisted in Table 1.5.

TABLE 1.5 Novel Approaches Developed For Phytoconstituents and Extracts.

Plant extract/ phytoconstituents	Novel approach developed	Application of novel approach	Activity reported	Reference
Colchicine	Niosomes	High encapsulation of colchicines with good stability, prolonged release, and less side effects	Cytotoxicity	[1, 78]
Curcumin	Nanoparticles	Specific targeted delivery by antibody-coupled particles	Anticancer	[27, 37, 42, 71]
Glycyrrhiza glabra L. leaf extract	Phytocomplexes	GIT exposure enhances indirect improvement of bioavailability	Antioxidant, anti-inflammatory, antigenotoxic	[100]
Plant extract of Sapinduse marginatus	Nanoparticles	Nanosaponin induces dose-dependent cancer cell death with lower toxicity on normal cells	Cytotoxicity to cancer cell line	[93]
Vincristine	Transferosomes	Increase entrapment efficiency and skin permeation	Anticancer	[14]
Zedoary turmeric oil from Rhizoma zedoariae	Microspheres	Sustained release and higher bioavailability	Hepatoprotective, anticancer	[99]

1.4 SUMMARY

Since time immemorial, nature has bestowed its benefits on mankind by providing food, fiber, shelter, and medicines in many different forms that are still being used in healthcare management in many underdeveloped and developing countries. Considering widespread use of herbal medicines, standardization and validation are necessary for maintaining quality of herbal drugs. Quality control and the standardization of herbal extracts are important to protect the integrity of the herbal extracts for pharmaceutical quality. Several qualitative tests such as morphology, macroscopy, and microscopy have been used routinely for monitoring the quality of raw materials along with processed extracts during herbal drug development. Chemical evaluation of marker compounds and total fingerprint analysis are considered quantitative and semiquantitative measures, respectively, for monitoring the quality of several herbal drugs. Marker compounds suggest identification and quantitative measurement of known marker compounds in the raw material, processed extracts, and finished products using chromatographic techniques such as HPTLC, HPLC, and GC.

In fingerprint analysis, raw material or processed extracts are considered as drug in its entirety. Chromatographic fingerprint with known and unknown components can be developed, validated, and used as quality control tool for raw, processed, and finished products. The fate of herbal drugs especially the standardized extracts or the isolated active phytoconstituents from the extracts after their oral administration have not been explored much yet. Pharmacokinetic parameters (such as ADME of phytoconstituents after oral administration of extracts or isolated compounds) provide valuable information on bioavailability in plasma and/or targeted tissues and therefore, clinical utility of herbal drugs. New emerging categories may offer new opportunities to scientists, practitioners, and industries to develop research-based drugs from botanical sources, which can get global acceptance from modern medicine.

KEYWORDS

- absorption
- bioavailability
- DNA fingerprinting
- efficacy
- EMEA
- excretion
- GACP
- GMP
- niosomes
- pharmacodynamic interaction
- pharmacokinetic
- phytoconstituent
- phytopharmaceuticals
- plant extract
- poor bioavailability
- safety pharmacology
- standardization
- transferosomes

REFERENCES

1. Abe, E. A Novel LC-ESI-MS-MS Method for Sensitive Quantification of Colchicine in Human Plasma: Application to Two Case Reports. *J. Anal. Toxicol.* **2006,** *30*(3), 210–215.
2. Andrade, A. S. Pharmacokinetic and Metabolic Effects of American Ginseng (*Panax quinquefolius*) in Healthy Volunteers Receiving the HIV Protease Inhibitor Indinavir. *BMC Complementary Altern. Med.* **2008,** *8*(1), 50.
3. Arora, D.; Sharma, A. Pharmacognostic and Phytochemical Studies of *Stellaria media* Linn. *J. Pharmcological Sci. Res.* **2012,** *4*(5), 1819–1822.
4. Bansal, A. Chemometrics: A New Scenario in Herbal Drug Standardization. *J. Pharm. Anal.* **2014,** *4*(4), 223–233.
5. Bauer, R. Quality Criteria and Standardization of Phytopharmaceuticals: Can Acceptable Drug Standards be Achieved? *Drug Inf. J.* **1998,** *32*(1), 101–110.
6. Biber, A. Oral Bioavailability of Hyperforin from Hypericum Extracts in Rats and Human Volunteers. *Pharmacopsychiatry* **1998,** *31*(S1), 36–43.
7. Borins, M. The Dangers of Using Herbs: What Your Patients Need to Know. *Postgrad. Med.* **1998,** *104*(1), 91–100.
8. Boullata, J. I.; Nace, A. M. Safety Issues with Herbal Medicine. *Pharmacother: J. Human Pharmacol. Drug Ther.* **2000,** *20*(3), 257-269.
9. Cai, Z.; Sinhababu, A. K.; Harrelson, S. Simultaneous Quantitative Cassette Analysis of Drugs and Detection of Their Metabolites by High Performance Liquid Chromatography/ion Trap Mass Spectrometry. *Rapid Commun. Mass Spectrom.* **2000,** *14*(18), 1637–1643.
10. Calixto, J. Efficacy, Safety, Quality Control, Marketing and Regulatory Guidelines for Herbal Medicines (Phytotherapeutic Agents). *Braz. J. Med. Biol. Res.* **2000,** *33*(2), 179–189.

11. Chen, L. C. Pharmacokinetic Study of Paeoniflorin in Mice After Oral Administration of Paeoniae Radix Extract. *J. Chromatogr. B Biomed. Sci. Appl.* **1999,** *735*(1), 33–40.

12. Chen, J. Quality Control of a Herbal Medicinal Preparation Using High-Performance Liquid Chromatographic and Capillary Electrophoretic Methods. *J. Pharm. Biomed. Anal.* **2011,** *55*(1), 206–210.

13. Choudhary, N.; Sekho, B. S. An Overview of Advances in the Standardization of Herbal Drugs. *J. Pharm. Educ. Res.* **2011,** *2*(2), 55.

14. Colalto, C. Herbal Interactions on Absorption of Drugs: Mechanisms of Action and Clinical Risk Assessment. *Pharmacol. Res.* **2010,** *62*(3), 207–227.

15. Collett, A. Comparison of P-Glycoprotein-Mediated Drug–Digoxin Interactions in Caco-2 with Human and Rodent Intestine: Relevance to in vivo Prediction. *Eur. J. Pharm. Sci.* **2005,** *26*(5), 386–393.

16. Cooks, R. G. Ambient Mass Spectrometry. *Science* **2006,** *311*(5767), 1566–1570.

17. Cragg, G. M.; Newman, D. J. Biodiversity: A Continuing Source of Novel Drug Leads. *Pure Appl. Chem.* **2005,** *77*(1), 7–24.

18. De Smet, P. A.; Brouwers, J. R. Pharmacokinetic Evaluation of Herbal Remedies. *Clin. Pharmacokinet.* **1997,** *32*(6), 427–436.

19. Dettmer, K.; Aronav, P. A.; Hammock, B. D. Mass Spectrometry—Based Metabolomics. *Mass Spectrom. Rev.* **2007,** *26*(1), 51–78.

20. Dubey, N.; Kumar, R.; Tripathi, P. Global Promotion of Herbal Medicine: India's Opportunity. *Curr. Sci.* **2004,** *86*(1), 37–41.

21. Eid, H. M.; Haddad, P. S. Mechanisms of Action of Indigenous Antidiabetic Plants from the Boreal Forest of Northeastern Canada. *Adv. Endocrinol.* **2014,** *2014,* 1–11.

22. Ernst, E. The Efficacy of Herbal Medicine—An Overview. *Fundam. Clin. Pharmacol.* **2005,** *19*(4), 405–409.

23. Felgines, C. Blackberry Anthocyanins are Mainly Recovered from Urine as Methylated and Glucuronidated Conjugates in Humans. *J. Agric. Food Chem.* **2005,** *53*(20), 7721–7727.

24. Folashade, O.; Omoregie, H.; Ochogu, P. Standardization of Herbal Medicines—A Review. *Int. J. Biodiversity Conserv.* **2012,** *4*(3), 101–112.

25. Fugh-Berman, A. Herb-Drug Interactions. *Lancet* **2000,** *355*(9198), 134–138.

26. Fugh-Berman, A.; Ernst, E. Herb–Drug Interactions: Review and Assessment of Report Reliability. *Br. J. Clin. Pharmacol.* **2001,** *52*(5), 587–595.

27. Goel, A.; Kunnumakkara, A. B.; Aggarwal, B. B. Curcumin as "Curecumin": from Kitchen to Clinic. *Biochem. pharmacol.* **2008,** *75*(4), 787–809.

28. Gokhale, S.; Kokate, C.; Purohit, A. *Text Book of Pharmacognosy;* Nirali Prakshan: Pune, India, 1993; pp 345-348.

29. Guo, J.; Xu, Q.; Chen, T. Quantitative Determination of Astilbin in Rabbit Plasma by Liquid Chromatography. *J. Chromatogr. B* **2004,** *805*(2), 357–360.

30. Gurib-Fakim, A. Medicinal Plants: Traditions of Yesterday and Drugs of Tomorrow. *Mol. Aspects Med.* **2006,** *27*(1), 1–93.

31. Gurley, B. J. In vivo Effects of Goldenseal, Kava Kava, Black Cohosh, and Valerian on Human Cytochrome P450 1A2, 2D6, 2E1, and 3A4/5 Phenotypes. *Clin. Pharmacol. Ther.* **2005,** *77*(5), 415–426.

32. Gurley, B. J. Effect of Goldenseal (*Hydrastis canadensis*) and Kava Kava (*Piper methysticum*) Supplementation on Digoxin Pharmacokinetics in Humans. *Drug Metab. Dispos.* **2007**, *35*(2), 240–245.

33. Handa, S. Medicinal Plants-Priorities in Indian Medicines Diverse Studies and Implications. *Suppl. Cultiv. Util. Med. Plants* **1996,** 33–51.

34. Hao, X. Validation of an HPLC Method for the Determination of Scutellarin in Rat Plasma and its Pharmacokinetics. *J. Pharm. Biomed. Anal.* **2005,** *38*(2), 360–363.

35. Hattori, M. Metabolism of Glycyrrhizin by Human Intestinal Flora, II: Isolation and Characterization of Human Intestinal Bacteria Capable of Metabolizing Glycyrrhizin and Related Compounds. *Chem. Pharm. Bull.* **1985,** *33*(1), 210–217.

36. Heyman, H. M.; Meyer, J. J. M. NMR-Based Metabolomics as a Quality Control Tool for Herbal Products. *S. Afr. J. Bot.* **2012,** *82,* 21–32.

37. Holder, G. M.; Plummer, J. L.; Ryan, A. J. The Metabolism and Excretion of Curcumin (1,7-bis-(4-Hydroxy-3-Methoxyphenyl)-1,6-Heptadiene-3,5-Dione) in the Rat. *Xenobiotica* **1978,** *8*(12), 761–768.

38. Hua, X. Determination of Pinostrobin in Rat Plasma by LC–MS/MS: Application to Pharmacokinetics. *J. Pharm. Biomed. Anal.* **2011,** *56*(4), 841–845.

39. Hua-Bin, Z. Quality Control Methodology and Their Application in Analysis on HPLC Fingerprint Spectra of Herbal Medicines. *Chromatogr. Res. Int.* **2011,** *2012,* 1–12.

40. Hussain, K. Bioactive Markers Based Pharmacokinetic Evaluation of Extracts of a Traditional Medicinal Plant, Piper Sarmentosum. *Evidence Based Complementary Altern. Med.* **2011,** *2011,* 1–7.

41. Iosio, T. Oral Bioavailability of Silymarin Phytocomplex Formulated as Self-Emulsifying Pellets. *Phytomedicine* **2011,** *18*(6), 505–512.

42. Jachak, S. M.; Saklani, A. Challenges and Opportunities in Drug Discovery from Plants. *Curr. Sci. Bangalore* **2007,** *92*(9), 1251.

43. Joo, K. M. Pharmacokinetic Study of Ginsenoside Re with Pure Ginsenoside Re and Ginseng Berry Extracts in Mouse Using Ultra Performance Liquid Chromatography/mass Spectrometric Method. *J. Pharm. Biomed. Anal.* **2010,** *51*(1), 278–283.

44. Jordan, S. A.; Cunningham, D. G.; Marles, K. J. Assessment of Herbal Medicinal Products: Challenges, and Opportunities to Increase the Knowledge Base for Safety Assessment. *Toxicol. Appl. Pharmacol.* **2010,** *243*(2), 198–216.

45. Kamboj, V. P. Herbal Medicine. *Curr. Sci. Bangalore* **2000,** *78*(1), 35–38.

46. Kanaze, F. I.; Bounartzi, M. I.; Niopas, I. Validated HPLC Determination of the Favone Aglycone Diosmetin in Human Plasma. *Biomed. Chromatogr.* **2004,** *18*(10), 800–804.

47. Khandelwal, K. R. *Practical Pharmacognosy;* Pragati Books Pvt. Ltd.: New Delhi—India, 2008; pp 146–148.

48. Kumar, V. An overview of herbal medicine. *Int. J. Ph. Sci.* **2009,** *1*(1), 1–20.

49. Kumar, D. Pharmacognostic Evaluation of Leaf and Root Bark of Holoptelea Integrifolia Roxb. *Asian Pac. J. Trop. Biomed.* **2012,** *2*(3), 169–175.

50. Lahlou, M. The Success of Natural Products in Drug Discovery. *Pharmacol. Pharm.* **2013,** *4*(3A), 17.

51. Lee, L. S. Possible Differential Induction of Phase 2 Enzyme and Antioxidant Pathways by American Ginseng, *Panax quinquefolius*. *J. Clin. Pharmacol.* **2008,** *48*(5), 599–609.

52. Lee, J. H.; Kim, J. M.; Kim, C. Pharmacokinetic Analysis of Rhein in *Rheum undulatum* L. *J. Ethnopharmacol.* **2003**, *84*(1), 5-9.

53. Lei, F. Pharmacokinetic Study of Ellagic Acid in rat After Oral Administration of Pomegranate Leaf Extract. *J. Chromatogr. B* **2003**, *796*(1), 189–194.

54. Li, H. L. Bioavailabilty and Pharmacokinetics of Four Active Alkaloids of Traditional Chinese Medicine Yanhuanglian in Rats Following Intravenous and Oral Administration. *J. Pharm. Biomed. Anal.* **2006**, *41*(4), 1342–1346.

55. Liang, Y. Z.; Xie, P.; Chan, L. Quality Control of Herbal Medicines. *J. Chromatogr. B* **2004**, *812*(1), 53–70.

56. Markowitz, J. S. Multiple-Dose Administration of Ginkgo Biloba did not Affect Cytochrome P-450 2D6 or 3A4 Activity in Normal Volunteers. *J. Clin. Psychopharmacol.* **2003**, *23*(6), 576–581.

57. Michelitsch, A. Determination of Hypericin in Herbal Medicine Products by Differential Pulse Polarography. *Phytochem. Anal.* **2000**, *11*(1), 41–44.

58. Mills, S.; Bone, K. *Principles and Practice of Phytotherapy: Modern Herbal Medicine;* Churchill Livingstone: London, United Kingdom, 1999; pp 945–948.

59. Molavi, O. Development of a Sensitive and Specific Liquid Chromatography/Mass Spectrometry Method for the Quantification of Cucurbitacin I (JSI-124) in Rat Plasma. *J. Pharm. Pharm. Sci.* **2006**, *9*(2), 158–164.

60. Moltó, J. Herb-Drug Interaction Between *Echinacea purpurea* and Darunavir-Ritonavir in HIV-Infected Patients. *Antimicrob. Agents Chemother.* **2011**, *55*(1), 326–330.

61. Mukherjee, P. K. *Quality Control of Herbal Drugs: An Approach to Evaluation of Botanicals;* Business Horizons Publication: New Delhi, 2002; pp 112–120.

62. Naik, H. Development and Validation of a High-Performance Liquid Chromatography–Mass Spectroscopy Assay for Determination of Artesunate and Dihydroartemisinin in Human Plasma. *J. Chromatogr. B* **2005**, *816*(1), 233–242.

63. Ogbole, E. A. Phytochemical Screening and in vivo Antiplasmodial Sensitivity Study of Locally Cultivated *Artemisia annua* Leaf Extract Against *Plasmodium berghei*. *Am. J. Ethnomed.* **2014**, *1*(1), 42–49.

64. Olivier, M. T.; Muganza, F. M. Phytochemical Screening, Antioxidant and Antibacterial Activities of Ethanol Extracts of *Asparagus suaveolens* Aerial Parts. *S. Afr. J. Bot.* **2017**, *108,* 41-46.

65. Pandey, A.; Tripathi, S. Concept of Standardization, Extraction and Pre-Phytochemical Screening Strategies for Herbal Drug. *J. Pharmacogn. Phytochem.* **2014**, *2*(5), 115–119.

66. Patil, D. Physicochemical Stability and Biological Activity of *Withania somnifera* Extract Under Real-Time and Accelerated Storage Conditions. *Planta Med.* **2010**, *76*(05), 481–488.

67. Qian, T.; Jiang, Z. H.; Cai, Z. High-Performance Liquid Chromatography Coupled with Tandem Mass Spectrometry Applied for Metabolic Study of Ginsenoside Rb 1 on Rat. *Anal. Biochem.* **2006**, *352*(1), 87–96.

68. Rajnarayana, K. Study on the Influence of Silymarin Pretreatment on Metabolism and Disposition of Metronidazole. *Arzneimittelforschung* **2004**, *54*(02), 109–113.

69. Rasheed, A.; Sravya-Reddy, B.; Rosa, C. A Review on Standardization of Herbal Formulation. *Int. J. Phytother.* **2012**, *2,* 74–88.

70. Razak, N. A. Effect of Initial Leaf Moisture Content on the Herbal Quality Parameter of *Orthosiphon stamineus* Dried Leaf During Storage. *Int. J. Agric. Innovations Res.* **2014,** *2*(6), 1131–1136.

71. Rejinold, N. S. Saponin-Loaded Chitosan Nanoparticles and Their Cytotoxicity to Cancer Cell Lines in vitro. *Carbohydr. Polym.* **2011,** *84*(1), 407–416.

72. Sahoo, N.; Manchikanti, P.; Dey, S. Herbal Drugs: Standards and Regulation. *Fitoterapia* **2010,** *81*(6), 462–471.

73. Saraf, S. Applications of Novel Drug Delivery System for Herbal Formulations. *Fitoterapia* **2010,** *81*(7), 680–689.

74. Schubert, S. Y.; Lansky, E. P.; Neeman, I. Antioxidant and Eicosanoid Enzyme Inhibition Properties of Pomegranate Seed Oil and Fermented Juice Flavonoids. *J. Ethnopharmacol.* **1999,** *66*(1), 11–17.

75. Sen, S.; Chakraborty, R.; De, B. Challenges and Opportunities in the Advancement of Herbal Medicine: India's Position and Role in a Global Context. *Herb. Med.* **2011a,** *1*(3), 67–75.

76. Sen, S.; Chakraborty, R.; De, B. Ethnobotanical Survey of Medicinal Plants Used by Ethnic People in West and South District of Tripura, India. *J. For. Res.* **2011b,** *22*(3), 417–426

77. Shan, J. J. Challenges in Natural Health Product Research: The Importance of Standardization. *West. Pharmacol. Soc.* **1998,** *54,* 24–30.

78. Siracusa, L. Phytocomplexes from Liquorice (*Glycyrrhiza glabra* L.) Leaves: Chemical Characterization and Evaluation of Their Antioxidant, Anti-Genotoxic and Anti-Inflammatory Activity. *Fitoterapia* **2011,** *82*(4), 546–556.

79. Smita, L.; Chougule, A. Need of Herbal Drug Standardzation. *Int. Ayurvedic Med. J.* **2015,** *3*(3), 574–577.

80. Staffeldt, B. Pharmacokinetics of Hypericin and Pseudohypericin After Oral Intake of the *Hypericum perforatum* Extract LI 160 in Healthy Volunteers. *J. Geriatr. Gsychiatry Neurol.* **1994,** *7*(1), S47–53.

81. Suryawanshi, S.; Asthana, R.; Gupta, R. Simultaneous Estimation of Mangiferin and Four Secoiridoid Glycosides in Rat Plasma Using Liquid Chromatography Tandem Mass Spectrometry and its Application to Pharmacokinetic Study of Herbal Preparation. *J. Chromatogr. B* **2007,** *858*(1), 211–219.

82. Thakkar, K. Recent Advances in Herbal Drug Standardization—A Review. *Int. J. Adv. Pharm. Res.* **2013,** *4*(8), 2130–2138.

83. Thillaivanan, S.; Samraj, K. Challenges, Constraints and Opportunities in Herbal Medicines—A Review. *Int. J. Herb. Med.* **2014,** *2*(1), 21–24.

84. Tilburt, J. C.; Kaptchuk, T. J. Herbal Medicine Research and Global Health: An Ethical Analysis. *Bull. W. H. O.* **2008,** *86*(8), 594–599.

85. Vibha, J. Study on Pharmacokinetics and Therapeutic Efficacy of *Glycyrrhiza glabra*: A Miracle Medicinal Herb. *Bot. Res. Int.* **2009,** *2*(3), 157–163.

86. Wang, S. Simultaneous Determination of Oxymatrine and its Active Metabolite Matrine in Dog Plasma by Liquid Chromatography–Mass Spectrometry and its Application to Pharmacokinetic Studies. *J. Chromatogr. B* **2005,** *817*(2), 319–325.

87. Warude, D.; Patwardhan B. Botanicals: Quality and Regulatory Issues. *J. Sci. Ind. Res.* **2005,** *64,* 83–92.

88. Wiesner, J. Challenges of Safety Evaluation. *J. Ethnopharmacol.* **2014,** *158,* 467–470.

89. World Health Organization (WHO). Traditional Medicine Strategy, 2002–2005.

90. Xie, J. Simultaneous Determination of Ginkgolides A, B, C and Bilobalide in Plasma by LC–MS/MS and its Application to the Pharmacokinetic Study of Ginkgo Biloba Extract in Rats. *J. Chromatogr. B* **2008,** *864*(1), 87–94.

91. Xing, J.; Xie, C.; Lou, H. Recent Applications of Liquid Chromatography–Mass Spectrometry in Natural Products Bioanalysis. *J. Pharm. Biomed. Anal.* **2007,** *44*(2), 368–378.

92. Yadav, N.; Dixit, V. Recent Approaches in Herbal Drug Standardization. *Int. J Integr. Biol.* **2008,** *2*(3), 195–203.

93. You, J. Study of the Preparation of Sustained-Release Microspheres Containing Zedoary Turmeric Oil by the Emulsion–Solvent-Diffusion Method and Evaluation of the Self-Emulsification and Bioavailability of the Oil. *Colloids Surf. B* **2006,** *48*(1), 35–41.

94. Zeng, Z. D. Mass Spectral Profiling: An Effective Tool for Quality Control of Herbal Medicines. *Anal. Chim. Acta* **2007,** *604*(2), 89–98.

95. Zhang, X. *General Guidelines for Methodologies on Research and Evaluation of Traditional Medicine*. Report WHO/EDM/TRM/2000.1; WHO (World Health Organization), Geneva; **2000,** pages 1–71.

96. Zhang, H. G. Separation and Identification of Aconitum Alkaloids and Their Metabolites in Human Urine. *Toxicon* **2005,** *46*(5), 500–506.

97. Zhang, J. Quality of Herbal Medicines: Challenges and Solutions. *Complementary Ther. Med.* **2012,** *20*(1), 100–106.

98. Zhao, Z. Z. Application of Microscopy in Authentication of Chinese Patent Medicine—Bo Ying compound. *Microsc. Res. Tech.* **2005,** *67*(6), 305–311.

99. Zhaowu, Z. Preparation of Matrine Ethosome: Its Percutaneous Permeation in vitro and Anti-Inflammatory Activity in vivo in Rats. *J. Liposome Res.* **2009,** *19*(2), 155–162.

100. Zheng, Y. Preparation and Characterization of Transfersomes of Three Drugs in vitro. Zhongguo Zhong yao za zhi Zhongguo zhongyao zazhi. *China J. Chin. Mater. Med.* **2006,** *31*(9), 728–731.

CHAPTER 2

ANTIMICROBIAL PROPERTIES OF TRADITIONAL MEDICINAL PLANTS: STATUS AND POTENTIAL

V. DURAIPANDIYAN, T. WILLIAM RAJA,
NAIF ABDULLAH AL-DHABI, and
IGNACIMUTHU SAVARIMUTHU

CONTENTS

ABSTRACT

Local healers and benefactors of natural medicines believe that traditional medicines (TMs) are effective against various diseases and provide complete cure for diseases. In recent years, the demand for TMs has been rapidly growing, and people have realized their importance. The study of medicinal plants as antimicrobial agents is necessary to gain insight into the medicinal flora and their real value. Many studies have confirmed the antimicrobial activity of plants.

2.1   INTRODUCTION

Traditional medicine (TM) is a widely-used phrase that describes original medicine derived from nature. It includes various forms, such as traditional Chinese medicine (TCM), Indian Ayurveda, and Arabic Unani medicine.[40] TMs play an essential role in human health care in many developing and developed countries, providing a better route for drug discovery with enhanced characteristics in terms of cost, socioeconomic relations, safety, and efficacy.[28] TM success stories are included in many research articles. Researchers and the drug industry in China have endeavored to improve TCM branded products based on its comprehensive features.[41] TMs are used to find "new entity drugs" and can be considered an attractive strategic option.[21] At present, we follow a "one-drug-fits-all" approach; however, this approach may not be sustainable in the future. The concept of "traditional herbal formulations" has been created by researchers to identify multiple targets, risk factors, or symptoms. Traditional herbal formulations are safe and effective, while offering a faster and more economical alternative.[22]

"Reverse pharmacology" is an innovative approach that mixes living traditional knowledge such as Ayurveda with modern technology, providing better and safer lead drug candidates. In this process, safety is the starting point and markers of efficacy are used for validation purposes. This approach has been utilized in the process of preclinical and clinical research.[38] Practical guidelines for plant-based "drug" development have been framed by the Food and Drug Administration (FDA), as well as other agencies, to cure existing diseases in the world.[16] The importance of TMs has recently been recognized. TMs include 25,000–75,000 plant species that are used to treat various diseases, including microbial diseases and several products are accepted for commercial purposes.[1] Plant-based drugs and traditional herbal formulations can be used to fight against infectious pathogens, effectively killing microorganisms, and inhibiting their growth. Thus, they are helpful for improving human health.

Plants have been shown to have many biologically active compounds, many of which are very effective against microbial diseases.[4] Lemon balm (*Melissa officinalis*), garlic (*Allium sativum*), and tea tree (*Melaleuca alternifolia*) contain antimicrobial agents.[17] The essential oils from these plants can greatly impact the treatment of contagious pathogens found on the skin, in the respiratory system, urinary tract, gastrointestinal tract, and biliary systems. Essential oils from *M. alternifolia* can be used to treat acne and skin infections.[39] Diterpenoid alkaloids are isolated from the *Ranunculaceae* family of plants and have antimicrobial properties.[27] The antibacterial properties of volatile oils from black pepper (*Piper nigrum*), geranium (*Pelargonium graveolens*), oregano (*Origanum vulgare*), cloves (*Syzygium aromaticum*), thyme (*Thymus vulgaris*), and nutmeg (*Myristica fragrans*) have been tested against 25 different genera of bacteria, including plant pathogens, animal pathogens, bacteria known to cause food spoilage, and bacteria associated with food poisoning. Considerable inhibitory effects were observed against all organisms.[7]

Antimicrobial compounds possess properties that are effective against microbial growth and several methods are used to assess the antimicrobial activity. Various tribes have depended upon plant-based drugs for their basic medical needs and diseases have been cured using different parts of medicinal plants. Administration of TMs does not cause side effects. Consequently, TMs are playing a pivotal role in protecting human beings from microorganisms.

This chapter focuses on antimicrobial properties of traditionally-used medicinal plants against pathogenic microbes. The active extracts and pure compounds have been screened by many researchers. This information will help in drug-discovery programs.

2.2 MATERIALS AND METHODS

2.2.1 *ISOLATION OF PLANT COMPOUNDS*

Researchers typically approach locals to gain information before collecting TM plants with antimicrobial activity. The plant parts are chopped into small pieces, dried, and maintained under dark conditions. The dried plant parts are thoroughly ground and then extracted with various solvents. The plant extracts are then used to determine antibacterial activity of the plant. Effective extracts are separated using column chromatography; the active fractions and crystalline compounds are further tested for antimicrobial activity. Based on the results, researchers report plants that are potentially effective against various microorganisms.

2.2.2 VARIOUS METHODS FOR ANTIMICROBIAL ACTIVITY

At present, researchers have developed an interest in deriving antimicrobial agents from various plant extracts and compounds to treat microbial infections. Bioassays such as disk diffusion, well diffusion, agar plug diffusion, broth or agar dilution, and cross-streaking are used to estimate inhibitory zones and to check the effectiveness of plant extracts and compounds.

2.2.3 AGAR DISC DIFFUSION METHOD

Antimicrobial activity can be determined using the disc diffusion method,[26] which is an antimicrobial susceptibility test. Sterile petri dishes containing 20 mL Mueller-Hinton agar (MHA) for bacteria and 20 mL Sabouraud dextrose agar (SDA) for fungi are used in many clinical microbiology laboratories. Certain fastidious bacterial pathogens, such as *Streptococcus* sp., *Haemophilus influenzae*, *Haemophilus parainfluenzae*, *Neisseria gonorrhoeae,* and *Neisseria meningitidis* can be analyzed using this method. Test cultures (104 CFU/mL) are applied using swab sticks on top of gelatinized media and permitted to dry for 10 min; filter paper discs are approximately 6 mm in diameter. The initial concentration of the extract is 1 mg/mL; the initial concentration of the test compound is 250 µg/mL. Tests can be conducted with various concentrations of the crude extract (5, 2.5, and 1.25 mg/disc) in triplicates. Discs having the preferred drug concentrations are maintained on the surface of the medium for 30 min at 37°C for drug diffusion. Negative controls are prepared using the respective solvents and streptomycin (10 µg/disc) is used as a positive control. The petri dishes are incubated for 24 h at 37°C for bacteria and for 48 h at 27°C for fungi. After incubating, inhibition zones are formed around the discs and are measured with transparent rulers (mL). The experiment is repeated twice. This is a simple and low-cost method used to screen various plant extracts, essential oils, and other drugs for antimicrobial activity.[5]

2.2.4 AGAR WELL DIFFUSION METHOD

The agar well diffusion method resembles the agar disc diffusion method in its simplicity. Plant extracts or compounds are used to assess the antimicrobial activity by applying microbial inocula to the surface of an agar plate. Then, a hole of 6–8 mm in diameter is made aseptically with a sterile cork to load the drug (20–100 mL) into the respective wells at a given concentration. Agar plates are then incubated under suitable conditions. The antimicrobial agent binds in the agar medium and inhibits the microbial growth.

2.2.5 AGAR PLUG DIFFUSION METHOD

The agar plug diffusion method is similar to that of the disc diffusion method and includes making an agar culture of the microbial strain and applying tight streaks in a suitable culture medium. Microbial cells express their own molecules, which diffuse into the agar medium. After incubating, a sterile cork-borer is used to cut an agar plot, which is then deposited on another agar plate inoculated with test microorganisms. The substances then diffuse from the plug to the agar medium, forming zones of inhibition.

2.2.6 BROTH DILUTION METHOD

Broth micro- and macro-dilutions are fundamental methods for testing antimicrobial susceptibility. A twofold dilution of the antimicrobial agent (e.g., 1, 2, 4, 8, 16, and 32 mg/mL) is prepared per standard procedures. The microorganisms along with test antimicrobial agents can then be grown in a liquid medium dispensed in tubes containing a minimum volume of 2 mL (macro-dilution) or with smaller volumes using 96-well microtitration plates (micro-dilution). A minimum inhibitory concentration (MIC) is determined by the unaided eye and considered to be the lowest concentration of antimicrobial agent that fully inhibits microbial growth in the tubes or micro-dilution wells.

To determine a MIC end point, viewing devices can facilitate reading the micro-dilution tests and recording of results, increasing the ability to discern growth in the wells. Tetrazolium salts, 3-(4, 5-dimethyl thiazol-2-yl)-2, 5-diphenyl tetrazolium bromide (MTT), and 2, 3-bis {2-methoxy-4-nitro-5-[(sulfenyl amino) carbonyl]-2H-tetrazolium-hydroxide} (XTT), are often used to determine MIC endpoints in both antifungal and anti-bacterial micro-dilution assays. The alamar blue dye (resazurin) is also an effective indicator of microbial growth that can be used for this purpose.

It is well-known that MIC values can be influenced by the inoculum size, type of growth medium, incubation time, and inoculum preparation method. As shown in a number of studies, the importance of preparing the inoculum using a hemocytometer for reproducibility and under suitable preparation conditions is independent of the color and the size of the conidia. The most common estimation of bactericidal or fungicidal activity is a determination of minimum bactericidal concentration (MBC) or minimum fungicidal concentration (MFC), also known as minimum lethal concentration (MLC). The bacterial endpoint can also be subjectively explained as the lowest concentration, wherein 99.9% of the final inoculum is killed. MFC can also be defined as the lowest concentration of drug with the capacity to kill 98–99.9% of the microorganisms included in the initial inoculum. Several studies have evaluated different test parameters to determine the MFC of various drugs against fungi.

2.2.7   CROSS-STREAK METHOD

The cross-streak method is helpful when rapidly screening microorganisms. A single streak of the microbial strain is placed in the middle of the agar plate and applied vertically to the petri dish. Inhibition zones are observed after the incubation period. Analysis includes evaluating the size of the inhibition zone.

2.3   RESULTS AND DISCUSSION

Tribes and local healers have a great deal of exposure and experience with traditional plants and their medicinal values; these people have been known to cure diseases with TM plants and to be healthy over the course of a lifetime. At present, researchers are looking to design drugs based on natural plant resources that do not adversely affect human health. Tribal communities use TM plants to cure various diseases.

2.3.1 TRIBAL COMMUNITIES

Tribal communities in various parts of the world use TM plants for their health care; diseases that are sometimes severe can be controlled by ingesting TM plants. Several tribes from India and Iran are highlighted here. China and other countries also have rich natural resources and commonly use TM plants. Indigenous tribes from the Andaman and Nicobar Islands also depend upon plants to recover from a disease and regain their health.

2.3.1.1 ASSAM TRIBES

Assam tribes use TM plants for their primary health care needs. A rich floristic diversity is available to them and they are fully dependent on TM plants for treating their health problems. Assam tribes use the rhizome of *Acorus calamus* to treat rheumatic pain, diarrhea, flatulence, pneumonia, bronchitis, and cough. *Ageratum conyzoides* roots are used to treat ophthalmic problems and issues with appetite. Also, the leaves of *A. conyzoides* have been shown to stop bleeding. *Alpinia allughas* rhizomes can be used to treat gout and colic. Sores can be controlled by ingesting *Alpinia malaccensis* rhizomes. *Ananas comosus* leaf extract is used against amebic dysentery and intestinal worms. The leaves of *Clerodendrum colebrookianum* are useful to eradicate intestinal worms; the tender leaves can also be boiled and the soup can be used to reduce blood pressure. *Dillenia indica* is used to treat stomach disorders. The fruits of this TM plant can control dandruff and thinning hair. The latex of *Ficus racemosa* is useful for piles and diarrhea; the powder of which is mixed with honey. The leaves of *Gomphostemma parviflora* have the potential to cure malaria. Tubers of *Kaempferia rotunda* are used for swellings, tumors, wounds, and ulcers. *Musa paradisiaca* fruits are ingested for the treatment of chronic dysentery. Rice (*Oryza sativa*) wash water and *Sarcochlamys pulcherrima* are also used to treat dysentery and diarrhea. *Pueraria tuberosa* tubers and *Scoparia dulcis* leaves are useful to cure fever. The inflorescence of *Spilanthes paniculata* plays a pivotal role in relieving toothache, bronchial pathologies, and mouth ulcers. The stem bark from *Zanthoxylum nitidum* has been used to treat toothache as well as problems with dental gums. The rhizome juice from *Thelypteris angustifolia* is effective for treating stomach problems; Assam tribes take four teaspoons of rhizome juice, thrice daily. As mentioned, they use the above plants for various treatments.[36]

2.3.1.2 JAINTIA TRIBES

Jaintia tribes live in the North Cachar Hills district of Assam, Northeast India. They use different TM plants for their ill-health, including boiled flowers and leaves of *Adhatoda vasica* with water. This decoction is consumed once daily to stop nosebleed, dysentery, and the vomiting of blood. Jaintia tribes use 2–3 drops of the water extract of *Barleria cristata* for skin infections. *Cassia tora* leaves, bark, and roots are used externally against skin diseases and leprosy. This plant destroys ringworms on the affected skin. *Nicotiana tabaeccum* is applied thrice daily to treat skin infections. Crushed leaves of *Plantago major* and raw milk are mixed equally and taken on an empty stomach for a week to cure jaundice. The plant leaf extract also cures toothache, earache, and bleeding gums.[34]

2.3.1.3 KANI TRIBES

Kani tribes are situated in Kouthalai of Tirunelveli hills, Tamil Nadu, India. They have practical knowledge about their TM plants. The combination of *A. conyzoides* leaf juice and *Cocculus hirsutus* is used to treat diarrhea. A mixture of *Borreria ocymoides* leaf juice, *Garcinia pictoria* leaf juice, and *Syzigium cumini* stem bark, along with gingerly oil is prepared as a paste and applied to wounds. *Carmona retusa* leaf powder is primarily useful for cleaning teeth. The whole plant is mixed with *Acacia nilotica* leaves and the seeds of *Areca catechu* and then used to cure toothache and to fortify teeth. Skin diseases are easily cured by a formulation of *Crotalaria pallida* leaf powder and root bark along with *Tragia involucrata* and *Wrightia tinctoria* leaves. This paste can be used externally to cure skin diseases. *Scleropyrum pentandrum* stem bark and leaves are also used to treat skin diseases. Extracted oil from *Pongamia pinnata* and the leaf paste from *Eupatorium odoratum* are mixed thoroughly and applied to wounds.

Hemionitis arifolia leaf paste, the root bark of *P. pinnata, Achyranthes aspera,* and *Datura metel* leaves are mixed together and applied externally for the treatment of rabies. Two formulations can be taken internally to destroy stomach worms and to cure ulcers. Specifically, equal amounts of leaves of *Zehneria maysorensis, Erythrina variegata, P. pinnata,* and *Ricinus communis* leaves that have been dried and stored in powder form in the dark are mixed with honey and used to kill stomach worms. Ulcers are cured by ingesting *Zornia diphylla, Madhuca longifolia* stem bark, *Begonia malabarica* rootstock, and *Hybanthes enneaspermus* leaves. A paste is made from these plants, which is then used to treat ulcers. Kani tribes are well versed in TM plants and their medicinal values. They use medicinal plants to treat their health issues, resulting in overall good health.[2]

2.3.1.4   KORKU TRIBES

Korku tribes belong to the Amravati district of Maharashtra, India. They obtain natural medicines from various TM plants. Chewing *Cardiospermum helicacabum* cures stomachache. *Curcuma pseudomontana* completely purifies blood. Raw tubers from *Leea macrophylla* are consumed for the treatment of diarrhea. Crushed *Curcuma amada* tubers are used to treat muscular, non-bleeding injuries. *Ougeinia oojeinensis* bark powder is used to cure wounds quickly. Ash is prepared from the dried leaves of *Tridax procumbens* and salt is added; *Cocus nucifera* (coconut) oil is mixed with the prepared ash. This formulation is helpful for eradicating ringworms. *Pueraria tuberosa* root powder is applied along with coconut oil to assist with wound healing. This paste should be applied daily, twice in a day, on the wound. The fresh leaves of *D. metel,* jaggery, and lime are mixed thoroughly into a paste and used externally for the treatment of a swollen throat. Juice from the fresh leaves of *D. metel,* mixed with 50 mL cow milk can be used to treat severe stomachache; this medication must be taken thrice daily for seven days to be completely effective.[19]

2.3.1.5   PALIYAR TRIBES

Paliyar tribes use TM plants from the Palani hills of Tamil Nadu, India. These plants are used in folk medicine to treat skin diseases, dysentery, stomachache, wounds, diarrhea, toothache, and headache. *Cassia alata* root powder as well as *Cassia fistula* stem bark and leaves are useful for treating skin diseases. However, they are prepared somewhat differently from each other. *C. alata* root powder with lime juice heals skin diseases. Stem bark and leaves from *C. fistula* are ground into a paste and coconut oil is added; this paste is then used to cure skin diseases. *Acalypha fruticosa* root and leaves are mixed with water and ground thoroughly. The water extract of this plant can be taken internally to cure dysentery. Water extracts of *C. fistula* and *Toddalia asiatica* leaves can be used to cure stomachache. Stomachache may also be cured with *Syzygium cumini* fruits, which are taken orally. The coat from the dried fruit of *Punica granatum* is ground and mixed with water and then taken internally to control stomachache and diarrhea. Wounds are treated with the paste of *Peltophorum pterocarpum* stem bark. *T. asiatica* stem bark powder is used for brushing teeth and to treat toothache. Burned and dried *Solanum xanthocarpum* fruits are inhaled to treat toothache as well as headache.[9]

2.3.1.6   KURD TRIBES

The Ilam province is situated in Iran. The Kurd tribe belongs to the district of Dehloran and Abdanan, Iran. The flowers and leaves of *Adiantum capillus* have been used as antiseptic, anticalculus, and analgesic agents. Some TM plants are useful in the treatment of septic wounds and injuries including, *Teucrium polium, Opoponax hispidus, Allium Akaka, Ferulago angulata, Achillea biebersteinii, Myrtus communis, Allium ampeloprasum, Ferula haussknechtii, Peganum harmala,* and *Alhagi persarum.* The stems and leaves of *A. persarum* have been used to treat the urinary tract infections. *A. ampeloprasum* leaves and bulbs are also useful to treat urinary infections. Toothache is treated with the flowers and seeds of *P. harmala* as well as the flowers and leaves of *A. biebersteinii. T. polium* is an antiseptic agent for gastric and mouth odors. Certain diseases caused by pathogens can be cured by TM plants. Further, microbes present in wounds can degrade cells and various plant-based drugs are used to protect cells and destroy these microorganisms. For example, *M. communis* is effective against microorganisms that are present in wounds. Skin diseases, burns, wounds, and cut wounds are very well treated with *Scrophularia deserti, Scrophularia striata, Alcea angulate, Capparis spinosa,* and *Solanum nigrum.* The different plant parts such as leaves, seeds, flowers, fruits, stem, stem bark, root, and root bark are used to eradicate microorganisms in various diseases. Based on the need, it is advisable to take plant-based drugs internally or externally.[37]

2.3.2 ANTIMICROBIAL COMPOUNDS FROM TRADITIONAL MEDICINAL PLANTS

Clerodane diterpene [(5R, 10R)-4R, 8 dihydrox-2S, 3R: 15, 16-diepoxy-cleroda-13 (16), 17, 12S, 18, 1S-dilactone] is isolated from an ethanolic extract of *Tinospora cordifolia*.[10] This compound, depicted in Figure 2.1a, has antimicrobial activity against bacteria and fungi. Friedelin (Fig. 2.1b) is fractionated from hexane, ethyl acetate, and methanoicl extracts of *Azima tetracantha* leaves.

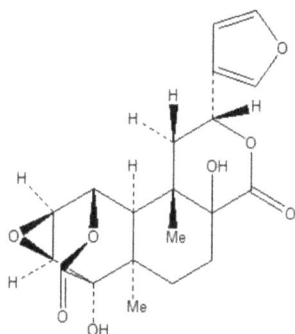

(a) Clerodane diterpene (or clerodanoids)

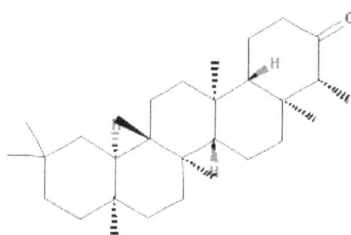

(b) Friedelin, $C_{30}H_{50}O$
[<By Jatlas, Own work: CC BY-SA 3.0, https://commons.wikimedia.org/w/index. php? curid=28,977,101>]

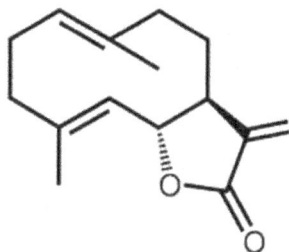

(c) Costunolide, $C_{15}H_{20}O_2$ [<https:// en.wikipedia.org/wiki/Costunolide>]

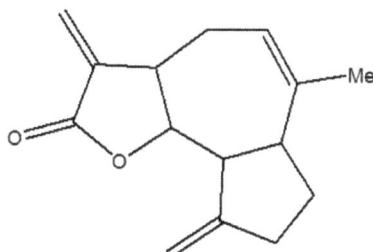

(d) Eremanthin $(C_{15}H_{18}O_2)$

FIGURE 2.1 Structural formula of four antimicrobial compounds, (a) Clerodane diterpene (or clerodanoids), (b) Friedelin, $C_{30}H_{50}O$ [<By Jatlas, Own work: CC BY-SA 3.0, https://commons.wikimedia.org/w/index.php? curid=28,977,101>], (c) Costunolide, $C_{15}H_{20}O_2$ [<https://en.wikipedia.org/wiki/Costunolide>], (d) Eremanthin $(C_{15}H_{18}O_2)$.

The leaves of this plant have been used to keratinize tissue infected with *Candida albicans*, which are dermatophytes responsible for creating infections. Further, this compound has been shown to inhibit certain fungi.[11] Costunolide (Fig. 2.1c) and Eremanthin (Fig. 2.1d) are compounds that are isolated from the hexane extract of *Costus speciosus* rhizomes.[29] They have similar antimicrobial activities against the tested fungi;[11] both compounds are used for micro-broth dilutions.

The compound 4-hydroxybenzoic acid hydrate (Fig. 2.2a) was derived from *C. fistula*, which possesses antidermatophytic, antifungal, and antibacterial properties. Skin infections, diarrhea, and fever can be treated with this plant-based compound.[14] Quinone 6-(4, 7-dihydroxy-heptyl) and cyclohex-2-ene-1, 4-dione have antibacterial activity and are isolated from *Pergularia daemia* (Fig. 2.2b). Further, it inhibits the growth of *Bacillus subtilis*, *Staphylococcus aureus,* and *Proteus vulgaris*.[18] Thymol and carvacrol were derived from thymol oil and oregano oil, respectively. Both compounds have antimicrobial activity. Thymol (Fig. 2.2c) is active against *C. albicans* and *Escherichia coli*.[37] Carvacrol (Fig. 2.2d) is active against *B. subtilis* and *Saccharomyces cerevisiae*.[24]

Epicatechin (Fig. 2.2e) was obtained from *Pappea capensis* leaves. It is very active, with a low MIC (7.81 μg/mL) against *Klebsiella pneumoniae* and *E. coli*.[6] Confertifolin (Fig. 2.2f) has been tested using the broth micro-dilution method and has antimicrobial activity against bacteria and fungi.[8] The compound can be dissolved in water with 2% dimethyl sulfoxide (DMSO). An initial concentration of 0.5 mg/mL has been shown to have antimicrobial activity. The initial test application was serially-diluted twofold and each well was loaded with a 5 μL suspension, holding 10^8 CFU/mL bacteria and 10^4 spores/mL (5 μL) fungi, respectively. Fluconazole and ketoconazole for fungi, and strep-tomycin for bacteria were used in the assays as positive controls. For fungi, the plates were incubated for 24, 48, or 72 h at 27°C, whereas for bacteria, the plates were kept for 24 h at 37°C. After the incubation period, researchers observed that a low concentration of the compound inhibited bacterial and fungal growth. Table 2.1 shows that a low MIC is effective against *Enterococcus faecalis*; however, higher concentrations are required against *Staphylococcus epidermidis*, *Pseudomonas aerugi-nosa*, *P. vulgaris*, *B. subtilis*, *S. aureus*, *E. coli*, *K. pneumoniae*, *Erwinia* sp., and *Salmonella typhi*.[12]

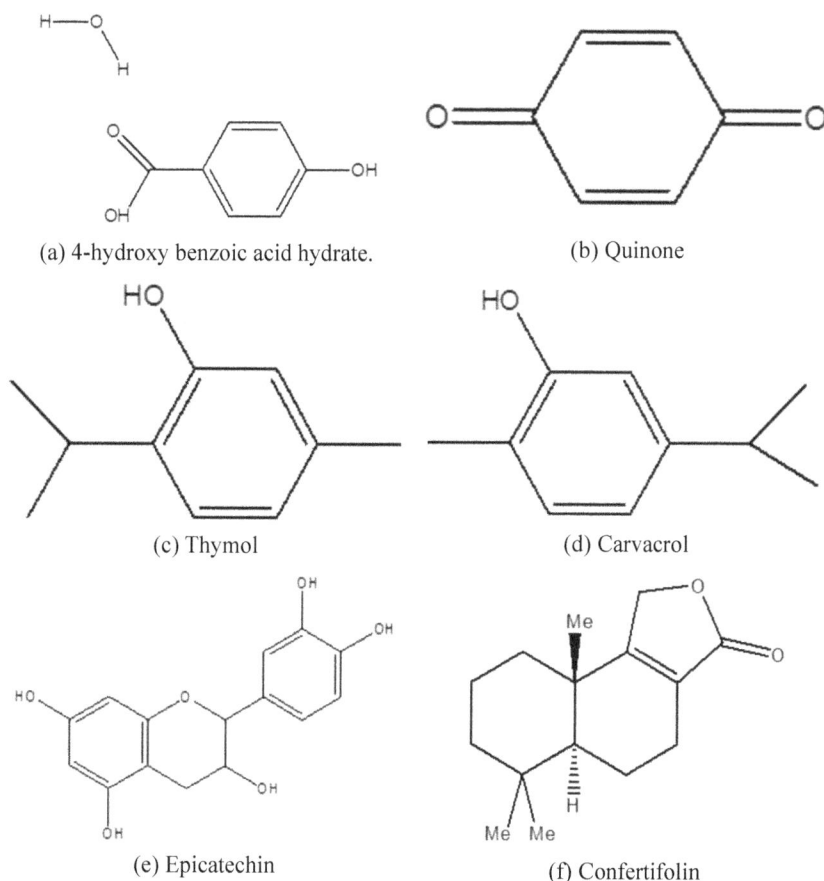

(a) 4-hydroxy benzoic acid hydrate.

(b) Quinone

(c) Thymol

(d) Carvacrol

(e) Epicatechin

(f) Confertifolin

FIGURE 2.2 Bioactive compounds from traditional medicinal plants, (a) 4-hydroxybenzoic acid hydrate, (b) quinone, (c) thymol, (d) carvacrol, (e) epicatechin, (f) confertifolin.

A single X-ray diffraction (XRD) was used to determine the structure of confertifolin (6,6,9a-trimethyl-4,5,5a,6,7,8,9,9aoctahydronaphtho[1,2-c] furan-3 (1H)-one) isolated from *Polygonum hydropiper* L (*Polygonaceae*) leaves. It is effective against pathogenic fungi. Duraipandiyan et al. observed that this compound has significant antifungal activity. MIC values vary depending on the least resistant to the most resistant species, including *Epidermophyton floccosum, Cochliobolus lunata, Trichophyton mentagrophytes, Trichophyton rubrum* MTCC 296, *Botrytis cinerea, Aspergillus niger, Magnaporthe grisea, T. rubrum* (clinical isolate), *Scopulariopsis* sp., and *Trichophyton simii*.

TABLE 2.1   Antibacterial Activity of Confertifolin from *Polygonum hydropiper*.

Compounds from traditional medicinal plants	Test microorganism	Minimum inhibitory concentration (MIC) μg/mL	Streptomycin having MIC (μg/mL)	Reference
Confertifolin	*Bacillus subtilis*	>500	25	[12]
	Staphylococcus aureus	>500	6.25	
	Staphylococcus epidermidis	500	>50	
	Enterococcus faecalis	31.25	25	
	Escherichia coil	>500	12.5	
	Pseudomonas aeruginosa	500	25	
	Klebsiella pneumonia	>500	6.56	

The antimicrobial properties of TMs described here were chosen from previous published literature. Other preferred methods for testing antimicrobial activity are described elsewhere. Based on ethnobotanical methods, researchers gain information regarding the importance of TM plants from local people and then conduct studies in the laboratories. Researchers produce plant-based compounds and drugs against microorganisms and various diseases. This report is useful for designing new drugs and will be useful for assisting marginalized people.

Plants are a source of inspiration and contribute to improving the welfare of humans. The plants described here have been chosen from published literature and exhibit antimicrobial activity, including *M. officinalis, A. sativum, M. alternifolia, P. nigrum, P. graveolens, O. vulgare, S. aromaticum, T. vulgaris, M. fragrans, P. hydropiper, Satureja bachtiarica, Satureja calamintha, Satureja hortensis, Satureja intermedia, Satureja khuzestanica, Satureja montana, Satureja wiedemanniana, C. speciosus, Anisomeles malabarica* and *Terminalia arjuna*.

2.3.3   *MINIMUM INHIBITORY CONCENTRATION (MIC)*

T. asiatica (L.) Lam. leaves and roots have been shown to have antimicrobial activity. Flindersine is isolated from this plant and is an effective antimicrobial agent at low concentrations (Fig. 2.3).

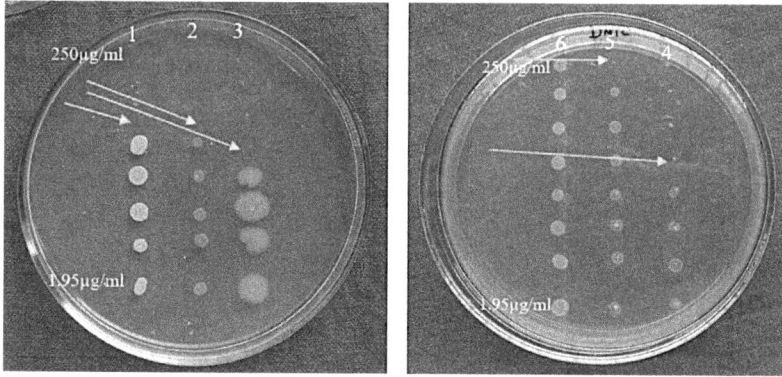

FIGURE 2.3 (See color insert.) Minimum inhibitory concentration (MIC) of flindersine against tested bacteria: 1-*Staphylococcus aureus* (62.5 µg/mL); 2-*Staphylococcus epidermidis* (62.5 µg/mL); 3-*Bacillus subtilis* (31.2 µg/mL); 4-*Enterococcus faecalis* (31.2 µg/mL); 5-*Pseudomonas aeruginosa* (250 µg/mL); and 6-*Escherichia coli* (>250 µg/mL).

Rhein is derived from the leaves of *Cassia reticulata* and has been tested against bacteria such as *S. aureus* (62.5 µg/mL), *S. epidermidis* (62.5 µg/mL), *B. subtilis* (31.2 µg/mL), *E. faecalis* (62.5 µg/mL), *P. aeruginosa* (125 µg/mL), and *E. coli* (>250 µg/mL). The MIC of this compound is effective against these bacteria (Fig. 2 4).

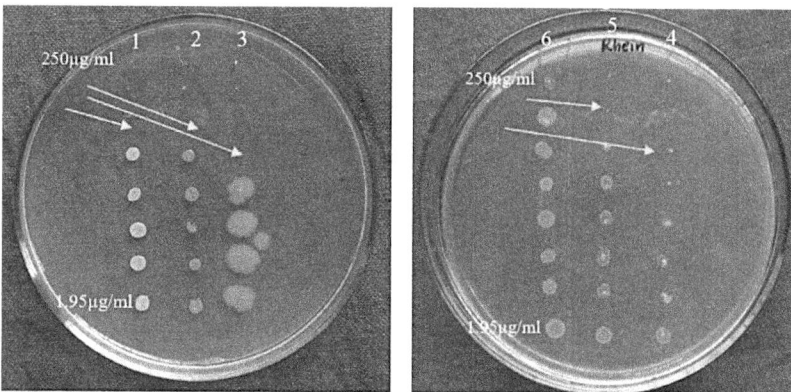

FIGURE 2.4 (See color insert.) MIC of rhein against tested bacteria: 1-*Staphylococcus aureus* (62.5 µg/mL); 2-*Staphylococcus epidermidis* (62.5 µg/mL); 3-*Bacillus subtilis* (31.2 µg/mL); 4-*Enterococcus faecalis* (62.5 µg/mL); 5-*Pseudomonas aeruginosa* (125 µg/mL); 6-*Escherichia coli* (>250 µg/mL).

2.3.4 BROTH MICRO-DILUTION

Flindersine has been effective in the treatment of some fungal infections, with gradually increasing concentrations. The compound, along with test fungi, has been tested using the broth micro-dilution method. The results and test fungi are displayed below (Fig. 2.5). Rhein has been used to kill bacteria. The micro-dilution method was preferred for determining the inhibition of bacterial growth and antimicrobial activity (Fig. 2.6).

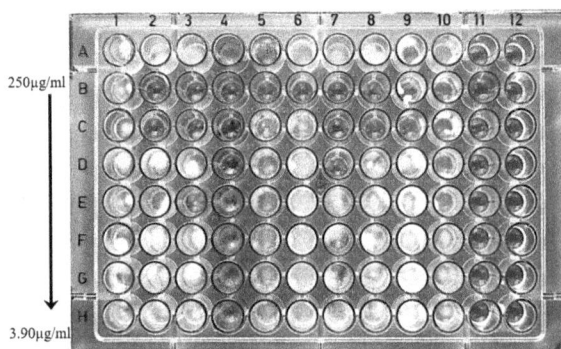

FIGURE 2.5 (See color insert.) Antifungal activity of Flindersine: B1 to H1-*Trichophyton mentagrophytes* (>250 µg/mL); B2 to H2-*Epidermophyton flocossum* (125 µg/mL); B3 to H3-*Trichophyton simii* 125 µg/mL); B4 to H4-*Curvularia lunata* (250 µg/mL); B5 to H5-*Aspergillus niger* (>250 µg/mL); B6 to H6-*Botrytis cinerea* (>250 µg/mL); B7 to H7-*Trichophyton rubrum* (296) (62.5 µg/mL); B8 to H8-*Magnoporthe grisea* (125 µg/mL); B9 to H9-*T. rubrum* (57) (>250 µg/mL); B10 toH10-*Scopulariopsis* sp (>250 µg/mL); A1-A10: Control (culture and broth only); A\11to H11 and A12 to H12—Blank.

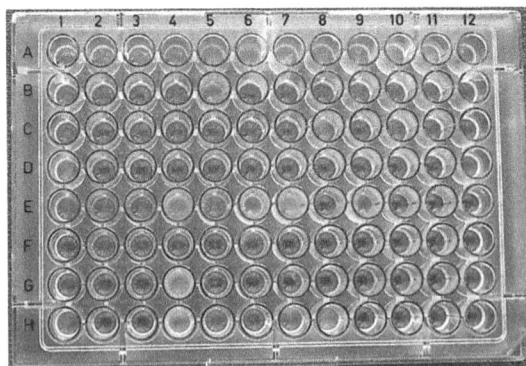

FIGURE 2.6 (See color insert.) MIC of rhein as determined using the micro-dilution method against bacteria.

2.3.5 ANTIMICROBIAL ACTIVITY OF SATUREJA SPECIES AGAINST TEST MICROORGANISMS

Essential oils, ethanolic extracts, and hydroalcoholic extracts from various *Satureja* species have antimicrobial activity against many test microorganisms. The essential oil of *S. bachtiarica* is very active against *P. aeruginosa*.[30] The ethanolic extract of *S. bachtiarica* is also active against various harmful microorganisms such as *E. coli, K. pneumoniae, S. aureus,* and *Streptococcus agalactiae.*[43] The hydroalcoholic extract of *S. khuzestanica* is effective against *Cryptococcus neoformans.*[3] Various other extracts from different *Satureja* species also have antimicrobial activity, as shown in Table 2.2.

2.3.6 MEDICINAL PROPERTIES AND PHARMACOLOGICAL ACTIONS OF TRADITIONAL MEDICINE (TM) PLANTS

The hexane extract of *C. speciosus* (Koen ex. Retz.) Sm has better antimicrobial activity than that of other extracts from the same plant. The inhibitory zone was shown against *S. aureus* and *S. epidermidis.*[13] Hexane and ethyl acetate extracts of *A. malabarica* (Linn.) (*Lamiaceae*) showed substantial antimicrobial activity and maximum inhibitory zones against bacteria. These extracts are very effective against *Shigella flexneri*, forming a 14 mm zone of inhibition against this gram negative bacterium. The hexane extract of the *A. malabarica* leaf and inflorescence inhibited *Enterobacter aerogenes*, a gram-negative rod-shaped bacteria, with an inhibition zone of 13 mm.[15] The methanolic extract of *T. arjuna* bark has been shown to have antimicrobial activity against some gram-positive and gram-negative bacteria. The largest inhibition zone was observed to be 12 mm against *S. aureus*. Further, among the gram-positive and gram-negative bacteria, the methanolic extract is more effective against *S. aureus*; however, good antimicrobial activity was observed against all bacteria tested. The methanolic extract of *Butea monosperma* has a favorable antimicrobial activity against fungi, with an inhibition zone of 14 mm against *C. albicans.*[20] All other plant extracts and their corresponding inhibition zones are displayed in Table 2.3.

Dichloromethane and methanolic extracts of *Centella asiatica* showed antimicrobial activity (Table 2.4) against numerous bacteria such as *B. subtilis, Bacillus cereus, E. coli, Klebsiella aerogenes,* and *P. vulgaris*. The largest zone of inhibition (23 mm) was found against *S. aureus* using

TABLE 2.2 Extracts of *Satureja* Species Against Test Microorganisms.

Genus	Botanical name	Usage of plant parts	Name of the microorganism	Reference
Satureja species	*Satureja bachtiarica*	Essential oil	*P. aeruginosa*	[30]
Satureja species	*S. bachtiarica*	Ethanol extract	Some gram-positive bacteria (*Escherichia coli, Klebsiella pneumoniae, S. aureus,* and *Streptococcus agalactiae*)	[43]
Satureja species	*Satureja calamintha*	Essential oil	*E. coli, P. aeruginosa, S. aureus, K. pneumoniae, Enterobacter aerogenes, Proteus mirabilis and Streptococcus,* and *Enterococcus.*	[3]
Satureja species	*Satureja hortensis*	Essential oil	various gram-positive and gram-negative microorganisms	[25]
Satureja species	*Satureja intermedia*	Essential oil	*E. coli, E. faecalis, S. aureus, S. epidermidis, B. subtilis, Bacillus pumulis, Candida albicans,* and *Saccharomyces cerevisiae.*	[32]
Satureja species	*Satureja khuzestanica*	Fresh leaves ethanol extract	Some saprophytic fungi	[33]
Satureja species	*S. khuzestanica*	Hydroalcoholic extract	*Cryptococcus neoformans*	[44]
Satureja species	*Satureja montana*	Ethanol extract	*S. aureus, E. coli, Proteus morgani, Candida tropicalis,* and *Trichophyton mentagrophytes*	[29]
Satureja species	*S. wiedemanniana*	Essential oil	Antibacterial effects against *Bacillus* species isolated from raw meat samples	[42]

TABLE 2.3 Various Plant Extracts Against Test Microorganisms.

Botanical name	Extract type	Inhibition zone	Test microorganism	Reference
Costus speciosus (Koen ex. Retz.) Sm	Hexane extract	15 mm	*S. aureus*	[13]
		15 mm	*S. epidermidis*	
		12 mm	*B. subtilis*	
	Chloroform extract	12 mm	*S. aureus*	[13]
		13 mm	*S. epidermidis*	
		9 mm	*B. subtilis*	
	Ethyl acetate extract	10 mm	*S. aureus*	[13]
		9 mm	*S. epidermidis*	
		9 mm	*B. subtilis*	
	Methanol extract	14 mm	*S. aureus*	[13]
		12 mm	*S. epidermidis*	
		10 mm	*B. subtilis*	
Anisomeles malabarica (inflorescence)	Ethyl acetate extract	14 mm	*Shigella flexneri*	[35]
A. marabalica (inflorescence)	Hexane extract	14 mm	*S. flexneri*	[35]
		13 mm	*E. aerogenes*	[35]
Terminalia arjuna	Methanol extract	12, 11 and 11 mm respectively	Gram-positive bacteria: *S. aureus, Micrococcus luteus,* and *S. epidermidis*	[20]
		11 mm, 11 mm, 10 mm and 10 mm respectively	Gram-negative bacteria: *E. aerogenes, Vibrio parahaemolyticus, K. pneumoniae,* and *Yersinia enterocolitica*	[20]

TABLE 2.3 *(Continued)*

Botanical name	Extract type	Inhibition zone	Test microorganism	Reference
B. monosperma	Methanol extract	14 mm	C. albicans	[20]
C. speciosus rhizome	Methanol extract	12 mm	S. aureus	[13]
		9 mm	S. epidermidis	
		14 mm	B. subtilis	
	Water extract	12 mm	S. aureus	

TABLE 2.4 Antibacterial Activity of Different Crude Extracts from Leaves of *Centella asiatica.*

Bacteria	Concentration of dichloromethane (mg/mL) extract	Zone of inhibition (mm)	Concentration of methanol (mg/mL) extract	Zone of inhibition (mm)	Reference drug	
					Optimum concentration (36 mg/mL) of chloramphenicol drug (30 µg/disc) with zone of inhibition	Optimum concentration (31 mg/mL) of streptomycin drug (30 µg/disc) with zone of inhibition
B. subtilis	3.5	10	3.5	12	28	33
	7.25	12	7.25	15		
	15	15	15	18		
	30	18	30	20		
	60	20	60	25		
Bacillus cereus	3.5	11	3.5	10	24	22
	7.25	13	7.25	12		
	15	17	15	15		
	30	20	30	17		
	60	21	60	26		
E. coli	3.5	–	3.5	10	31	34
	7.25	12	7.25	12		
	15	14	15	12		
	30	15	30	14		
	60	18	60	26		
Klebsiella aerogenes	3.5	14	3.5	12	32	36

TABLE 2.4 *(Continued)*

Bacteria	Concentration of dichloromethane (mg/mL) extract	Zone of inhibition (mm)	Concentration of methanol (mg/mL) extract	Zone of inhibition (mm)	Reference drug	
					Optimum concentration (36 mg/mL) of chloramphenicol drug (30 µg/disc) with zone of inhibition	Optimum concentration (31 mg/mL) of streptomycin drug (30 µg/disc) with zone of inhibition
	7.25	17	7.25	15		
	15	20	15	19		
	30	22	30	20		
	60	23	60	20		
Proteus vulgaris	3.5	12	3.5	10	24	23
	7.25	14	7.25	11		
	15	17	15	13		
	30	20	30	16		
	60	21	60	25		
P. mirabilis	3.5	13	3.5	12	27	29
	7.25	16	7.25	15		
	15	18	15	18		
	30	20	30	19		
	60	20	60	23		
P. aeruginosa	3.5	—	3.5	10	30	32

TABLE 2.4 *(Continued)*

Bacteria	Concentration of dichloromethane extract (mg/mL)	Zone of inhibition (mm)	Concentration of methanol extract (mg/mL)	Zone of inhibition (mm)	Reference drug	
					Optimum concentration (36 mg/mL) of chloramphenicol drug (30 µg/disc) with zone of inhibition	Optimum concentration (31 mg/mL) of streptomycin drug (30 µg/disc) with zone of inhibition
	7.25	13	7.25	14		
	15	14	15	16		
	30	18	30	18		
	60	19	60	22		
S. aureus	3.5	11	3.5	16	34	29
	7.25	13	7.25	19		
	15	14	15	20		
	30	16	30	23		
	60	23	60	28		
Salmonella typhi	3.5	10	3.5	15	26	30
	7.25	15	7.25	17		
	15	17	15	20		
	30	18	30	22		
	60	22	60	27		

the dichloromethane extract (60 mg/mL). The highest concentration of methanolic extract used was 60 mg/mL, which resulted in the largest zone of inhibition against *S. typhi* (27 mm). Reference drugs were chloramphenicol and streptomycin.[34]

2.3.7 SOCIOECONOMIC IMPACT AND INNOVATIVE TECHNOLOGIES

TMs are easily available and cost-effective. Tribes from different parts of the world familiar with the medicinal value of various plants are well protected against infections from different pathogens. We need to ingest plant-based foods and vegetables to live a healthy life. There are many innovative technologies and techniques that have improved plant-based drugs. TMs cure many diseases; however, they sometimes require long periods of time to be effective. The new technologies and techniques available may accelerate the time required for TMs to be effective in the future.

2.3.8 CASE STUDY

TMs enhance health in human beings. Case studies have been conducted by investigators to determine the efficacy of plant-based drugs. Many case studies have proven that nature-based compounds have good activity against various diseases, including microbial diseases.

2.4 SUMMARY

TMs play a pivotal role in human health care in numerous developing and developed countries, with increasing commercial value. Local healers and people take medicinal plants to manage illnesses. TMs offer better routes for drug discovery, with enriched presentation in terms of cost, safety, and efficacy.[28] Though medicinal plants have been used in various treatments during the past centuries, the demand has increased in recent years, attributable to the ill-effects of modern medicines. Nearly half of all small molecules from natural products have been approved and used for drug discovery. They are also being used to find "new entity drugs".[23] The current "one drug fits all" approach may not be sustainable in the future.[22] Researchers are looking to find additional traditional herbal formulations that are effective against infectious pathogens. A recent review by Rios et al. showed that there were 115 published articles on the "antimicro-bial activity of medicinal plants" in PubMed from 1966–1994; however, this number rose to more than 307 during the period between 1995 and 2004. Various plant extracts and compounds have antimicrobial proper-ties[31] and extensive research has been conducted since 2005. Essential oils from lemon balm (*M. officinalis*), garlic (*A. sativum*), and tea tree (*M. alternifolia*) have antimicrobial properties that are used to treat infec-tious pathogens found on our skin, in the urinary tract, respiratory system, gastrointestinal tract, and biliary systems. Various antimicrobial methods are used to test the antimicrobial activity of these compounds. Tribal communities from different parts of the world depend upon TM plants for their primary health care. Most of the known effective antimicrobial compounds are highlighted in this chapter. Hence, TM plants have many antimicrobial properties that are effective for curing microbial diseases.

This chapter focuses on the antimicrobial properties of traditionally-used medicinal plants against pathogenic microbes. The active extracts and pure compounds have been screened by many researchers. This infor-mation will help in drug discovery programs.

2.5 ACKNOWLEDGMENT

The authors are grateful to the Deanship of Scientific Research, King Saud University for funding through Vice Deanship of Scientific Research Chairs.

KEYWORDS

- antimicrobial activity of *Satureja* species against test microorganisms
- antimicrobial compounds from traditional medicinal plants
- antimicrobial properties
- drug discovery
- efficacy
- medicinal properties
- pharmacological actions of traditional medicinal plants
- safety
- traditional herbal formulation
- traditional medicine
- tribal communities

REFERENCES

1. Aguilar, G. Access to Genetic Resources and Protection of Traditional Knowledge in the Territories of Indigenous Peoples. *Environ. Sci. Policy* **2001**, *4,* 241–256.
2. Ayyanar, M.; Ignacimuthu, S. Traditional Knowledge of Kani Tribals in Kouthalai of Tirunelveli Hills, Tamil Nadu, India. *J. Ethnopharmacol.* **2005**, *10* (2), 246–255.
3. Bensouici, C.; Benmerache, A.; Chibani, S.; Kabouche, A.; Abuhamdah, S.; Semra, Z. Antibacterial Activity and Chemical Composition of the Essential oil of *Satureja Calamintha* sp. Sylvatica from Jijel, Alger. *Der. Pharm. Lett.* **2013**, *5,* 224–227.
4. Cowan, M. M. Plant Products as Antimicrobial Agents. *Clin. Microbial. Rev.* **1999**, *12*(4), 564–582.
5. Das, K.; Tiwari, K. S.; Shrivastava, D. K. Techniques for Evaluation of Medicinal Plant Products as Antimicrobial Agents: Current Methods and Future Trends. *J. Med. Plants Res.* **2010**, *4*(2), 104–111.
6. Devienne, K. F.; Raddi, M. S. G. Screening for Antimicrobial Activity of Natural Products. *Braz. J. Microbiol.* **2002**, *33*(2), 97–105.
7. Dorman, H. J. D.; Deans, S. G. Antimicrobial Agents from Plants: Antibacterial Activity of Plant Volatile Oils. *J. Appl. Microbiol.* **2000**, *88*(2), 308–316.
8. Duraipandiyan, V.; Ignacimuthu, S. Antibacterial and Antifungal Activity of *Cassia fistula* L.: An Ethnomedicinal Plant. *J. Ethnopharmacol.* **2007**, *112,* 590–594.
9. Duraipandiyan, V.; Ayyanar, M.; Ignacimuthu, S. Antimicrobial Activity of Some Ethnomedicinal Plants Used by Paliyar Tribe from Tamil Nadu, India. *BMC Complementary Altern. Med.* **2006**, *6*(1), 354.
10. Duraipandiyan, V.; Ignacimuthu, S.; Valanarasu, M. Antibacterial and Antifungal Activity of *Syzygium lineare* Wall. *Int. J. Integr. Biol.* **2008**, *3*(3), 159–162.
11. Duraipandiyan, V.; Gnanasekar, M.; Ignacimuthu, S. Antifungal Activity of Triterpenoid Isolated from *Azima tetracantha* Leaves. *Folia Histochem. Cytobiol.* **2010a**, *48*(2), 311–313.
12. Duraipandiyan, V.; Indwar, F.; Ignacimuthu, S. Antimicrobial Activity of Confertifolin from *Polygonum hydropiper*. *Pharm. Biol.* **2010b**, *48,* 187–190.
13. Duraipandiyan, V.; Al-Harbi, N. A.; Ignacimuthu, S.; Muthukumar, C. Antimicrobial Activity of Sesquiterpene Lactones Isolated from Traditional Medicinal Plant, *Costus speciosus* (Koen ex. Retz.) Sm. *BMC complementary Altern. Med.* **2012a**, *12,* 13.
14. Duraipandiyan, V.; Ignacimuthu, S.; Balakrishna, K.; Al-Harbi, N. A. Antimicrobial Activity of *Tinospora cordifolia*: An Ethnomedicinal Plant. *Asian J. Tradit. Med.* **2012b**, *7,* 59–65.
15. Ghasemi, P. A.; Momeni, M.; Bahmani, M. Ethnobotanical Study of Medicinal Plants Used by Kurd Tribe in Dehloran and Abdanan Districts, Ilam Province, Iran. *Afr. J. Tradit. Complementary Altern. Med.* **2013**, *10*(2), 368–385.
16. Guidance for Industry on Botanical Drug Products. U.S. Department of Health and Human Services, Food and Drug Administration, Center for Drug Evaluation and Research (CDER); June 2014; <http://www.fda.gov/cder/guidance/index.htm>.
17. Heinrich, M.; Barnes, J.; Gibbons, S.; Williamson, E. M. *Fundamentals of Pharmacognosy and Phytotherapy;* Churchill Livingstone: Edinbrugh, 2004; pp 245–252.

18. Ignacimuthu, S.; Pavunraj, M.; Duraipandiyan, V.; Raja, N.; Muthu, C. Antibacterial Activity of a Novel Quinone from the Leaves of *Pergularia daemia* (Forsk.), a Traditional Medicinal Plant. *Asian J. Tradit. Med.* **2009,** *4*(1), 36–40.

19. Jagtap, S. D.; Deokule, S. S.; Bhosle, S. V. Some Unique Ethnomedicinal Uses of Plants Used by the Korku Tribe of Amravati District of Maharashtra, India. *J. Ethnopharmacol.* **2006,** *107*(3), 463–469.

20. Kinsalin, V. A.; Kumar, P. S.; Duraipandiyan, V.; Ignacimuthu, S.; Al-Dhabi, N. A. Antimicrobial Activity of Methanol Extracts of Some Traditional Medicinal Plants from Tamil Nadu, India. *Asian J. Pharm. Clin. Res.* **2014,** *7*(2), 36–40.

21. Kong, D. X.; Li X. J,; Zhang, H. Y. Where is the Hope for Drug Discovery? Let History Tell the Future. *Drug Discovery Today* **2009,** *14*(3–4), 115–119.

22. Kumar, V. Pharmaceutical Issues in the Development of a Polypill for the Treatment of Cardiovascular Diseases. *Drug Discov. Today: Ther. Strategies* **2008,** *5,* 63–71.

23. Labadie, R. P. Problems and Possibilities in the use of Traditional Drugs. *J. Ethnopharmacol.* **1986,** *15,* 221–230.

24. Lambert, R. J. W.; Skandamis, P. N.; Coote, P. J.; Nychas, G. J. E. A Study of the Minimum Inhibitory Concentration and Mode of Action of Oregano Essential Oil, Thymol and Carvacrol. *J. Appl. Micrbiol.* **2001,** *91*(3), 453–462.

25. Mahboubi, M.; Kazempour, N. Chemical Composition and Antimicrobial Activity of *Satureja hortensis* and *Trachyspermum copticum* Essential Oil. *Iran J. Microbiol.* **2011,** *3,* 194–200.

26. Murray, P. R.; Baron, E. J.; Pfaller, M. A.; Tenover, F. C.; Yolke, R. H., Eds. *Manual of Clinical Microbiology,* 6th ed.; ASM (Am. Soc. of Microbiol.) Press: Washington, D.C., 1995; E-book.

27. Omulokoli, E.; Khan, B.; Chhabra, S. C. Antiplasmodial Activity of Four Kenyan Medicinal Plants. *J. Ethnopharmacol.* **1997,** *56,* 133–137.

28. Patwardhan, B. Ayurveda and Natural Products Drug Discovery. *Curr. Sci.* **2004,** *86,* 789–799.

29. Pepeljnjak, S.; Stanic, G.; Potocki P. Antimicrobial Activity of the Ethanolic Extract of *Satureja montana* ssp. Montana. *Acta Pharm.* **1999,** *49,* 65–69.

30. Pirbalouti, A. G.; Dadfar, S. Chemical Constituents and Antibacterial Activity of Essential oil of *Satureja bachtiarica* (Lamiaceae). *Acta Pol. Pharm.* **2013,** *70,* 933–938.

31. Rios, J. L.; Recio M. C. Medicinal Plants and Antimicrobial Activity. *J. Ethnopharmacol.* **2005,** *100*(1–2), 80–84.

32. Sadeghi, I.; Yousefzadi, M.; Behmanesh, M.; Sharifi, M.; Moradi, A. In vitro Cytotoxic and Antimicrobial Activity of Essential Oil from *Satureja intermedia*. *Iran. Red Crescent Med. J.* **2013,** *15,* 70–74.

33. Sadeghi-Nejad, B.; Shiravi, F.; Ghanbari, S.; Alinejadi, M.; Zarrin, M. Antifungal Activity of *Satureja khuzestanica* (Jamzad) Leaves Extracts. *Jundishapur J. Microbiol.* **2007,** *3,* 36–40.

34. Sajem, A. L.; Gosai, K. Traditional use of Medicinal Plants by the Jaintia Tribes in North Cachar Hills District of Assam, Northeast India. *J. Ethnobiol. Ethnomed.* **2006,** *2*(1), 1.

35. Saravana Kumar, P.; Al-Dhabi, N. A.; Duraipandiyan, V; Ignachimuthu, S. In vitro Antimicrobial Activity and Phytochemical Analysis of *Anisomeles malabarica* (Linn.) (Lamiaceae) Leaf and Inflorescences. *Int. J. Pharm. Phytopharmacol. Res.* **2014,** *3,* 323–326.

36. Sharma, U. K.; Pegu, S. Ethnobotany of Religious and Supernatural Beliefs of the Mising Tribes of Assam with Special Reference to the 'Dobur Uie'. *J. Ethnobiol. Ethnomed.* **2011,** *7*(1), 16.

37. Topuz, O. K.; Özvural, E. B.; Zhao, Q.; Huang, Q.; Chikindas, M.; Gölükçü, M. Physical and Antimicrobial Properties of Anise Oil Loaded Nanoemulsions on the Survival of Foodborne Pathogens. *Food Chem.* **2016,** *203,* 117–123.

38. Vaidya, A. D. B.; Devasagayam, T. P. A. Current Status of Herbal Drugs in India: An Overview. *J. Clin. Biochem. Nutr.* **2007, *41,*** 1–11.

39. Vanaclocha, B. V.; Folcara, S. C., Eds. *Fitoterapia: vademécum de prescripción (Phytotherapy: Prescription Formulary),* 4th ed.; Editorial Masson: Barcelona, 2003; pp 153–154.

40. WHO (World Health Organization). Traditional Medicine Strategy 2002–2005. [http://whqlibdoc.who.int/hq/2002/WHO_EDM_TRM_2002.1.pdf], (accessed May 4, 2017).

41. Xie, P. S.; Wong, E. Chinese Medicine—Modern Practice. In *Annals of Traditional Chinese Medicine;* World Scientific Publishing: Singapore, 2005; p 99.

42. Yucel, N.; Aslim, B.; Ozdogan H. In vitro Antimicrobial Effect of *S. wiedemanniana* Against Bacillus Species Isolated from Raw Meat Samples. *J. Med. Food* **2009,** *12,* 919–923.

43. Zareii, B.; Seyfi, T.; Movahedi, R.; Cheraghi, J.; Ebrahimi, S. Antibacterial Effects of Plant Extracts of *Alcea digitata* L., *Satureja bachtiarica* L. and *Ferulago angulata* L. *J. Babol Univ. Med. Sci.* **2014,** *16,* 31–37.

44. Zarrin, M.; Amirrajab, N.; Sadeghi-Nejad, B. In vitro Antifungal Activity of *Satureja khuzestanica* Jamzad against *Cryptococcus neoformans. Pak. J. Med. Sci.* **2010,** *26,* 880.

CHAPTER 3

MARINE BIOACTIVE COMPOUNDS: INNOVATIVE TRENDS IN FOOD AND MEDICINE

MUNAWAR ABBAS, FARHAN SAEED, and
HAFIZ ANSAR RASUL SULERIA

CONTENTS

ABSTRACT

Marine organisms are surplus sources of biologically active and available compounds. The presence of these bioactive molecules in marine sources puts them at very important position in diet-based regimens. Thus, the trend to consume food as a "medicine" is gradually becoming more common among people because of the awareness of the role of bioactive or nutraceutical in human health beyond the basic nutrition. Accordingly, marine lives are major source of these bioactive compounds such as polysaccharides, peptides, polyunsaturated fatty acids (PUFAs), and phlorotannins. These bioactive compounds have the ability to minimize the effect of lifestyle-related disorders. Hence, the functional and nutraceutical potential of these bioactive components from marine sources play a significant role against various aliments such as antioxidant, antidiabetic, anticancer, antihypertensive, and antibacterial effects, and these are the limelight of this chapter.

3.1   INTRODUCTION

In these millennia, the natural products are gaining popularity in the domain of nutrition to cure various maladies and to improve the quality of life. Earlier, many of these were consumed more often as a complete food or a part of diet to earn additional health benefits beyond their basic nutrition.[78] Although exact biological consequences of such diets were not clearly identified, yet evolutionary experiences have brought sufficient knowledge regarding the medicinal properties of natural products.[14] The inspiration from ancient diet and its promising health effect result from the innovation of a number of natural products based on commercial products.[75] Among them, functional food and nutraceutical, the most dynamic natural product-based segments in the food industry, create a revolution in food industries and make them more research oriented similarly to pharmaceutical industries.[12,61] Both concepts are industrially emerged few decades ago, nevertheless, consumer demand and the market share are tremendously increasing while it has been estimated that global nutraceutical market was about US$ 243 billion in 2015.[61]

By definition, functional and nutraceutical foods are extracted from food or its food ingredient.[71] However, it seems no longer applicable for modern nutraceuticals; a number of nonfood-derived components that are known as safe, have been used in nutraceutical product formation targeted to avoid chronic diseases such as cancer and neurodegenerative diseases. Various kinds of natural plants,[70] animals,[31] and marine sources[68] have been explored for the formulation of functional and nutraceutical foods by using their biologically active by-products. Among those sources, utilization of microorganisms for the production of nutraceutical and the functional food is emerging as an interesting and economically viable concept.[22] In addition to conventional ingredients, many novel metabolites with advanced functional and health benefits have been identified from microorganisms;[21,48] and such metabolites from food grade microorganisms are principal compounds for developing modern nutraceuticals, as microbial fermentation generally offers many advantages such as low energy requirements, lower CO_2 emissions, low toxic waste, and simpler purification scheme.[41]

Nowadays, science is playing a strategic role for human beings via curing various diseases by utilizing marine bioactives in pharmaceutical industries[72] as the old saying *"Life begins from sea by* Marika Reinke <https://marikareinke.com/2015/03/life-begins-at-sea/>." In this century, research studies are being focused on the marine life and their importance in medicinal corporations, utilizing these biologically active compounds by using previous investigations of marine resources. Marine life and environment consist of wide range of the living organisms that provide vast resources to the humanitarian system as compared to the land organisms.[25] Hence, marine biota has nutraceutical importance due to the presence of a number of bioactive compounds, for example, carbohydrates, proteins, terpenoids, and polyphenols for the remedy of many lifestyle-related disorders.[74]

Consumers show more interest in natural bioactive compounds, which are used by cosmetics companies, medicinal companies, and most importantly as a functional constituent in the daily diet. Especially those bioactive compounds, which are derived from marine organisms, are richest source of health promoters.[72] These extracted components include peptides, vitamins, fatty acids (FAs), phytochemicals, and polysaccharides,[12] which may derived from natural or chemical extraction.[73] About 71% of the earth surface is occupied by oceans having a large population of biological organisms.[3] These different groups of species contain bacteria, seaweeds (micro- and macroalgae), fish species, cyanobacteria, and crustaceans, which are used for secondary metabolites to change their friendly marine environment.[48] Secondary metabolites have certain health-related benefits as they can be used and utilized in nutraceutical products while many sources of these metabolites are not in access or utilized to specify their uses for mankind.[80] As the marine resources are gaining importance recently, research studies are being conducted on marine-derived substance and for finding many bioactive components (Fig. 3.1) having biological characteristics that enhance their capability as nutraceutical in food industry.[7]

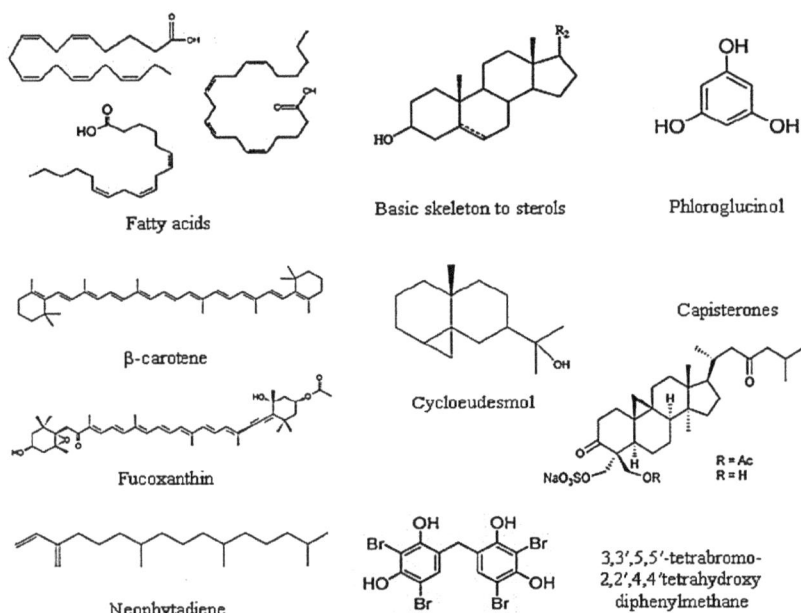

FIGURE 3.1   Chemical structure of bioactive compounds from marine food sources.

3.2   MARINE MICROORGANISMS

Oceanic microflora has remained largely unexplored recently, as marine science has focused only on either large charismatic creatures or economically important food species in the marine environment. As an ecological group, marine microorganisms also possess health promoting characteristics together with additional unique properties. For example, the diversity of marine microbes is substantially high and they exhibit diverse physiological adaptations for the marine environment.[87] The adaptation to such extreme environments has made marine microorganisms capable of producing unique microbial metabolites.[29] Furthermore, marine microorganisms are metabolically efficient and have evolved with effective strategies to use limited dissolved organic matters to produce more metabolites than the energy consumed. Energy-yielding carbon-monoxide-oxidation cycle, broadly distributed bacteriochlorophyll- and proteorhodopsin-based light energy harvesting systems are typical examples of oceanic microbial energy-saving cycle.[2] Moreover, during the past few decades, marine microbiology has gained a remarkable progress in terms of biodiversity and ecological role of marine microbes. The genome sequences of the most abundant inhabitants in the marine microbial world have been completely revealed.[13] Thus, marine microbes would be an ideal source to explore novel metabolites and biotechnological potentials that could be used in the food industry.

Nevertheless, it is a well-known fact that the marine microorganisms make up more than 90% of the marine biomass.[13] These tiny microscopic organisms are the front line of the marine food chains and act as living lungs for the planet by producing more than half of the world's oxygen. All microscopic organisms in salt water are referred as marine microorganisms.

With the identification of those constraints associated with marine animal- and plant-derived biologically active metabolites, marine microbes have received a growing attention as an alternative source of marine natural products. Tremendously diverse marine microbial community produces compounds with unique structural properties, and these compounds possess a broad spectrum of pharmaceutical properties such as antimicrobial, antituberculosis, anticoagulant, antiviral, antiparasitic, anti-inflammatory, antidiabetic, antimalarial, antiprotozoal, anthelmintic, antiplatelet, antiparasitic, and antitumor effects.[29,48] Hundreds of research articles are being published every year to reveal the potentials of the marine microbial metabolites in pharmaceutical applications. Waters et al.[87] have highlighted that more than half of the molecules currently in the marine drug development pipeline are highly likely to be produced by microorganisms. In addition, a number of food-grade metabolites, which possess promising pharmaceutical properties, have been isolated from marine microbes, and there is a clear potential to develop those active ingredients as modern nutraceuticals and functional food.[12]

Marine life contains mostly invertebrates which fall under different taxonomic groups that are categorized into various phyla, that is, mollusca, bryozoan, annelida, echinodermata, porifera, and arthropoda. Their habitat starts from intertidal to the deep-sea environment.[77] Hippocrates (father of modern medicine) explained the uses of different marine organisms, their components, and their diseases related cure for human beings.[85] In this regard, 40% trade of fishery and sea products is due to their beneficial health outcomes in the presence of bioactive compounds, that is, taurine and carotenoids while the other ingredients are minerals (selenium and iodine), polyunsaturated fatty acids (PUFAs), polysaccharides, and peptides.[5]

This chapter focuses on the current knowledge that demonstrates the suitability of marine bioactive compounds in drug discovery to treat and prevent various chronic diseases. The antioxidant, antidiabetic, anticancer, antihypertensive, and antibacterial effects are the limelight of this chapter.

3.3 MARINE-DERIVED COMPOUNDS

3.3.1 POLYSACCHARIDES

Polysaccharides are helpful in the structural formation of marine life and save them from the external environment. Brown and red algae contain alginates and carrageenan, which are linear biopolymers and are abundantly present in these polysaccharides.[35,84] Marine brown algae have more complicated structure, that is polysaccharides of sulfated matrix known as fucoidans, than red algae.[7] Complex structure of the fucoidans extracted from different marine species varies in sugar composition, sequence of sugar residue, linkage mode, and position of sulfate groups.[12,40,48] Sulfate linkage in the structure enhances the biological properties of fucoidans that is beneficial for the use in dairy industry as a nutraceutical component. Table 3.1 indicates polysaccharides in macro- and microalgae.

TABLE 3.1   Some of the Polysaccharides Found in Macro- and Microalgae.

Name	Main sources	Reference
Agar	Red algae (*Gelidiella acerosa*)	[57]
Alginate	Brown alage (*Laminaria digitat, Macrocystis pyrifera*)	[18]
Chitin	Crabs, shrimps, lobsters, prawns, krill	[58]
Laminarin	Brown algae (*Eisenia bicyclis*)	[46]

Microalgae are photosynthetic microorganisms as they can manufacture their food. They are native to marine and freshwater. The main characteristics of these organisms are presence of biochemical molecules having a great ability for various industrial activities such as the production of biofuels, pharmaceuticals, and cosmetics.[47] Dietary fibers can also be obtained from various marine sources. These dietary fibers are enriched with different mixtures of the polysaccharides that exhibit unique chemical composition and physical structure, whatever the source is.[38] That is why the species of marine algae that are taken in by man or animals as food exhibit physiological and metabolic outcomes. Some properties such as viscosity, fermentation, water holding capacity, binding ability of bile acid, cation exchange ability, and fecal bulking properties are due to the physiological changes instigated by dietary fibers.[16]

Marine bioactives are responsible to minimize the risk factors associated with cardiovascular diseases. It was found that low-density lipoprotein (LDL) cholesterol has lower levels in rats when they are fed with a diet containing dried *Ulva rigida*.[81]

3.3.2 PROTEINS AND ESSENTIAL FATTY ACIDS (PUFAS)

Protein plays a vital role like polysaccharides in structural integrity of marine species. The enzymatic breakdown of marine proteins gives collagen and gelatin that have the potential to act like bioactive peptides which is beneficial for nutraceutical purposes.[80] Recent studies examined the biochemical composition of microalgae and revealed the fact that they have high levels of organic molecules such as proteins, polysaccharides, lipids, phytosterols, and pigments including mineral salts.[7] High levels of PUFAs are found in marine microalgae unusually from the n-3 series including eicosapentaenoic (EPA) and docosahexaenoic acids (DHA). The FAs are integrated as neutral lipids known as triacyclglycerols, acting as storage compounds. They also act as polar lipids, that is, phospholipids and galactolipids as the essential component of chloroplast and endoplasmic reticulum compartments, respectively.[42] Table 3.2 lists PUFAs in marine organisms.

During the gastrointestinal digestion, bioactive peptides come out from the protein.[50] The arrangement of bioactive peptides of gelatin and collagen is too much strong and is used in nutraceutical companies to increase the product value. Particularly, fishes and marine algae are the major source of PUFAs especially omega 3 (ω-3) and omega 6 (ω-6) FAs and their precursor which are linoleic acid and linolenic acids,[42] while DHA and EPA play a beneficial role to improve the human health (Table 3.2).

TABLE 3.2   Proteins and Essential Fatty Acids (PUFAs) Found in Some Marine Organisms.

Name	Main sources	Reference
Arachidonic acid (AA)	*Mortierella alpina*	[90]
Docosahexaenoic acid (DHA)	Herring, mackerel, salmon, spirulina	[42]
Eicosapentaenoic acid (EPA)	Herring, mackerel, sardine, salmon	[65]
Hexadecatetraenoic acid	*Ulva pertusa*	[42]
γ-Linolenic acid (GLA)	*Mortierella, Spirulina*	[24]

3.3.3 POLYPHENOLS, CAROTENOIDS, AND PREBIOTICS

In marine algae, phenolic compounds are present as a result of oxidative stress. Green algae contain a significant proportion of flavonoids while the red algae have phlorotannins in abundance. Brown algae contain phlorotannins compounds including dieckol, eckol, and phloroglucinol.[64] Phlorotannins represent antioxidant behavior which allows its constituents to act as nutraceutical agents.[1] Carotenoid is a natural pigment consisting of 40 carbon skeletons and show lipophilic behavior.[39] Marine organisms synthesize various carotenoids, that is, fucoxanthin, beta-carotene, and astaxanthin (Table 3.3). *Dunaliella* sp. are used for the commercial preparation of beta-carotene while *Haematococcus* sp. are used for astaxanthin production on industrial level.[83]

TABLE 3.3   Antioxidants Found in Algae.

Name	Main sources	References
Astaxanthin	*Haematococcus pluvialis*	67
Phlorotannin	Brown algae (Fucaceae, Cystoseiraceae, Laminariaceae)	63
β-Carotene	*Dunaliella salina*	23

Prebiotics are the compounds, which are nondigestible, specifically fermented, and initiate the gut microbiota activities to provide certain health benefits to the host. They are oligosaccharides; prebiotic behavior also had been shown by algal polysaccharides.[51,80] Exopolysaccharide produced by lactic acid bacteria (LAB) of marine water that exhibits bifidogenic attributes. Moreover, *Spirulina platensis*, species of cyanobacteria, have the capacity to activate *Bifidobacterium* and *Lactobacillus* species by the enhancement of their prebiotics properties.[26,48]

3.3.4 TERPENOIDS

Antioxidants protect human body from the attack of reactive oxygen species (ROS). Food deterioration begins when lipid peroxidation, undesirable by-products and rancidity, occurs in the food product. Hydroxyl radicals, superoxide anion, and H_2O_2 are the ROS which reduce product safety and nutritional value by producing lipid peroxidation compounds.[76] To avoid these losses; many synthetic commercial antioxidants are being produced, for example, butylated hydroxy anisole, propyl gallate, tert-butylhydroquinone, and butylated hydroxytoluene that stops the oxidation and peroxidation reactions. However, to use these, synthetic antioxidants have been avoided as much as possible because of their potential health hazards. Otherwise, several studies reveal that terpenoids extracted from marine fungi, that is, torulene, neurosporaxanthin, beta-carotene, and gamma-carotene also exhibit antioxidant potential.[17]

All the terpenoids are formed by 2- and 5-carbon building units. On the basis of a number of terpenoid building blocks, they are categorized into sesterterpenes (C25), diterpenes (C20), sesquiterpenes (C15), and monoterpenes (C10). These compounds are utilized in the manufacturing of perfumes, cosmetics, herbicides, phytohormones, drugs, insecticides, and have many other applications. Current studies on terpenoids led to methylerythritol phosphate cycle and the discovery of a second biosynthetic cycle that prevails in plants and bacteria. Terpenoid carotenoids antioxidant activity is estimated by different methods that include: ferric reducing antioxidant power, lipid peroxide inhibition, nitric oxide scavenging, 1,1-diphenyl-2-picrylhydrazyl radical scavenging (DPPH), hydroxyl radical scavenging assays, and 2,2'-azino-bis-3-ethylbenzthiazoline-6-sulphonic acid—ABTS radical scavenging assays. Terpenoids derived from marine fungi are also useful in cosmetics, medicinal- and food industries. Moreover, these terpenoids show antiproliferative behavior on human cancerous cell lines in vitro while inhibitory behavior of cancer cell growth in mice.[79]

3.4 REMEDIES VIA MARINE FOODS

3.4.1 ANTICANCER EFFECTS

Cancer cells show uncontrolled growth and this inversion is life-threatening. The external contributing factors toward cancer include infectious organisms, radiation, chemical agents, tobacco, and poor nutrition while the internal factor includes inherited changes in genes. The detrimental effects of these external factors can be prevented by removing precancer lesions and early detection of the cancerous cells.[30] In developing countries, tobacco plays a major role in cancer prevalence, that is, 22% global cancer deaths while 71% deaths are due to lungs cancer and 20% deaths due to viral infection hepatitis B or C virus and human papillomavirus. Stomach, breast, lung, colon, and liver cancers contributed mostly to cancer deaths every year, and worldwide cancer deaths are increasing day by day, with an estimated 13.1 million deaths in 2030.[89] To prevent and treat these cancers, we need a safe, cheap, and potent medicine. Bioactive components-derived from marine biota behave as antimicrobial, antioxidant, anti-inflammatory, hypocholesterolemic, and antiviral agents.[69]

Since the 17th century, marine algae have been proven beneficial in manufacturing of drug due to the presence of phytochemical ingredients and vast diversification of algae groups. Categorized into seaweeds (macroalgae) and phytoplankton (microalgae), phytoplankton (microalgae) have more than 5000 species while macroalgae (seaweeds) have 6000 species. Phytoplankton includes dinoflagellates (Dinophyta), Chlorophyta, green and yellow-brown flagellates, (Chrysophyta, Prasinophyta, Prymnesiophyta, Cryptophyta, and Raphidophyta) diatoms (Bacillariophyta), and blue-green algae (Cyanophyta). Several studies found hydrophilic anticancerous components from marine algae. The side effects of these compounds are low, although various antitumor agents express their side effects on normal cells.[12] These bioactive agents may kill the cancerous cell by inducing apoptosis or to damage the cell signaling process by the stimulation of signaling enzymes, that is, protein kinase C family.[66]

3.4.2 ANTIHYPERTENSIVE EFFECTS

In developing countries, the major prevailing disease is hypertension (high blood pressure) that is a major risk factor for congestive heart failure, stroke, myocardial infarction, end-stage renal disease, and arteriosclerosis, angiotensin-converting enzyme (ACE, a dipeptidyl carboxypeptidase) plays a vital role in blood pressure regulation and also in cardiovascular function by changing the decapeptide angiotensin-I to the vasoconstricting octapeptide angiotensin II. In addition to it, angiotensin II inactivates bradykinin, which causes the increase in blood pressure in the arteries.[62,80] Therefore, the ACE inhibitor is a powerful agent to treat heart failure, myocardial infarction, stroke, and high blood pressure. Three types of ACE inhibitors are produced to control blood pressure, such as captopril having sulfhydryl moiety, fosinopril having a phosphorous group, and on the basis of ligand on the active site on ACE.

The first representative of this category is captopril, which has sulfhydryl moiety while the fosinopril has a phosphorus group, and lisinopril and enalapril have carboxyl moiety. Many studies have been conducted to investigate the manufacturing of ACE inhibitor that is still used for clinical antihypertensive drugs.[48,88] Otherwise, antitumor, immune modulatory, antihypertensive, antimicrobial, anticoagulant, and antioxidant behavior are physiological characteristic of marine fish-derived bioactive peptides.[37] Marine fish-derived ACE inhibitory peptides have been purified from enzymatic digestion of different fish materials derived from salmon,[53] bonito,[19] tuna,[28,53] sardine,[54] and Alaska pollack.[49]

3.4.3   CARDIOPROTECTION THROUGH MARINE FOODS

Cardiovascular disease (CVD) is a major disorder that affects blood vessels and heart causing heart failure, cerebrovascular disease (stroke), peripheral vascular disease, heart disease (heart attack), and hypertension (high blood pressure).[48] In 1999, only CVD contributed to about 1/3 of world deaths and, in 2010, it will be major reason of death in the developing countries.[84] Researches have revealed the preventive role of high fruits- and vegetable diet against CVD.[27,84] In addition, ω-3 PUFAs, antioxidants, vitamins and minerals, and dietary fibers with physical exercise are important to control and cure CVD. The risk of CVD can be lowered by glucose-lowering agents. Glucagon-like peptide-1 receptor (GLP-1R) belongs to B1 family of G protein, a paired receptor[36] and has been found to lower blood sugar level. Control of heart rate and blood pressure is very complex and is species specific due to GLP-1.

The production of nitric oxide (NO) and endothelial nitric oxide synthase phosphorylation increases by GLP-1R via the 5-AMP-activated protein kinase dependent pathway (AMPK). Some studies have suggested the potential clinical benefits of GLP-1 agonists.[86] The agonists may lower unfavorable circumstances of cerebrovascular and cardiovascular events that include stroke, acute coronary syndrome, cardiac mortality, and procedure of revascularization.[36]

Glucagon-like peptide-1 is delivered into the bloodstream as a result of ingestion of food. However, the amount of GLP-1 is reduced rapidly by the activation of dipeptidyl peptidase-4 (DPP-4) enzyme. DPP-4 also slows down the activation of postprandial response of GLP-1 that is compromised with the pathogenesis of type 2 diabetes.[82] It was examined that individuals on freshwater algae diet exhibit a low level of total cholesterol, LDL cholesterol and triglyceride levels, and high-level of high-density lipoprotein cholesterol than those given a water placebo. It has been shown that intake of about 8 ounces of seafood per week minimizes the risk of cardiac death among individuals either having or preexisting CVD.[52]

3.4.4   ANTIOXIDANT PROPERTY

Positive health effects are found in the presence of antioxidants. Antioxidants can protect the human body against descent due to the free radicals and ROS. This is because of the presence of reactive oxygen, hydrogen peroxide, and super oxidation and hydroxyl radicals. When ROS and free radicals attack macro molecules such as DNA, proteins, and lipids, they may increase the incidence of cancer, diabetes mellitus, and inflammatory diseases with tissues.[6]

Marine algae are the richest source of antioxidants among natural sources.[44] The function of sulfated polysaccharides (SPs) from algae not only provides dietary fiber, but it also subsidizes as antioxidant in marine life. Fucoidans, laminaran, and alginic acid have high antioxidant activities.[59] The antioxidant activity of SPs has been investigated by different methods such as DPPH radical scavenging, lipid peroxide inhibition, NO scavenging, super oxide radical, and hydroxyl scavenging assays. By the function, many marine-derived SPs have their antioxidant properties in phosphatidylcholine-liposomal suspension and an organic solvent such as acetone, methyl acetate, and toluene.[4] The dependence of antioxidants is on structural features as the degree of sulfating, molecular weight, and the type of major sugar. For example, the lower molecular weighted SPs show potent antioxidant activities than that of the higher weighted SPs.

β-carotene is the most important food colorant used in every part of the world. It is used in food and beverage industries for improvement of the appearance of products.[15] It has strong antioxidant properties. The area of attraction for the researchers is to know the microalgal production of carotenoids such as β-carotene and their valuable bioactive ingredients that are found in algal cell in high concentration. The formation of ROS can be prevented by the protective role of carotenoids.[4] The specific scavenging of oxygen containing compounds or metal chelating ability during peroxidation is due to the antioxidant activity. Moreover, greater antioxidant properties are found in peptides isolated from marine fish proteins than α-tocopherol in different oxidative systems.[32]

3.4.5 ANTIBACTERIAL ACTIVITY

Phlorotannins comprise 1–15% of the thallus dry mass and occur in fucosan granules, which are called physodes within algal cells.[81] The antibacterial activity of phlorotannins has been described to be due to the reservation of oxidative phosphorylation, and special capability to cause cell lysis by binding with enzymes and cell membranes in bacteria. The phenolic rings of aromatic compounds and OH groups of the phloroglucinol units fix to the-NH groups of proteins of bacteria by H-bond and hydrophobic interactions.[60] Kamei and Isnansetyo[33] stated that the bacteriolytic activity of phloroglucinol compounds to fight with *Vibrio* species is increased when tertiary structures like methyl-or acetyl-vinyl are present. However, for the penetration of the gram-negative species *Vibrio parahaemolyticus*, a greater minimum inhibitory concentration (MIC) is required as compared with the gram-positive methicillin-resistant *Staphylococcus aureus* (MRSA). Due to less penetrable nature and physiological differences in β-lactamase mechanisms, the outer lipopolysaccharide membrane of gram-negative species is the case for most antibacterial actions in comparison with the peptidoglycan gram-positive layer.[56]

Phlorotannins extracted from a seaweed *Sargassum thunbergii* also can injure the cell wall and cell membrane of *Vibrio parahaemolyticus* which cause leakage and deconstruction of permeability of membrane.[56] Therefore, phlorotannis collected from algae may potentially be used in aquaculture drugs as well as to control food safety.

3.4.6   ANTICOAGULANT PROPERTY

Marine brown algae hold blood anticoagulant properties.[34] The SPs that are derivatives of marine are alternative sources for the formation of novel anticoagulant drugs.[10] The most widely studied properties of SPs is the anticoagulant properties.[11] High anticoagulant activity has been identified by two types of SPs including sulfated galactans that are also known as carrageenan derived from marine algae[8] and sulfatedfucoidan that is derived from marine brown algae.[9] Moreover, other studies have revealed that six-fold higher activities are obtained from heparin contained in the *Monostroma nitidum*. Moreover, anticoagulant activity in the extracts of marine brown algae is higher than the red and green algae extracts.[55] Marine brown algae are built-up of phloroglucinol-based polyphenols and have blood anticoagulant characteristics.

3.4.7   IMMUNOMODULATORY EFFECTS

Infection and neoplasia can be avoided by immune system participation, while many approaches are introduced to stimulate the immune responses properly. Nutritional intervention includes balancing the intake of essential nutrients, and consuming functional foods is widely known method to introduce the health promoting benefits of immune system. Recently, interest has been arising due to the bioactive compound obtained from the marine-derived proteins.[43] These are certain protein fragments, which are easily absorbable and digestible, low in molecular weight and exhibiting beneficial response on body function to improve positive health effects, that is, antimicrobial, immunomodulatory, antihypertensive, and antioxidant properties.[45]

Immunomodulatory peptides come from dietary protein, and the breakdown occurs in the gut. However, the proportion of peptides is too low to induce effect on the immune system.[20] On the other hand, the breakdown of enzymatic proteins helps in the production of bioactive peptide fractions and possesses beneficial food constituent or nutraceutical ingredient.

3.5   SUMMARY

Marine organisms are a rich source of novel and promising bioactive molecules for a wide range of applications, including new therapeutics, cosmetics, and biotechnology. Biodiversity in the seas has only been partially explored, although marine organisms are excellent sources for many industrial products. Although marine invertebrates have been the focus in search for new marine natural products, yet marine algae contribute to the wealth of novel bioactives. It is possible to successfully collect, isolate, and classify marine organisms, such as bacteria, fungi, micro- and macroalgae, cyanobacteria, and marine invertebrates from the oceans and seas globally. Extracts and purified compounds of these organisms can be studied for the several therapeutically and industrially significant biological activities including anticancer, anti-inflammatory, antihypertensive, antibacterial, anticoagulant, and immunomodulatory activities by applying a wide variety of screening tools. Traditionally, the search for bioactivity from natural sources primarily involves target-driven approaches using single-target biochemical assays for primary screening. The research innovations can be targeted for the industrial product development to improve the growth and productivity of marine-based medicine. Marine research aims at a better understanding of environmentally conscious sourcing of marine-based medicine and food products and increased public awareness of marine biodiversity. Marine research is expected to offer novel marine-based lead compounds for industries and strengthen their product portfolios related to pharmaceutical, nutraceutical, cosmetic, agrochemical, food processing, and material applications.

This chapter focuses on the current knowledge that demonstrates the suitability of marine bioactive compounds in drug discovery to treat and prevent various chronic diseases. The antioxidant, antidiabetic, anticancer, antihypertensive, and antibacterial effects are the limelight of this chapter.

KEYWORDS

- anti-inflammation
- antioxidants
- Ayurveda medicines
- carcinogenesis
- cardiovascular disease
- carotenoids
- diet-based therapies
- fatty acids
- free radicals
- functional foods
- hypercholesterolemia
- hyperglycemia
- immune dysfunctions
- marine bioactives

- marine invertebrates
- marine metabolites
- metabolic syndromes
- nutraceuticals
- nutrition
- oxidative stress
- peptides

- pharmaceuticals
- phenolics
- phlorotannins
- phytochemicals
- polysaccharides
- proteins
- terpenoids

REFERENCES

1. Arct, J.; Pytkowska, K. Flavonoids as Components of Biologically Active Cosmeceuticals. *Clin. Dermatol.* **2008,** *26,* 347–357.
2. Azam, F.; Malfatti, F. Microbial Structuring of Marine Ecosystems. *Nat. Rev. Microbiol.* **2007,** *5*(10), 782–791.
3. Baharum, S. N.; Beng, E. K.; Mokhtar, M. A. A. Marine Microorganisms: Potential Application and Challenges. *J. Biol. Sci.* **2010,** *10,* 555–564.
4. Boisvert, C.; Beaulieu, L.; Bonnet, C.; Pelletier, E. Assessment of the Antioxidant and Antibacterial Activities of Three Species of Edible Seaweeds. *J. Food Biochem.* **2015,** *39,* 377–387.
5. Borresen, T. Seafood for Improved Health and Wellbeing. *Food Technol.* **2009,** *63,* 88.
6. Butterfield, D. A.; Castenga, A.; Pocernich, C. B.; Drake, J.; Scapagnini, G.; Calabrese, V. Nutritional Approaches to Combat Oxidative Stress in Alzheimer's Disease. *J. Nutr. Biochem.* **2002,** *13,* 444–461.
7. Cardoso, M. J.; Costa, R. R.; Mano, J. F. Marine Origin Polysaccharides in Drug Delivery Systems. *Mar. Drugs* **2016,** *14*(2), 34.
8. Carlucci, M. J.; Pujol, C. A.; Ciancia, M.; Noseda, M. D.; Matulewicz, M. C.; Damonte, E. B. Antiherpetic and Anticoagulant Properties of Carrageenans from the Red Seaweed Gigartina Skottsbergii and Their Cyclized Derivatives: Correlation Between Structure and Biological Activity. *Int. J. Biol. Macromol.* **1997,** *20,* 97–105.
9. Chevolot, L.; Foucault, A.; Chaubet, F.; Kervarec, N.; Sinquin, C.; Fisher, A. M. Further Data on the Structure of Brown Seaweed Fucans: Relationships with Anticoagulant Activity. *Carbohydr. Res.* **1999,** *319,* 154–165.
10. Church, F. C.; Meade, J. B.; Treanor, E. R.; Whinna, H. C. Antithrombin Activity of Fucoidan. The Interaction of Fucoidan with Heparin Cofactor II, Antithrombin III, and Thrombin. *J. Biol. Chem.* **1989,** *264,* 3618–3623.
11. Costa, L. S.; Fidelis, G. P.; Cordeiro, S. L.; Oliveira, R. M.; Sabry, D. A.; Camara, R. B. G. Biological Activities of Sulfated Polysaccharides from Tropical Seaweeds. *Biomed. Pharmacother.* **2010,** *64,* 21–28.

12. De-Jesus Raposo, M. F.; de Morais, A. M. B.; Santos Costa de Morais, R. M. Marine Polysaccharides from Algae with Potential Biomedical Applications. *Mar. Drugs* **2015**, *13*, 2967–3028.

13. Delong, E. F. Modern Microbial Seascapes. *Nat. Rev. Microbiol.* **2007**, *5*(10), 755–757.

14. Dias, D. A.; Urban, S; Roessner, U. A Historical Overview of Natural Products in 466 Drug Discovery. Metabolites **2012**, *2*(2), 303–336.

15. Dufossé, L.; Galaup, P.; Yaron, A.; Arad, S. M.; Blanc, P.; Chidambara Murthy, K. N.; Ravishankar, G. A. Microorganisms and Microalgae As Sources of Pigments for Food Use: A Scientific Oddity or An Industrial Reality. *Trends Food Sci. Technol.* **2005**, *16*, 389–406.

16. Eastwood, M. A.; Morris, E. R. Physical Properties of Dietary Fiber that Influence Physiological Function: A Model for Polymers Along the Gastrointestinal Tract. *Am. J. Clin. Nutr.* **1992**, *55*(2), 436–442.

17. Ebel, R. Terpenes from Marine-Derived Fungi. *Mar. Drugs* **2010**, *8*, 2340–2368.

18. Fertah, M., Belfkira, A., Dahmane, E. M., Taourirte, M., Brouillette, F. Extraction and Characterization of Sodium Alginate from Moroccan Laminaria Digitata Brown Seaweed. *Arabian J. Chem.* **2014**, 1–8. https://www.researchgate.net/deref/http%3A%2F%2Fdx.doi.org%2F10.1016%2Fj.arabjc.2014.05.003

19. Fujita, H.; Yokoyama, K.; Yoshikawa, M. Classification and Antihypertensive Activity of Angiotensin I-Converting Enzyme Inhibitory Peptides Derived from Food Proteins. *J. Food Sci.* **2000**, *65*, 564–569.

20. Gauthier, S. F.; Pouliot, Y.; Saint-Sauveur, D. Immunomodulatory Peptides Obtained by the Enzymatic Hydrolysis of Whey Proteins. *Int. Dairy J.* **2006**, *16*, 1315–1323.

21. Ghorai, S.; Banik, S. P.; Verma, D.; Chowdhury, S.; Mukherjee, S.; Khowala, S. Fungal Biotechnology in Food and Feed Processing. *Food Res. Int.* **2009**, *42*, 577–587.

22. Gobbetti, M.; Di Cagno, R.; De Angelis, M. Functional Microorganisms for Functional Food Quality. *Crit. Rev. Food Sci. Nutr.* **2010**, *50*(8), 716–727.

23. Guedes, A. C.; Amaro, H. M.; Malcata F. X. Microalgae as Sources of Carotenoids. *Mar. Drugs* **2011**, *9*, 625–44.

24. Hamed, I.; Özogul, F.; Özogul, Y.; Regenstein, J. M. Marine Bioactive Compounds and Their Health Benefits: A Review. *Compr. Rev. Food Sci. Food Saf.* **2015**, *14*(4), 446–465.

25. Hill, R. T.; Fenical, W. Pharmaceuticals from Marine Natural Products: Surge or Ebb? *Curr. Opin. Biotechnol.* **2010**, *21*, 777–779.

26. Honypattarakere, T.; Cherntong, N.; Wickienchot, S.; Kolida, S.; Rastall, R. A. In vitro Prebiotic Evaluation of Exopolysaccharides Produced by Marine Isolated Lactic Acid Bacteria. *J. Biotechnol.* **2010**, S1–S57.

27. Hu, F. B.; Willett, W. C. Optimal Diets for Prevention of Coronary Heart Disease. *J. Am. Med. Assoc.* **2002**, *288*, 2569–2578.

28. Hwang, J. S. Impact of Processing on Stability of Angiotensin I-Converting Enzyme (ACE) Inhibitory Peptides Obtained from Tuna Cooking Juice. *Food Res. Int.* **2010**, *43*, 902–906.

29. Imoff, J. F.; Labes, A.; Wise, J. Bio-Mining the Microbial Treasures of the Ocean: New Natural Products. *Biotechnol. Adv.* **2011**, *29*, 468–482.

30. Imran, M.; Saeed, F.; Nadeem, M.; Arshad, M. U.; Ullah, A.; Suleria, H. A. R. Cucurmin; Anticancer and Antitumor Perspectives—A Comprehensive Review. *Crit. Rev. Food Sci. Nutr.* (just-accepted), **2016,** 00–00.

31. Johanningsmeier, S. D.; Harris, G. K. Pomegranate as a Functional Food and Nutra-ceutical Source. *Ann. Rev. Food Sci. Technol.* **2011,** *2,* 181–201.

32. Jung, W. K.; Je, J. Y.; Kim, S. K. A Novel Anticoagulant Protein from *Scapharca broughtonii. J. Biochem. Mol. Biol.* **2001,** *35,* 199–205.

33. Kamei, Y.; Isnansetyo, A. Lysis of Methicillin-Resistant *Staphylococcus aureus* by 2,4-Diacetylphloroglucinol Produced by *Pseudomonas sp.* AMSN Isolated from A Marine Alga. *Int. J. Antimicrob. Agents* **2003,** *21,* 71–74.

34. Killing, H. Zur biochemie der meersalgen (On Biochemistry of Seaweed). *Zeitschrift fur Physiologische Chemie* **1913,** *83,* 171–197.

35. Kim, S. K.; Ravichandran, Y. D.; Khan, S. B.; Kim, Y. Prospective of the Cosme-ceuticals Derived from Marine Organisms. *Biotechnol. Bioprocess Eng.* **2008,** *13,* 511–523.

36. Kirkpatrick, A.; Heo, J.; Abrol. R.; Goddard, W. A. Predicted Structure of Agonist-Bound Glucagon-Like Peptide 1 Receptor, a Class B G Protein Coupled Receptor. *Proc. Natl. Acad. Sci. U. S. A.* **2012,** *109*(49), 19988–19993.

37. Kobayashi, Y.; Yamauchi, T.; Katsuda, T.; Yamaji, H.; Katoh, S. Angiotensin-I Converting Enzyme (ACE) Inhibitory Mechanism of Tripeptides Containing Aromatic Residues. *J. Biosci. Bioeng.* **2008,** *106,* 310–312.

38. Lahaye, M.; Kaeffer, B. Seaweed Dietary Fibers: Structure, Physico-Chemical and Biological Properties Relevant to Intestinal Physiology. *Sci. Aliments* **1997,** *17*(6), 563–584.

39. Lee, J. H.; Seo, Y. B.; Jeong, S. Y.; Nam, S. W.; Kim, Y. T. Functional Analysis of Combinations in Astaxanthin Biosynthesis Genes from *Paracoccus haeundaensis. Biotechnol. Bioprocess Eng.* **2007,** *12,* 312–317.

40. Li, L. Y.; Sattler, I.; Deng, Z. W.; Groth, I.; Walther, G.; Menzel, K. D.; Peschel, G.; Grabley, S.; Lin, W. H. Seco-Oleane-Type Triterpenes from *Phomopsis sp.* (Strain HK10458) Isolated from the Mangrove Plant *Hibiscus tiliaceus.* Phytochemistry **2008,** *69,* 511–517.

41. Lin, C.; Koval, A.; Tishchenko, S.; Gabdulkhakov, A.; Tin, U.; Solis, G.P.; Katanaev, V.L. Double Suppression of the gα Protein Activity by RGS Proteins. *Mol. Cell* **2014,** *53*(4), 663–671.

42. Lordan, S.; Ross, R. P.; Stanton, C. Marine Bioactives as Functional Food Ingredi-ents: Potential to Reduce the Incidence of Chronic Diseases. *Mar. Drugs* **2011,** *9,* 1056–100.

43. Ma, M. S.; Bae, I. Y.; Lee, H. G.; Yang, C. B. Purification and Identification of Angio-tensin I-Converting Enzyme Inhibitory Peptide from Buckwheat (*Fagopyrum escul-entum* Moench). *Food Chem.* **2006,** 96(1), 36–42.

44. Mayer, A. M. S.; Hamann, M. T. Marine Pharmacology in 1999: Compounds with Antibacterial, Anticoagulant, Antifungal, Anthelmintic, Anti-Inflammatory, Anti-platelet, Antiprotozoal and Antiviral Activities Affecting the Cardiovascular, Endo-crine, Immune and Nervous Systems, and Other Miscellaneous Mechanisms of Action. *Comp. Biochem. Physiol. Part C* **2002,** *132,* 315–339.

45. Meisel, H. Food-Derived Bioactive Proteins and Peptides as Potential Components of Nutraceuticals. *Curr. Pharm. Des.* **2007,** *13*(9), 873–874.
46. Menshova, R. V.; Ermakova, S. P.; Anastyuk, S. D.; Isakov, V. V.; Dubrovskaya, Y. V.; Kusaykin, M. I.; Um, B. H.; Zvyagintseva, T. N. Structure, Enzymatic Transformation and Anticancer Activity of Branched High-Molecular-Weight Laminaran from Brown Alga *Eisenia bicyclis*. *Carbohydr. Polym.* **2014,** *99,* 101–109.
47. Mimouni, V.; Ulmann, L.; Pasquet, V.; Mathieu, M.; Picot, L.; Bougaran, G.; Cadoret, J. P.; Morant-Manceau, A.; Schoefs, B. The Potential of Microalgae for the Production of Bioactive Molecules of Pharmaceutical Interest. *Curr. Pharm. Biotechnol.* **2012,** 13(15), 2733–50.
48. Nadia, R.; Susan, C.; Stefano, G.; Maria, C. Polysaccharides from the Marine Environment with Pharmacological, Cosmeceutical and Nutraceutical Potential. *Molecules* **2016,** *21,* 551.
49. Nakajima, K.; Yoshie-Stark, Y; Ogushi, M. Comparison of ACE Inhibitory and DPPH Radical Scavenging Activities of Fish Muscle Hydrolysates. *Food Chem.* **2009,** *114,* 844–851.
50. Ngo, D. H.; Vo, T. S.; Ngo, D. N.; Wijesekara, I.; Kim, S. K. Biological Activities and Potential Health Benefits of Bioactive Peptides Derived from Marine Organisms. *Int. J. Biol. Macromol.* **2012,** *51,* 4, 378–383.
51. O'Sullivan, L.; Murphy, B.; McLoughlin, P.; Duggan, P.; Lawlor, P. G.; Hughes, H.; Gardiner, G. E. Prebiotics from Marine Macroalgae for Human and Animal Health Applications. *Mar. Drugs* **2010,** *8,* 2038–2064.
52. Oben, J.; Enonchong, E.; Kuate, D.; Mbanya, D.; Thomas, T.; Hildreth, D.; Ingolia, T.; Tempesta, M. The Effects of Proalgazyme Novel Algae Infusion on Metabolic Syndrome and Markers of Cardiovascular Health. *Lipids Health Dis.* **2007,** *6,* 20.
53. Ohta, T.; Iwashita, A.; Sasaki, S.; Kawamura, Y. Antihypertensive Action of the Orally Administered Protease Hydrolysates of Chum Salmon Head and Their Angiotensin I-Converting Enzyme Inhibitory Peptides. *Food Sci. Technol. Int.* **1997,** *4,* 339–343.
54. Otani, L.; Ninomiya, T.; Murakami, M.; Osajima, K.; Kato, H.; Murakami, T. Sardine Peptide with Angiotensin I-Converting Enzyme Inhibitory Activity Improves Glucose Tolerance in Stroke-Prone Spontaneously Hypertensive Rats. *Biosci. Biotechnol. Biochem.* **2009,** *73,* 2203–2209.
55. Patankar, M. S.; Oehninger, S.; Barnett, T.; Williams, R. L.; Clerk, G. F. A Revised Structure for Fucoidan may Explain Some of its Biological Activities. *J. Biol. Chem.* **1993,** *268,* 21770–21776.
56. Poudyal, H.; Panchal, S. K.; Diwan, V.; Brown, L. Omega-3 Fatty Acids and Metabolic Syndrome: Effects and Emerging Mechanisms of Action. *Prog. Lipid Res.* **2011,** *50,* 372–387.
57. Prasad, K.; Goswami, A. M.; Meena, R.; Ramavat, B. K.; Ghosh, P. K.; Siddhanta, A. K. Superior Quality Agar from Red Alga *Gelidiella acerosa* (*Rhodophyta, Gelidiales*) from Gujarat Coast of India: An Evaluation. *Indian J. Mar. Sci.* **2006,** *35*(3), 268–274.
58. Rinaudo, M. Chitin and Chitosan: Properties and Applications. *Prog. Polym. Sci.* **2006,** *31,* 603–632.
59. Rocha de Souza, M. C.; Marques, C. T.; Dore, C. M. G.; Ferreira da Silva, F. R.; Rocha, H. A. O.; Leite, E. L. Antioxidant Activities of Sulphated Polysaccharides from Brown and Red Seaweeds. *J. Appl. Phycol.* **2007,** *19,* 153–160.

60. Sampath, N. S. K.; Nazeer; R. A.; Jaiganesh, R. Purification and Identification of Antioxidant Peptides from the Skin Protein Hydrolysate of Two Marine Fishes, Horse Mackerel (*Magalaspis cordyla*) and Croaker (*Otolithes ruber*). *Amino Acids* **2012,** *42,* 1641–1649.

61. Schieber, A. Functional Foods and Nutraceuticals. *Food Res. Int.* **2012,** *46*(2), 437.

62. Shahidi, F.; Zhong, Y. Bioactive Peptides. *J. AOAC Int.* **2008,** *91,* 914–931.

63. Shibata, T.; Kawaguchi, S.; Hama, Y.; Inagaki, M.; Yamaguchi, K.; Nakamura, T. Local and Chemical Distribution of Phlorotannins in *Brown algae*. *J. Appl. Phycol.* **2004,** *16,* 291–6.

64. Shibata, T.; Ishimaru, K.; Kawaguchi, S.; Yoshikawa, H.; Hama, Y. Antioxidant Activities of Phlorotannins Isolated from Japanese Laminariaceae. *J. Appl. Phycol.* **2008,** *20,* 705–711.

65. Sijtsma, L., de Swaaf, M. E. Biotechnological Production and Applications of the Omega-3 Polyunsaturated Fatty Acid Docosahexaenoic Acid. *Appl. Microbiol. Biotechnol.* **2004,** *64,* 146–53.

66. Sithranga Boopathy, N.; Kathiresan, K. Anticancer Drugs from *Marine flora*: An Overview. *J. Oncol.* **2010,** *2010,* 18.

67. Spiller, G. A.; Dewell, A. Safety of an Astaxanthin-Rich *Haematococcus pluvialis* Algal Extract: A Randomized Clinical Trial. *J. Med. Food* **2003,** *6,* 51–6.

68. Suleria, H. A. R. Bioactive Compounds from Australian Blacklip Abalone (*Haliotis rubra*) Processing Waste. Ph.D. Thesis, School of Medicine, The University of Queensland, Australia. 2016, DOI: 10.14264/uql.2016.814.

69. Suleria, H. A. R.; Addepalli, R.; Masci, P.; Gobe, G.; Osborne, S. A. In vitro Anti-Inflammatory Activities of Blacklip Abalone (*Haliotis rubra*) in RAW 264.7 Macrophages. *Food Agric. Immunol.* **2017,** 1–14.

70. Suleria, H. A. R.; Butt, M. S.; Anjum, F. M.; Saeed, F.; Khalid, N. Onion: Nature Protection Against Physiological Threats. *Crit. Rev. Food Sci. Nutr.* **2015,** *55*(1), 50–66.

71. Suleria, H. A. R.; Butt, M. S.; Khalid, N.; Sultan, S.; Raza, A.; Aleem, M.; Abbas, M. Garlic (*Allium sativum*): Diet Based Therapy of 21st Century–A Review. *Asian Pac. J. Trop. Dis.* **2015,** *5*(4), 271–278.

72. Suleria, H. A. R.; Gobe, G.; Masci, P.; Osborne, S. A. Marine Bioactive Compounds and Health Promoting Perspectives; Innovation Pathways for Drug Discovery. *Trends Food Sci. Technol.* **2016,** *50,* 44–55.

73. Suleria, H. A. R.; Hines, B. M.; Addepalli, R.; Chen, W.; Masci, P.; Gobe, G.; Osborne, S. A. In vitro Anti-Thrombotic Activity of Extracts from *Blacklip abalone* (*Haliotis rubra*) Processing Waste. *Mar. drugs* **2016,** *15*(1), 8.

74. Suleria, H. A. R.; Masci, P.; Gobe, G.; Osborne, S. Current and Potential Uses of Bioactive Molecules from Marine Processing Waste. *J. Sci. Food Agric.* **2016,** *96*(4), 1064–1067.

75. Suleria, H. A. R.; Osborne, S.; Masci, P.; Gobe, G. Marine-Based Nutraceuticals: An Innovative Trend in the Food and Supplement Industries. *Mar. drugs* **2015,** *13*(10), 6336–6351.

76. Suleria, H. A.; Sadiq Butt, M.; Muhammad Anjum, F.; Saeed, F.; Batool, R.; Nisar Ahmad, A. Aqueous Garlic Extract and its Phytochemical Profile; Special Reference to Antioxidant Status. *Int. J. Food Sci. Nutr.* **2012,** *63*(4), 431–439.

77. Suleria, H. R.; Masci, P.; Gobe, G.; Osborne, S. Therapeutic Potential of Abalone and Status of Bioactive Molecules: A Comprehensive Review. *Crit. Rev. Food Sci. Nutr.* **2017,** *57*(8), 1742–1748.

78. Suleria, H. A. R. Marine Processing Waste-In Search of Bioactive Molecules. *Nat. Prod. Chem. Res.* **2016,** *4,* e118.

79. Sultan, M. T.; Buttxs, M. S.; Qayyum, M. M. N.; Suleria, H. A. R. Immunity: Plants as Effective Mediators. *Crit. Rev. Food Sci. Nutr.* **2014,** *54*(10), 1298–1308.

80. Tabarsa, M.; Park, G. M.; Shin, I. S.; Lee, E.; Kim, J. K.; You, S. Structure-Activity Relationships of Sulphated Glycoproteins from *Codium fragile* on Nitric Oxide Releasing Capacity from RAW264.7 Cells. *Mar. Biotechnol.* **2015,** *17,* 266–276.

81. Taboada, C.; Millán, R.; Míguez, I. Composition, Nutritional Aspects and Effect on Serum Parameters of Marine Algae *Ulva rigida. J. Sci. Food Agric.* **2010,** *90,* 445–449.

82. Tella, S. H., Rendell, M. S. Glucagon-Like Polypeptide Agonists in Type 2 Diabetes Mellitus: Efficacy and Tolerability, A Balance. *Ther. Adv. Endocrinol. Metab.* **2015,** *6*(3), 109–134.

83. Vidanarachchi, J. K.; Kurukulasuriya, M. S.; Wijesundara, W. M. N. M. Biological and Biomedicinal Applications of Marine Nutraceuticals. In *Marine Cosmeceuticals: Trends and Prospects;* Kim, S. K., Ed.; CRC Press: Boca Raton, FL, USA, 2013; pp 345–392.

84. Vo, T.S.; Ngo, D.H.; Kang, K.H.; Jung, W.K.; Kim, S.K. The Beneficial Properties of Marine Polysaccharides in Alleviation of Allergic Responses. *Mol. Nutr. Food Res.* **2015,** *59,* 129–138.

85. Voultsiadou, E. Therapeutic Properties and Uses of Marine Invertebrates in the Ancient Greek World and Early Byzantum. *J. Ethnopharmacol.* **2010,** *130,* 237–247.

86. Wang, B.; Zhong, J.; Lin, H.; Zhao, Z.; Yan, Z.; He, H.; Blood Pressure-Lowering Effects of GLP-1 Receptor Agonists Exenatide and Liraglutide: A Meta-Analysis of Clinical Trials. *Diabetes Obes. Metab.* **2013,** *15,* 737–749.

87. Waters, A. L.; Hill, R. T.; Place, A.R.; Hamann, M. T. The Expanding Role of Marine Microbes in Pharmaceutical Development. *Curr. Opin. Biotechnol.* **2010,** *21*(6), 780–786.

88. Wijesekara, I.; Qian, Z. J.; Ryu, B. M.; Ngo, D. H.; Kim, S. K. Purification and Identification of Antihypertensive Peptides from Seaweed Pipefish (*Syngnathus schlegeli*) Muscle Protein Hydrolysate. *Food Res. Int.* **2010,** *44*(3), 703–707.

89. World Health Organization (WHO). *February Cancer Fact Sheet.* Factsheet 297, 2012.

90. Yap, C. Y.; Chen, F. Polyunsaturated Fatty Acids: Biological Significance, Biosynthesis, and Production by Microalgae and Microalgae-Like Organisms. In *Algae and Their Biotechnological Potential;* Chen, F., Jiang, Y., Eds.; Kluwer Academic Publishers: Dordrecht, The Netherland 2001; pp 1–32.

PART II
Plant-Based Pharmaceuticals in Human Health: Review

MONOTERPENES-BASED PHARMACEUTICALS: A REVIEW OF APPLICATIONS IN HUMAN HEALTH AND DRUG DELIVERY SYSTEMS

IRINA PEREIRA, ALEKSANDRA ZIELIŃSKA, FRANCISCO J. VEIGA, ANA C. SANTOS, IZABELA NOWAK, AMÉLIA M. SILVA, and ELIANA B. SOUTO

CONTENTS

ABSTRACT

Plant-derived monoterpenes (such as α-pinene, citral, limonene, or linalool) constitute a group of terpenes (also called terpenoids or isoprenoids) produced by several plant species. Among them, we mention pine fruit which contains of α-pinene (turpentine scent), (−)-limonene (scent of oranges), and bornyl acetate (camphor aroma). Furthermore, there is also orange fruit consisting of (+)-limonene (scent of oranges) and lemon fruit consisting of (+)-limonene and citral (scent of lemons). Monoterpenes are the most widespread group of naturally occurring organic chemicals being constituted by two isoprene units. These plant-derived substances are used as flavors and fragrances for the synthesis of pharmaceuticals and cosmetic products on account of their volatility. Moreover, monoterpenes show antioxidant, antimicrobial, analgesic, and anxiolytic properties. Nowadays, cancer frequency is increasing and treatments with available chemotherapeutic agents become less effective. Therefore, the search for plant products as anticancer agents has achieved great popularity. The latest in vitro and in vivo studies have proven that monoterpenes are efficient in reducing the tumor volume and tumor cell proliferation without side effects. Additionally, plant-derived monoterpenes induce autophagy and apoptosis by interfering with different intracellular signaling pathways. The available chemotherapeutic agents show lack of selectivity for tumor tissue and can be cytotoxic. New drug delivery systems are under investigation in order to produce an anticancer agent with the maximum therapeutic value. In this chapter, we focus on the chemical properties and pharmaceutical applications of the selected plant-derived monoterpenes, which are suitable candidates for these new drug delivery systems mainly due to their antiproliferative activities.

4.1   INTRODUCTION

The scientific field is currently focused mostly on innovative therapeutic agents obtained from natural products, which can have a vast application in medicine,[99] pharmacy as well as in cosmetic and fragrance industries.[94,132] Essential oils containing monoterpenes are used widely to prevent and treat human disease.[54] These volatile constituents show a wide range of bioactivities such as antibacterial, antiviral, or antioxidants.[9] Nowadays, monoterpenes are used as flavors and fragrances for the synthesis of pharmaceuticals and cosmetic products on account of their volatility.[84] According to the latest scientific reports, these plant-derived substances are efficient in reducing the tumor volume and tumor cell proliferation without side effects.[16,34] The turning point of pharmaceutical contemporary research may involve the use of nanoparticle systems, based on lipids and loaded plant-derived monoterpenes. Lipid nanoparticles such as solid lipid nanoparticles (SLNs) and nanostructured lipid carriers (NLCs) due to the nanometric size and control release of the active substance have attracted most attention of pharmacy and cosmetology industries and are efficient carriers for the selected plant-derived monoterpenes.[133]

Monoterpenes are a class of terpenes consisting of two isoprene units and having the molecular formula $C_{10}H_{16}$. Structural formula of monoterpenes maybe linear (acyclic) or containing rings (cyclic).[81] Biochemical modifications such as oxidation or rearrangement produce the related monoterpenoids. In addition to linear attachments, the isoprene units can make connections to form rings. The most common ring size in monoterpenes is a six-membered ring.[81] A classic example is the cyclization of geranyl pyrophosphate to form limonene. Monoterpenes are mainly emitted by forests and form aerosols that can serve as cloud condensation nuclei. Such aerosols can increase the brightness of clouds and cool the climate.[127] Several monoterpenes derivatives have antibacterial or even anesthetic activity, such as linalool.[99]

In this chapter, the possible medical uses of the selected plant-derived monoterpenes are discussed. In the first part, the chemical and physical properties of α-pinene, citral, linalool, and limonene are described. Then, authors focus on the toxicological profile of α-pinene, limonene, and linalool. Special attention is drawn to the antiproliferative and anticancer activities of selected monoterpenes. Thereupon, clinical advances of monoterpenes extracted from essential oils are presented with regard to possible administration routes and applications in the treatment of some acute and chronic diseases. In addition, the therapeutic properties of essential oils in aroma and massage therapy have been included. Lastly, the authors explore the potential of monoterpene-loaded drug delivery systems in the treatment of cancer as well as the application of plant-derived monoterpenes as natural skin penetration enhancers for transdermal drug delivery.

4.2　CHEMICAL ANALYSIS OF SELECTED PLANT-DERIVED MONOTERPENES

4.2.1　ALPHA-PINENE

α-pinene ($C_{10}H_{16}$), also designated as α-pinene or 2,6,6-trimethylbicyclo [3.1.1] hept-2-ene,[63] is a universal, available, and inexpensive bicyclic monoterpene.[43,113] α-pinene is present in the essential oil of several coniferous trees (genus *Pinus*),[25,63,122,134,140] lavender,[122] rosemary (*Rosmarinus officinalis* L.—*Lamiaceae* family),[63,122,135] and in many other different plant species.

There are two structural isomers of pinene found in nature: α-pinene (Fig. 4.1) and β-pinene (Fig. 4.2), which are main volatile components of the essential oil of turpentine—resulting product of pine tree resin hydrodistillation.[37,43,88,113] Beside α-pinene, β-pinene is also one of the most common terpenoids released by forest trees.[44] Additionally, α-pinene can also be extracted from mandarin peel oil (*Citrus reticulata, Rutaceae* family).[98]

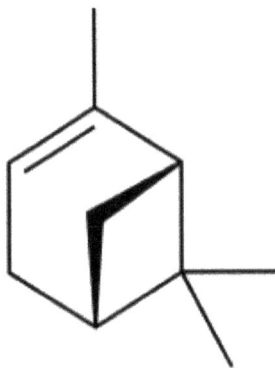

FIGURE 4.1　Structural formula of the (1R)-(+)-α-pinene.

α-pinene is a colorless liquid at room temperature, substantially insoluble in water. α-pinene is highly flammable and a power oxidizing agent. The main chemical and physical properties of α-pinene are shown in Table 4.1.

The reactivity of α-pinene derives from the strained four-membered ring which is prone to open thus leading to skeletal rearrangements.[32] α-pinene, including several monoterpenes and its isomer β-pinene, possesses two enantiomers: (−)-α-pinene (more common in European pines), and (+)-α-pinene (more common in North America). The racemate can be found in several plants and essential oils, namely eucalyptus oil.[37,122]

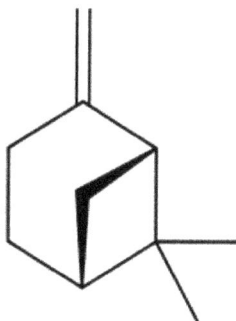

FIGURE 4.2   Structural formula of the (1R)-(+)-β-pinene.

Monoterpenes are extracted from the essential oils of plants. Nowadays, the interest in biosynthetic methods for monoterpene production is rising since the demand for these monoterpenes is increasing. Metabolic engineered microorganisms are being used to produce monoterpenes.[43] The metabolism of microorganisms can convert simple sugars (such as glucose) into acetyl coenzyme A and this last compound can be converted into isopentenyl diphosphate (IPP, also known as IDP[12]) and dimethylallyl diphosphate (DMAPP, also known as DMADP[12]) by mevalonate (MVA) pathway.[43,64,117] However, microorganisms do not possess efficient geranyl diphosphate synthases (GPPS) and adequate monoterpene synthases which make them unable to produce monoterpenes.[43] Although, this can be surpassed due to the development of genetic engineer that has enable the production of monoterpenes in modified microorganisms. For example, Yang, Jianming, et al.[141] succeeded in producing α-pinene in a genetically modified *Escherichia coli* strain, YJM28. The authors assembled a heterologous MVA pathway and inserted plasmids containing geranyl diphosphate synthases (GPPS2) and pinene synthases (PS; Pt30) genes from different plant species in *E. coli*.[141]

Sarria et al.[117] did a combinatory screening of GPPS and PS enzymes from different species of coniferous trees to improve flux of pinene production in engineered *E. coli*.[43] They constructed GPPS-PS protein fusions in order to reduce GPP inhibition of GPPS activity, and concluded that *Abies grandis* GPPS-PS fusion produced the highest amount of pinene (32 mg/L). Beside *E. coli*, other microorganisms such as *Corynebacterium glutamicum* were metabolically engineered to produce pinene.[64]

TABLE 4.1 Chemical and Physical Properties of Monoterpenes.

Property	α-pinene	Citral	Linalool	Limonene
CAS registry number	80–56–8	5392–40–5	78–70–6	138–86–3
Chemical formula	$C_{10}H_{16}$	$C_{10}H_{16}O$	$C_{10}H_{18}O$	$C_{10}H_{16}$
Appearance	Colorless liquid	Pale yellow liquid	Colorless liquid	Colorless liquid
Boiling point	156°C (at 760 mmHg), 429.15 K, 312.8°F	226°C, 499.15 K, 438.8°F	198°C, 471.15 K, 388.4°F	176°C, 449 K, 348.8°F
Density	0.8592 g/cm³ (at 20°C, 293.15 K, 68°F)	0.891–0.897 g/cm³ (at 15°C, 288.15 K, 59°F)	0.879 g/cm³ (at 20°C, 293.15 K, 68°F)	0.8411 g/cm³
Flash point	No data	91°C, 364 K, 196°F	71.11°C, 344.26 K, 160°F	50°C, 323 K, 122°F
Melting point	−62.5°C, 210.65 K, −80.5°F	<−10°C, 263.15 K, 14°F	−11.39°C, 261.76 K, 11.498°F	−74.35°C, 198. 80 K, −101.83°F
Molar mass	136.238 g/mol	152.24 g/mol	154.25 g/mol	136.24 g/mol
Odor	Pine tree aroma	Lemon aroma	Pleasant floral or woody scent, with a touch of spiciness	Orange aroma
Partition coefficient (log Pow)	No data	Neral: 2.8 (25°C, 298.15 K, 77°F), geranial: 3.0 (25°C, 298.15 K, 77°F)	No data	No data
Refractive index	1.4632 [–] (at 25°C, 298.15 K, 77°F)	1.488 [–]	1.462 [–] (at 20°C, 293.15 K, 68°F)	1.4727 [–]
Vapor pressure	No data	<130 Pa (at 40°C, 313.15 K, 104°F)	No data	No data
Viscosity	1.303 cP (at 25°C, 298.15 K, 77°F)	No data	No data	No data
Water solubility	No data	590 mg/L (at 25°C, 298.15 K, 77°F)	683.7 mg/L	Insoluble

α-pinene is used as a raw material for the synthesis of products with high commercial value being a widely used in the pharmaceutical, fragrance, and flavor industries.[43,113] Several products can be obtained by submitting α-pinene to different catalytic chemical processes. For example, α-pinene oxidation produces such as α-pinene oxide, verbenone, and verbenol, which are used in the production of artificial flavors, fragrances, and medicines.[96,130] Other terpenes used in industry such as β-pinene, tricyclene, camphene, limonene, p-cymene, terpinenes, or terpinolenes are result of α-pinene isomerization in the presence of acid catalysts.[27,134]

Acidic titanium dioxide (TiO_2) mesoporous molecular sieves[134] or praseodymium-incorporated aluminophosphate molecular sieves (PrAlPO-5)[130,142,143] are some of the catalysts used in α-pinene isomerization reaction due to their capacity to create acid sites (Lewis or Brönsted acid sites) and shape selectivity.[27,130,134] Moreover, α-pinene oxide can also undergo isomerization in order to produce *trans*-carveol, *trans*-sobrerol, and campholenic aldehyde.[130] Successively, campholenic aldehyde is one of the main components required to produce santalol (sandalwood fragrance).[96,130] α-terpineol has a distinctive lilac odor and is one of the many products of α-pinene hydration.[27]

α-pinene exhibits several biological as well as medical properties, such as antimicrobial, pesticidal, or antioxidant properties. Furthermore, it has anti-inflammatory, anti-stress, and anticonvulsive activities. It has also sedative effect and antitumor activity.[63,88]

4.2.2   CITRAL

Citral ($C_{10}H_{16}O$), also known as 3,7-dimethylocta-2,6-dienal, is a mixture of two geometric isomers: geraniol (Fig. 4.3), also called *trans* confirmation, *trans*-citral or citral A (approximately 55–70 wt.%); and neral (Fig. 4.4), also called *cis* confirmation, *cis*-citral or citral B (35–45 wt.%).[52] The natural form of citral consists citral A and citral B in the ratio 3:2. Citral is present in many essential oils and constitutes 70–85 wt.% of lemon grass oil.[16,105]

Citral is classified as α- and β-unsaturated aldehydes.[139] Therefore, it can be used as a model substance for heterogeneously catalyzed selective hydrogenations. The hydrogenation process of the carbonyl group is insignificant when palladium is put on as an active metal. Then, the conjugated

C=C bond is hydrogenated prior to the terminal isolated one. As a result of this reaction, citronellal (CAL) and dihydrocitronellal are obtained as main products.[139] The main chemical and physical properties of citral are shown in Table 4.1.[55] Physicochemical properties of citral are rare, in spite of that its hydrogenation products influence positively on chemicals and perfume industry.[84]

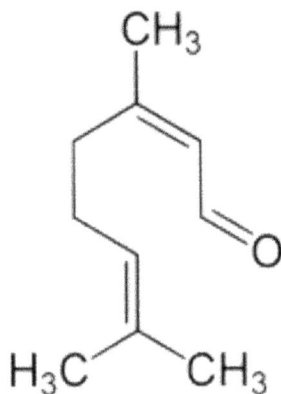

FIGURE 4.3   Structural formula of (Z)-3,7-dimethylocta-2,6-dienal (neral).

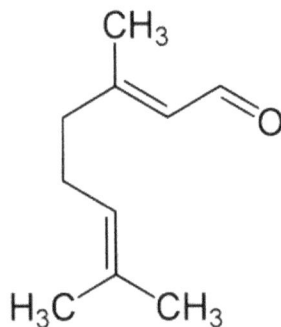

FIGURE 4.4   Structural formula of (E)-3,7-dimetylookta-2,6-dienal (geranial).

Citral exists in lemon oil and citrus flavors.[52] Owing to its strong, lemon-like fragrance, it can be successfully used in the food chemistry or perfumery industry,[84] as well as for the production of vitamin A. Citral is chemically unstable and readily biodegraded into geranic acid and 6-methyl-5-heptene-2-on over time in aqueous solutions. It has been confirmed that citral can be transformed into 2-formylmethy l-2-methyl-5-(1-hydroxy-1-methylethy)-tetrahydrofuran under oxygen atmosphere in aqueous acidic condition. According to Organization for Economic Co-operation and Development, almost 92 wt.% of citral can be degraded after 28 days. Furthermore, the degradation rate of citral is also dependent on pH. The half-life time for neral is 9.54 days at pH 4 and for geranial 9.81 days at the same pH. In turn, at pH 7, citral was slowly degraded and the half-life time of neral is 230 days, and geranial 106 days. Finally, in the alkaline solution (pH around 9) the half-life time of neral was 30.1 and 22.8 days.

Due to the presence of carbonyl group, conjugated C=C and isolated C=C bonds in the chemical structure of citral and by using selective hydrogenation with the right catalysts and reaction conditions. Citral can be obtained as an oxidation product from geraniol or nerol.[52] The aldehyde group causes the oxidative degradation of citral during its storage at low pH or oxidative environment. To the most frequent degradation products of citral belong: p-cymene, p-cresol, and p-methylacetophenone. Because of the high instability of citral and its short shelf life, there is a limited possibility to use it in citrus-flavored foods or alcoholic beverages. The main chemical and physical properties of citral are summarized in Table 4.1. Citral (neral and geranial) is believed to polymerize during the distillation process, whereas geranyl and neryl acetates seem to be the cause of α-terpineol increase, due to hydrolysis.[28] The essential oil of lemon balm (*Melissa officinalis* L.) containing neral or citral had strong antioxidant activity.[16]

4.2.3 LIMONENE

Limonene ($C_{10}H_{16}$), also known as 3,1-methyl-4-(1-methylethenyl)-cyclohexene or 4-isopropenyl-1-methylcyclohexene, is a chiral molecule as well as colorless liquid hydrocarbon classified as a cyclic terpene.[60] Limonene is the main component of the essential oils of *Citrus* (30–70 wt.% in different species). Apart from it, other significant components are: α- and β-pinene, γ-terpinene, terpinolene, or sabinene.[4,132] Racemic limonene is called dipentene. This compound has two isomers (Fig. 4.5):[41]

- **R-(+)-limonene or L-limonene**: Occurring mainly in mint oil and having turpentine-like aroma;

- **S-(−)-limonene or D-limonene**: More widespread than L-isomer and characterized by a strong orange aroma.[52] It can be used in chemical synthesis as a precursor to carvone and as a renewables-based solvent in cleaning products.

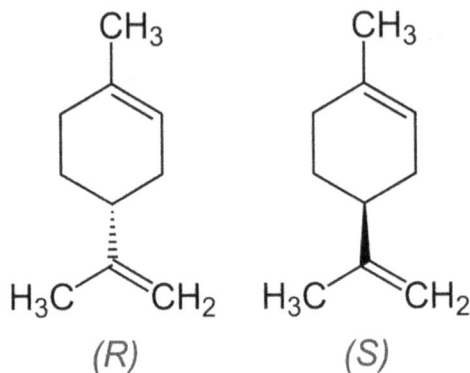

FIGURE 4.5 Chemical structures of both enantiomers of limonene: R-(+)-limonene (L-limonene; left) and S-(−)-limonene (D-limonene; right).

D-limonene is commonly produced from citrus fruits by using two kinds of technique such as centrifugal separation or steam distillation. Furthermore, the name of "limonene" comes from the lemon.[132] In addition, the rind of the lemon (like other citrus fruits) contains substantial amounts of this compound. Limonene is a chiral molecule, and biological sources produce one enantiomer: R-(+)-limonene.[60]

The main chemical and physical properties of limonene are presented in Table 4.1. It is worth mentioning that the main contaminants of limonene are most of all other monoterpenes such as myrcene, α-pinene, and sabinene. Limonene can be used as a substitute for hydrocarbon chloride, as a solvent or flavor and aroma of food in the food industry,[60] and also as a component of fragrance products in household chemistry (detergents, air fresheners), and finally, as a component of perfumes and cosmetics (concentration of 0.005–2 wt.%).[60–61,84] Limonene is a relatively stable terpene. It can be distilled without decomposition, but at the high temperatures it cracks to form of isoprene. D-limonene is the main component of the fragrance released by *Citrus* (plant family *Rutaceae*) and is commonly used in the production of medicines, as a flavoring agent in foods and as a fragrance in perfumery and cosmetic products. D-limonene has a lemon-orange scent. In turn, L-limonene has a pine-like scent. Additionally, D-limonene has also insecticidal properties and is frequently used in alternative medicine to relieve gastroesophageal reflux and heartburn.

4.2.4 LINALOOL

Linalool ($C_{10}H_{18}O$), also known as 3,7-dimethyl-1,6-octadien-3-ol[8], is a acyclic monoterpene tertiary alcohol. It is the most common monoterpene alcohol (also designated, monoterpenol[57]) in plants.[7,109] Linalool is a volatile, optically active compound present in the flower scent[57] of different plant families, but more prevalent in *Lamiaceae* family. The most common plant species containing linalool in their composition are summarized in Table 4.2.

TABLE 4.2 Most Common Plant Families and Plant Species Containing Linalool in Their Composition.

Plant family	Plants pecies	Reference
Apiaceae	*Coriandrum sativum* (coriander seeds)	[27, 42, 125]
Betulaceae	*Betula utilis*	[103]
Labiateae	*Thymus vulgaris* L.	[1, 124]
Lamiaceae	*Lavandula angustifolia* (true lavender), *Lavandula latifolia* (spike lavender), *Lavandula officinalis* (continental lavender), *Ocimum basilicum* (sweet basil), *Satureja Montana* (winter savory)	[21, 27, 35, 39, 51, 124]
Lauraceae	*Cinnamomum camphora*	[59]
Myrtaceae	*Myrtus communis* (myrtle)	[18, 70]
Rutaceae	*Citrus aurantium* L.	[33, 123]
Verbenaceae	*Lippia alba*	[131]

Linalool is a colorless liquid at room temperature but it may also be slightly yellowish and has a woody scent with a floral touch similar to French lavender or bergamot oil.[6] The main chemical and physical properties of linalool are presented in Table 4.1. The chemical structure of linalool (Fig. 4.6) presents alcohol functional group, which makes the compound not only more polar. Therefore, it is soluble in water, but also more chemically reactive. Linalool is also soluble in alcohol, chloroform, ether, fixed oils, and propylene glycol.[6,57]

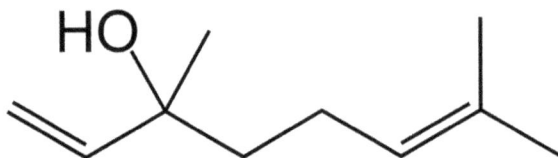

FIGURE 4.6   Structural formula of linalool.

The double bonds and hydroxyl group found in the chemical structure of linalool are responsible for the susceptibility of linalool to chemical modifications such as oxidation, glycosylation, esterification, and methylation.[57,109] Oxidation of linalool produces linalool oxides (furanoid and pyranoid) is common to wines, floral scents, and papaya fruit. Consequently, stereoisomeric aldehydes and alcohols ("lilac compound") can be produced from these linalool oxides.[109]

On account of the hydroxyl group in the third carbon (C3), linalool exhibit chiral properties.[7,109] The two linalool enantiomers, (3S)-(+)-linalool (coriandrol) and (3R)-(−)-linalool (licareol), (Fig. 4.7) are chemically, biosynthetically, electrophysiologically, and behaviorally distinct.[109] Coriandrol (Fig. 4.7) is the most common enantiomer found in nature.[7,57] Some fruits, such as passion fruit and apricots, contain in their composition the racemic mixture of these two enantiomers of linalool.[7]

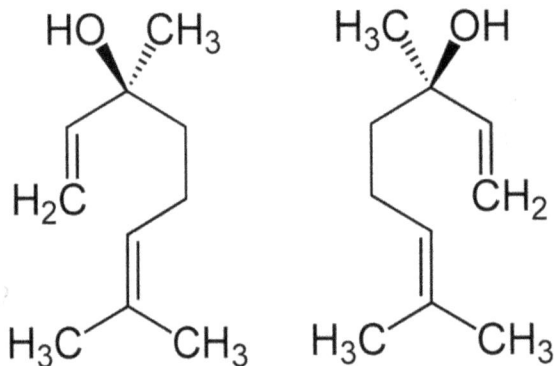

FIGURE 4.7   Structural formula of coriandrol (left) and licareol (right).

Linalool is a product of two different pathways present in higher plants: MVA and alternative non-MVA pathway, also known as 2-C-methyl-D-erythritol 4-phosphate (MEP) pathway. MVA and MEP pathways occur in cytoplasm and chloroplasts, respectively. Both MVA and MEP lead to the synthesis of IPP and DMAPP and the condensation of IPP or DMAPP generates geranyl diphosphate (GPP), which is a substrate for linalool synthases that belong to the terpene synthase (*tps*) gene family and catalyze the conversion of GPP to linalool.[3,7,12,109] Linalool can also be a by-product of geraniol and nerol biosynthesis.[7]

In order to produce linalool at the industrial scale, it is necessary to recreate efficient MVA or MEP pathways. Some metabolic engineering microorganisms, such as *Saccharomyces cerevisiae* are being used to produce linalool.[3] Amiri et al.[3] were able to produce linalool by inserting a plasmid in *S. cerevisiae* containing a linalool synthase (*LIS*) gene from *Lavender angustifolia* (lavender).

Increasing the production of linalool by using genetic engineered microorganisms is particularly important for pharmaceutical industry, as through linalool it is possible to produce other fragrance compounds such as geraniol, nerol, citral, farnesol, ionones, and citronellol.[7,39] Linalool is also main intermediate in the synthesis of vitamins A and E.[7,39]

Linalool is one of the ingredients usually present in cosmetic products such as perfumes, body lotions, shampoos, shower gels, soaps, hairsprays, creams, and antiperspirants.[3,7,73] In addition, linalool is also a common ingredient of house cleaning products.[7,73] In processed food and beverages, linalool is also present as a fragrance and flavor agent.[7]

In plants, linalool is responsible for pollinator attraction, but has also an insecticidal and repellent effects.[57,59]

Current reports indicate that linalool has a wide range of biological effects such as antioxidant, anti-inflammatory, antinociceptive, antimicrobial, anxiolytic, sedative, and anticancer effects.[8,38,39,77,93]

4.3 TOXICOLOGICAL PROPERTIES OF SELECTED MONOTERPENES

4.3.1 ALPHA-PINENE

α-pinene is a volatile organic compound widely used in household products as mentioned previously. In contact with ozone, α-pinene oxidizes forming air pollutants which impair indoor air quality. Air concentrations of α-pinene in indoor spaces vary from a few to 30 μg/m³ and the exposure limit (EL) value for humans is 450 μg/m³.[116]

α-pinene inhalation studies in humans have demonstrated that this compound is promptly metabolized.[2] The biotransformation of α-pinene is catalyzed by cytochrome P450 (CYP450) enzymes. In fact, daily intraperitoneal injection of α-pinene (40 mg/kg body weight [b.w.]) for 3 days in rats, followed by radioimmunoassay, concluded that α-pinene induced CYP2B subfamily in liver microsomes.[2,10] CYP450 enzymes catalyze oxidation of α-pinene and produce hydroxylation products that are either conjugated with glucuronic acid and excreted in the urine or undergone further oxidation. In a study with male albino rabbits, urine was analyzed after administration (with a stomach tube) of α-pinene in a water/polysorbate 80 suspension. Indeed, the major excreted products were glucuronic acid conjugates of hydroxylated terpene hydrocarbons. Verbenol was produced after further allylic oxidation of α-pinene.[2]

Toxicity studies focus mainly on the adverse effects that can be caused by inhalation of α-pinene. In a subchronic inhalation study in F344/N rats, females—that were exposed to 400 mg/L of α-pinene—showed a statistically significant decrease in the body weight gain and thymus weight, and increase in relative kidney weight and lung weight.[2] Some liver-related enzymes, such as alanine transaminase (ALT) and alkaline phosphatase, were significantly reduced after inhalation of α-pinene by female rats. In the same study, F344/N male rats liver weight was significantly increased at 200 mg/L and kidney weight was significantly increased in males exposed to 100 mg/L of α-pinene.[2] Males showed significant reductions in succinate dehydrogenase (SDH) activity at 400 mg/L, ALT activity at 50 mg/L, and alkaline phosphatase activity at 100 mg/L. Examination of the male kidneys at all dose levels (25, 50, 100, 200, and 400 mg/L) revealed lesions including granular casts and hyaline droplets. It was concluded that α-pinene inhalation in rats may cause adverse effects mainly on the liver and kidneys. The no-observed-adverse effect level (NOAEL) in this study with F344/N rats was 25 mg/L or 21 mg/kg b.w./day of α-pinene for male rats and the NOAEL for female rats was 200 mg/L or 170 mg/kg b.w./day of α-pinene.[2]

In another inhalation study, B6C3F1 mice were exposed to 0, 25, 50, 100, 200, or 400 mg/L of α-pinene for limited period of time during 14 weeks.[2] In both sexes at 400 and 200 mg/L, relative and absolute liver weights were increased, but no histological changes associated with liver weight were detected. However, microscopy examination of tissues collected from both sexes exposed to 100 mg/L of α-pinene revealed hyperplasia of the urinary bladder epithelium, which suggest that in mice α-pinene inhalation can also have harmful effects in bladder. In mice, the NOAEL for both sexes was 50 mg/L which corresponds approximately to 72 mg/kg b.w./day.[2]

In humans, exposure to high atmospheric concentrations of α-pinene (higher than 10,000 mg/m³) during a short period of time did not cause acute lung function alterations.[2]Not withstanding in the literature, it is described that inhalation of α-pinene by humans may cause palpitation, dizziness, nervous disturbances, chest pain, bronchitis, or nephritis.[116]

In a popliteal lymph node assay (PLNA) carried out in rats, α-pinene induced a clear immunostimulatory response due to its irritant properties but was unable to produce allergic and autoimmune-like reactions.[40] Hence, α-pinene may also cause irritation in humans when contact with the eyes or in case of ingestion.[116]

The mutagenicity of α-pinene was tested in vitro in different strains of *Salmonella typhimurium* in presence of metabolic activation and the results were negative for reverse mutation tests.[2,46–47] Similarly, mutagenicity was tested in rat hepatocytes and it was concluded that α-pinene (concentrations of up to 10 μg/mL) did not induce temporal modifications on deoxyribonucleic acid (DNA) synthesis.[2,46] Gminski et al.[46] tested the cytotoxicity and genotoxicity of α-pinene in A549 human lung cells in an air/liquid interface exposure system used for an acute exposure time of 1 h. A549 cells were exposed to different atmospheres that contained 1 up to 1800 mg/m³ of α-pinene. After staining with erythrosin-B and alkaline single-cell gel electrophoresis assay (also known as, comet assay), the authors concluded that α-pinene did not induce cytotoxicity neither genotoxicity in A549 cells.[46] Mutagenicity was also tested in vivo. After oral administration of 0.5 mL of α-pinene in Sprague–Dawley rats, urine was collected.[2] The urinary solutions (i.e., direct urine sample, urine–ether extract, and the aqueous fraction of the urine–ether extract) isolated from the rats administered with α-pinene did not show any evidence of mutagenicity.[2]

Regarding human exposure to α-pinene, further research is needed.[116]

4.3.2   LIMONENE

Similar to α-pinene, limonene oxidizes forming air pollutants, which impair indoor air quality when contact with ozone. Air concentrations of limonene in indoor spaces are less than 30 µg/m³ and the EL value for humans is 450 µg/m³.[116] The human occupational EL to limonene was estimated to be 17 mg/kg/day.[69]

When administered orally, D-limonene is rapidly and almost completely absorbed in the gastrointestinal tract of humans and animals and after metabolization. The excretion of D-limonene metabolites is made primarily through the urine.[69]

Limonene is quickly distributed throughout different tissues. The clearance from blood after 2 h inhalation exposure to 450 mg of limonene is 1.1 L/kg/h in male human body.[36,69] From the total limonene uptake, 1 wt.% was eliminated in the expired air after the end of exposure while approximately 0.003 wt.% was eliminated in the urine.[36] In humans as well as rats, metabolization of D-limonene generates oxygenated metabolites.[69] In rats, the two main oxygenated metabolites are perillic [4-(1-methylethenyl)-1-cyclohexene-1-carboxylic acid] and dihydroperillic acid. Humans also produce perillic acid, dihydroperillic acid, and one more metabolite, d-limonene-8, 9-diol, which is found in urine together with its glucuronide.[69]

Dermal toxicity of limonene, at high concentrations, is irritant but does not cause allergies.[69] In fact, dermal exposure of rabbits to D-limonene caused skin irritation, rash as well as eye irritation. Friedrich et al.[40] obtained negative results with the PLNA assay in rats, implying that limonene was unable to produce allergic and autoimmune-like reactions.[40] Oxidation products are the main reason while limonene is classified as a strong skin sensitizer.[69,85] Matura et al.[87] demonstrated that metabolites of limonene oxidation, such as limonene hydroperoxide, are main oxidation product responsible for the sensitization effect of limonene.[87] However, both R-(+)-limonene and S-(+)-limonene act as sensitizing agents.[87] The reaction of limonene with ozone or free radicals can produce other secondary organic aerosols either by multiphase oxidation or photooxidation.[83] This reaction products maybe one of the reasons for the irritability due to limonene.[69,83]

Oral intraperitoneal (ip), intravenous (iv), subcutaneous (sc), and dermal administration routes have low acute toxicity (Table 4.3) in rodents. In acute toxicity studies carried out in F344/N rats and B6C3F1 mice, the NOAEL of oral administered D-limonene was found to be 1650 mg/kg/day in both species and lowest-observed-adverse effect level (LOAEL) was 3300 mg/kg/day in both species.[69,100,108]

TABLE 4.3 Toxicity Studies on the Selected Plant-Derived Monoterpenes.

Plant-derived monoterpenes	Species	Toxicity study	Route	Results	Reference
α-pinene	F344/N rats	NOAEL	Inhalation	25 mg/L or 21 mg/kg·b.w./day for male rats 200 mg/L or 170 mg/kg·b.w./day for female rats	[2]
	B6C3F1 mice	NOAEL	Inhalation	50 mg/L which corresponds approximately to 72 mg/kg·b.w./day	[69]
Limonene	Rat	LD50	Oral	5 g/kg	[69]
	Mouse		Oral	6 g/kg	
	Rat		ip	3.6 g/kg for male, 4.5 g/kg for female	
	Rat		iv	0.125 g/kg for male; 0.11 g/kg for female	
	Mouse		sc	Mouse sc >41.5 g/kg	
	Rabbit		Dermal	>5 g/kg	
	F344/N rats	NOAEL LOAEL	Oral	1650 mg/kg/day 3300 mg/kg/day	[69, 108]
	B6C3F1 mice	NOAEL LOAEL	Oral	1650 mg/kg/day 3300 mg/kg/day	
	Fischer 344 rats	NOAEL LOAEL	Oral	5 mg/kg b.w./day for males 30 mg/kg b.w./day for males	[2, 136]
Linalool	F344/N rats	NOAEL	Oral	1200 mg/kg b.w./day for female rats	[2, 108]
	Fish	LC50	–	46.4 mg/L	[6]
	Scendesmus subspicatus	EμC50 E_BC50 NOEC	–	141.4 mg/L 86.0 mg/L 32.0 mg/L	[6]
	Daphnia	EC50	–	20 mg/L	[6]
	Humans	NOAEL LOAEL	Dermal	250 mg/kg b.w./day 1000 mg/kg b.w./day	[15, 7]

Short-term toxicity studies reveal that oral administration of D-limonene by gavage in male Fischer 344 rats has nephrotoxic effects specific to that sex and species.[2,136] The administration of 75 mg D-limonene/kg b.w./ day significantly increased liver and kidney weights in Fischer 344 rats. Acute administration of a single dose of D-limonene (200 mg/kg b.w./day) increased the formation of hyaline droplets in the corticomedullary junction of the kidneys in male but not in female Fischer 344 rats. Similarly, subchronic oral administration of D-limonene (75 mg D-limonene/kg b.w./ day) to male rats revealed modifications in histopathological features an increase in hyaline droplet formation, arising of granular casts at the corticomedullary junction, and multiple cortical changes collectively classified as chronic nephrosis.[136] In this study with Fischer 344 rats, the NOAEL and the LOAEL values were 5 and 30 mg D-limonene/kg body weight/ day, respectively.[2,136]

The nephrotoxic effects of oral administration were studied not only in Fischer 344 rats (F344/N rats) but also in Wistar rats.[2,69] In F344/N rats, histopathological features observed after administration of D-limonene in corn oil also showed signals of nephropathy. In this study with F344/N rats, the NOAEL value selected for the female rats was 1200 mg D-limonene/kg b.w./day.[2,108] In Wistar rats, oral administration of D-limonene at a dose of 400 mg/kg b.w for 30 days decreased the cholesterol levels, increased relative liver weight and the activity of liver enzymes (cytochrome b5, aminopyrine demethylase, and aniline hydroxylase).[69] Oral administration of D-limonene in rats caused nephrotoxicity and altered liver function. In dogs, diarrhea and emesis were some of the immediate symptoms after oral administration of high (1000 mg/kg b.w/day) and low-dose (100 mg/kg b.w./day) of D-limonene.[2,137] The highest dose group experienced an increase of 35% in serum cholesterol and a 2-fold increase in serum alkaline phosphatase. However, no nephrotoxic effects were noticed indicating that D-limonene is uniquely harmful to kidneys of specific male rats.[2]

In long term studies carried out in B6C3F1 mice, only oral administration of high dose (500 mg D-limonene/kg b.w./day) of D-limonene revealed morphological changes in hepatocytes (multinucleation and cytomegaly).[2] Nevertheless in B6C3F1 mice, evidence of carcinogenic effects in this prolonged study was not observed.[2] However, in another long term study, D-limonene (150 mg D-limonene/kg b.w./day) has shown carcinogenic activity in male F344/N rats.[2] At this high dose (150 mg

D-limonene/kg b.w./day), it was possible to observe increased tubular cell hyperplasia, adenomas, and adenocarcinomas of kidney of male rats.

In male rats, D-Limonene induced a unique form of nephropathy after subacute or chronic exposure.[69] The carcinogenic response in the kidney of male rats has been related to accumulation of aggregates of α2u-globulin (a low molecular-weight protein synthesized in the liver of male rats) and D-limonene metabolites.[2,69,74,136] The concentration of α2u-globulin was higher in male rats after oral administration of D-limonene and the increase of the protein is directly related to the arising of hyaline droplets and other above mentioned nephrological alterations in rats.[2,69,136,137] There is no known protein in humans that can act similarly to α2u-globulin. Therefore, the appearance of renal lesions in rats does not predict human carcinogenicity of oral administered D-limonene.[69,137]

Oral administration of limonene by humans (male adults) can cause severe symptoms such as diarrhea, painful constrictions, and proteinuria.[69] However, no significant modifications in liver enzymes such as serum aminotransferase and alkaline phosphatase occur after limonene ingestion. The estimated reference dose (RfD) for the ingestion of D-limonene by humans was 2.5 mg/kg/day.[69]

In studies carried out in bacteria and cells, it was possible to conclude that D-limonene did not induce cytogenetic damage.[2] In vivo studies also confirmed that limonene is not able to produce forward mutation, therefore, has no mutagenotoxicity.[2,69]

In rodent studies, the oral administration of limonene reduced motor activity.[69] Nevertheless, in humans short term exposure to limonene was unable to induce adverse effects on the central nervous system.[36]

In reproductive toxicity studies with pregnant rabbits, high dose (1000 mg/kg b.w./day) of D-limonene caused maternal toxicity which led to increased mortality.[2] In these studies, the NOEL value of D-limonene for maternal toxicity was 500 mg/kg b.w./day and the NOEL value for offspring toxicity was higher than 1000 mg/kg b.w/day. In fact, oral administration of limonene in rodents may cause maternal toxicity as well as teratogenic or embryo toxic effects.[69]

In safety evaluation and risk assessment study, Kim et al.[69] determined the values of RfD, the NOAEL, and the systemic exposure dose (SED) of D-limonene. Limonene estimated the values were RfD of 2.5 mg/kg/day, a NOAEL of 250 mg/kg/day, and a SED of 1.48 mg/kg/day. The estimated margin of exposure (MOE) was 169 and the hazard index (HI) for

D-limonene was 0.592. According to the estimated values, D-limonene maybe regarded as a safe ingredient despite its irritating properties, but needs regulation when used in cosmetics.[69]

4.3.3 LINALOOL

The global consumption of linalool per year exceeds 1000 mt.[7,73] Overall the International Fragrance Association (IFRA) estimated value for the total systemic exposure of 0.077 mg/kg/day. Specifically, the estimated values for inhalation exposure and dermal exposure were 0.03 and 0.3236 mg/kg/day, respectively.[6] Lapczynski et al.[73] indicated an estimated value of 6.3236 mg/kg for maximum daily exposure on the skin.[73] The acceptable daily intake (ADI) of linalool was estimated to be less than 0.5 mg/kg b.w./day.[7]

In mammals, linalool metabolism is rapid [half-life time ($t\frac{1}{2}$) for linalool = 11 min]. In liver, CYP450 catalyzes linalool oxidation. The metabolites of linalool oxidation are excreted as glucuronic acid conjugates in the urine. In a study with male Wistar rats after an oral administered single dose of 500 mg/kg b.w. radioactive linalool, urine analysis revealed 55% of the radiolabel linalool as a glucuronic acid conjugate.[85] In a period of 72 h after dose administration, CO_2 in the expired air was excreted as 23%, and 15% in the feces.

In the environment, linalool has presented risks in the aquatic compartment. Despite of linalool persistence, it does not have bioaccumulative effects in the environment.[6] In the aquatic compartment, linalool is toxic for fishes, algae, and *Daphnia*. In fishes, the highest reported median lethal concentration (LC50) was less than 46.4 mg/L. In algae (*Scendesmus subspicatus)*, a growth inhibition study concluded that the half-maximal effective concentration for the growth rate (EµC50) was 141.4 mg/L, the half-maximal effective concentration for the biomass (E_BC50) was 86.0 mg/L, and the no-observed-adverse effect concentration (NOEC) was 32.0 mg/L. In *Daphnia*, a crustacean, the reported half-maximal effective concentration (EC50) was 20 mg/L.[6]

Linalool has a weak allergenic potential per se.[6,7] Human studies confirmed that linalool has no sensitization effect.[6] Nevertheless, after atmospheric exposure, linalool forms oxides that are protein reactive. Autoxidation of linalool produces hydroperoxides, furan oxides, pyranoxides, alcohols, and linalyl aldehyde. Accordingly with the local lymph node assay[6–7,20] and repeated open application test studies.[5] Linalool oxides (mainly hydroperoxides) triggers skin sensitization and cause eczematous reactions in patients with allergic contact dermatitis. The storage temperature and the purity of linalool have influence in the autoxidation reaction.[7] Addition of antioxidants like, α-tocopherol or butylhydroxytoluene can prevent the formation of linalool oxides. It is hard to estimate the consumer risks to linalool auto-oxidation reaction because there is no knowledge of its extent.[20] To IFRA, only peroxide levels below or equal to 20 mmol/L are acceptable when using linalool as a fragrance ingredient.[7] For humans, the estimated NOAEL and LOAEL values for linalool dermal exposure are 250 and 1000 mg/kg b.w./day, respectively.[7,15]

Linalool does not absorb wavelengths in the ultraviolet (UV) region and therefore, does not present phototoxicity.[6]

Linalool does not display signs of mutagenotoxicity.[6–7] The mutagenotoxicity of linalool has been assessed in vitro studies in bacteria (*S. typhimurium* strains, *E. coli* and *Bacillus subtilis*), mouse lymphoma cells; rat primary hepatocytes and Chinese hamster ovary, and fibroblast cells.[6,85] In addition, linalool mutagenotoxicity was also studied in vivo in a mouse micronucleus assay and in Sprague—Dawley rats.[85] Previously mentioned in vitro and in vivo studies support the lack of mutagenotoxicity of linalool.

Linalool reproductive and developmental toxicity have been evaluated mainly in rats.[73,85] In Sprague—Dawley rats, the maternal NOAEL of linalool was 500 mg/kg b.w./day and the developmental NOAEL was 1000 mg/kg b.w./day.[85]

Since there are limited studies on reproductive and developmental toxicity, the adverse effects in human reproductive function are unknown.[7]

4.4 ANTIPROLIFERATIVE AND ANTICANCER ACTIVITIES OF SELECTED MONOTERPENES

In the past few years phytochemicals, in particular monoterpenes, have emerged as new candidate compounds to be used as antitumor drugs. According to several recent studies, monoterpenes can induce autophagy, cell death or apoptosis, and can also interfere with different intracellular signaling pathways such as mitogen-activated protein kinases (MAPK) pathway and factor nuclear kappa B (NF-κB) pathway. Moreover, monoterpenes can induce cytogenetic changes since these natural compounds are able to influence telomerase activity.[38]

4.4.1 ALPHA-PINENE

α-pinene is a monoterpene isolated from pine needle oil that has demonstrated antiproliferative activity against a wide range of cancer cells types and subtypes.

Recent studies demonstrate that α-pinene inhibits liver cancer cell proliferation.[25,26] Chen et al.[25] demonstrated through in vitro and in vivo studies that α-pinene has inhibitory effect on hepatocellular carcinoma. Chen et al.[25] concluded that α-pinene induces cell cycle arrest from G2 phase to M phase in BEL-7402 cells (hepatocellular carcinoma cell line). Consequently, since α-pinene induces cell cycle arrest, cell proliferation is suppressed and, therefore, tumor size is reduced. Chen et al.[25] observed that the combination between α-pinene and other anticancer drugs may be beneficial in late-stage hepatocellular carcinoma.

The interaction of cancer cells with stromal cells creates a tumor microenvironment that promotes cancer progression. In order to promote invasion and metastasis, tumor cells and tumor-associated stromal cells can release inflammatory cytokines such as tumor necrosis factor alpha (TNFα). In cancer cells, the expression of matrix metalloproteinases (MMPs) is stimulated by TNFα. MMPs are extracellular matrix zinc-containing endopeptidases responsible for the migration and invasion of tumor cells through the disruption of basement membranes.[45] Kang et al.[63] used several molecular genetics approaches, such as reverse transcription polymerase chain reaction, western blot, and immunofluorescence microscopy analysis to investigate the effect of α-pinene on highly metastatic MDA-MB-231 human breast cancer cells. The study conducted by Kang et al.[63] revealed that α-pinene inhibits NF-κB signaling pathway by suppressing IκB kinase complex.[65] Consequently, TNFα-induced MMP-9 gene expression was downregulated in MDA-MB-231 cells. Therefore, α-pinene is a potential chemotherapeutic drug due to the antimetastatic action demonstrated in highly metastatic breast cancer cells.

Apoptosis evasion is a hallmark of cancer. Tumor cells can circumvent apoptosis through multi-antiapoptotic mechanisms and therefore is crucial to discover an agent that can alter cancer cell's resistance to apoptosis. Matsuo et al.[86] analyzed α-pinene antiapoptosis activity using a metastatic melanoma model (B16F10-Nex2 cells). α-pinene was able to induce different biochemical reactions associated with apoptosis, such as release of pro-apoptotic factors from the mitochondrial intermembrane space, reactive oxygen species (ROS) production, exposure of phosphatidyl-serine on cell surface, activation of caspase-3, and DNA fragmentation.[86]

According to several studies on the cytotoxic effect on cancer cells exerted by α-pinene and doxorubicin (DOX), a widely used chemothera-peutic agent, is similar.[123]

Nowadays, our society lives under a stressful environment that has prejudicial consequences to health. Stress can alter glucocorticoid and noradrenaline values which may have a direct influence in carcinogen-esis.[72] The principal objective of Kusuhara et al.[72] work was to submit in vivo cancer models to a peaceful forest environment through the use of α-pinene since this volatile compound is present in the scent of pine trees. The results of this study revealed that α-pinene inhalation signifi-cantly reduced tumor growth in mice. Thus, the authors foresee that the recreation of a forest environment may have beneficial effects in cancer patients.

4.4.2   CITRAL

Bhalla et al.[16] mentioned that monoterpenes, including citral, are signifi-cant in cancer prevention and treatment. The researchers mention the active ingredients obtained from aromatic plants, such as D-limonene, CAL, or citral, which are modulated hepatic monooxygenase activity by interacting with procarcinogen xenobiotic biotransformation. Recent studies have demonstrated that essential oils can exhibit antimutagenicity toward mutation caused by UV light.[16] Hepatocarcinogenesis in rats is inhibited by lemon grass oil due to the high content of citral.[16]

4.4.3   LIMONENE

According to several studies, limonene has demonstrated chemopreven-tive and antiproliferative properties.[90,91] Oral administration of limonene inhibited the progression of dimethylbenz[α]anthracene- and N-nitroso-N-methylurea-induced mammary carcinomas.[50] Limonene is a skin penetra-tion enhancer and therefore, it is capable of promoting the absorption of anticancer drugs such as tamoxifen through the skin.[89,145] Rouanet et al.[114] assessed that topical administration of hydroxytamoxifen, a tamoxifen metabolite, yielded a significant decrease in mammary tissue proliferation.

Similarly, Miller et al.[89] aimed to evaluate if limonene could be used topically without causing adverse effects in normal mammary tissues of mice and women. In fact, limonene can be applied on the mammary tissue with no concerns since it does not cause alterations in the morphology and function of normal mammary gland.[91] The authors indicated that the efficacy of topical administration of limonene in mammary cancer models should be assessed in order to develop new anticancer topical formulations.

4.4.4 LINALOOL

The essential oil of *Lavandula angustofolia* demonstrated to decrease cell viability in human cervix carcinoma cells (HeLa) and lung adenocarcinoma cells (A549).[97] Linalool is one of the main components present in the essential oil of *L. angustofolia* and the anticancer properties of the plant are mainly attributed to this monoterpene.

Sun et al.[129] proved that linalool can be an effective anticancer agent to prostate cancer. Using 3-(4,5-dimethylthiazol-2-yl)-2,5-diphenyltetrazolium bromide (MTT) assay, they ascertained that linalool exerted a strong cytotoxicity in human prostate cancer cells (DU145). In DU145 cells, linalool induced sub-G1 cell cycle arrest. DNA damage induced after treatment with various concentrations of linalool (0, 20, 40, and 80 μM); cell shrinkage and membrane blebbing observed by inverted light microscopy revealed that linalool induced apoptosis in DU145 cells. The findings of Sun et al.[129] suggested that linalool exerted an anticancer activity in prostate cancer cells by inducing cell cycle arrest and apoptosis in a dose and time dependent manner. In fact in a more recent paper, the activity of *L. angustifolia* essential oil and linalool on prostate cancer cell models was also researched.[146] The MTT assay performed in (Zhao et al.[146]) study confirmed that linalool displays a potent cytotoxic effect against prostate cancer cells (PC-3 and DU145). Linalool also exerted its inhibitory effects in the migration of DU145 and PC-3 cells. Mice xenograft models treated with *L. angustifolia* essential oil, linalool or linalyl acetate showed a significant reduction in tumor size. Zhao et al.[146] findings indicate that this phytochemicals have slightly different effects in DU145 and PC-3 cell lines since cell cycle is stopped in the S phase in DU145; and cell cycle stops in the G2/M phase in PC-3, which makes this cell line more sensitive to these tested plant-based products.

Furthermore, linalool is also the major constituent of *Lindera umbellata* essential oil, a broadly used traditional eastern medicine.[82] In human leukemia HL-60 cells, both essential oil of *L.umbellata* and linalool at the same dose (5 or 50 μg/mL) inhibited cell proliferation, differentiation and induced apoptosis.[82,123]

Cerchiara et al.[22] revealed that linalool has antiproliferative activity against RPMI 7932 human melanoma cell line. The ultrastructural changes observed through scanning electron microscopy and transmission electron microscopy and the activation of caspase-3 detected by fluorescent immunostaining followed by confocal microscopy indicated that linalool maybe capable of inducing apoptosis in RPMI 7932 cell line. Thus, linalool is a potential therapeutic agent against melanoma due to its proapoptotic effect.

Skin exposure to UV radiation emitted by the sun or by artificial tanning beds is the main environmental risk factor associated with melanomagenesis.[62] UV radiation can cause DNA damage by the formation of cyclobutane pyrimidine dimers and 6–4 photoproducts.[112] Gunaseelan et al.[49] assessed linalool action on the skin of mice exposed to acute (7 days of exposure) and chronic (210 days of exposure) ultraviolet-B (UVB) radiation. Acute UVB radiation decreased the antioxidant content in mice skin. Topical and intraperitoneal administration of linalool prevented acute UVB-induced skin inflammatory response (hyperplasia, edema formation, and lipid peroxidation). Chronic UVB radiation induced carcinogenic process in mice skin as it triggers an increase in proliferative proteins expression such as NF-κB, TNF-α, interleukin 6 (IL-6), cyclooxygenase 2, vascular endothelial growth factor (VEGF), transforming growth factor beta 1 (TGF-β1), B-cell lymphoma 2 (Bcl-2), and mutated p53. In in vivo study, Gunaseelan et al.[49] revealed that pretreatment with linalool before chronic UVB-exposure significantly prevented the expression of proliferative markers and subsequently decreased the tumor incidence in mice skin. Besides that, linalool prevented the over expression of angiogenic factors such as VEGF and TGF-β1 in mice skin exposed to chronic UVB radiation. The topical and intraperitoneal administration of linalool in mice exposed to chronic UVB radiation promoted apoptosis through the inhibition of mutated p53 and simultaneously reduction of antiapoptotic factors (Bcl-2). Their results indicated that linalool is a potential agent against skin photocarcinogenesis.[49]

ROS promote the progression step in the carcinogenic process.[79] Mitić-Ćulafić et al.[92] investigated linalool effect against DNA damage caused by ROS-inducing agent t-butyl hydroperoxide (t-BOOH) in bacteria and human cell lines. In bacteria, the antimutagenic effects of linalool were assessed in reverse mutation assays with *E. coli* WP2 IC185 and oxyR mutant IC202 strain (that has downregulated antioxidant enzymes). Linalool demonstrated an antimutagenic effect only in IC202 *E. coli* strain. In human hepatoma cells (HepG2) and B lymphoblastoid (NC-NC), linalool potential to suppress t-BOOH induced DNA damage was ascertained using comet assay. Linalool (1.0 µg/mL) reduced DNA damage by 34 wt.% in hepatoma HepG2 cells. In NC-NC cells at the lowest concentration (0.01 µg/mL), co-treatment with linalool reduced DNA damage by 33 wt.%. In turn, pretreatment with linalool reduced DNA damage by 47 wt.% in NC–NC cell line. In this study,[92] they concluded that linalool and the other tested monoterpenes (eucalyptol and myrcene) had protective effect against ROS mutagenic action essentially due to its antioxidant activity.

Jana et al.[58] confirmed the apoptotic and antiproliferative effects of linalool in sarcoma-180 cells (S-180). In vivo studies using S-180 tumor-bearing mice, linalool is quickly metabolized and increases ROS production in cancer cells and a reduction in the activity of antioxidant enzymes. Linalool is less cytotoxic to normal cells (liver and bone narrow) than cyclophosphamide, a conventional anticancer agent. In turn, in normal cells (liver and bone narrow) linalool exerts antioxidant activity (increases antioxidant enzymes). Therefore, linalool has a selective activity since it induces antiproliferative mechanisms against cancer cells and protective mechanisms in normal cells.

Linalool activity on hematological malignancies was studied by Gu et al.[48] The authors used a variety of different leukemia cell lines some of them containing wild type p53 tumor suppressor and others containing mutant or null p53 gene. The cell lines used were Kasumi-1, HL-60, THP-1, U937, KG-1, NB4 (acute myeloid leukemia), K562 (blast crisis of chronic myeloid leukemia), Molt-4, H-9, Jurkat (acute T-lymphoblastic leukemia), Raji (Human Burkitt's lymphoma), and L428 (Hodgkin's lymphoma). Cell growth was analyzed using MTT assay. The authors concluded that linalool preferentially has cytotoxic activity on leukemia cells with wild type p53 while sparing normal hematopoietic cells.[38,48] Additionally, they[48] used quiescent leukemia cells (Molt-4 cells) since they are relatively resistant to conventional cytotoxic agents. In quiescent leukemia cells, linalool induced apoptosis through c-Jun N-terminal

kinase (JNK) signaling pathway, which makes this monoterpene a potential therapeutic agent for leukemia.

Linalool activity against leukemia cancer cells was also discussed
by Chang et al.[24] The water-soluble tetrazolium salts (WST-1) analysis in their study concluded that linalool is cytotoxic in human acute
myeloid leukemia cell line U937 (half-maximal inhibitory concentration
(IC_{50})=2.59 µM) and HeLa cervical cancer cells (IC_{50}=11.02 µM). In
U937 cell line, linalool induced cell cycle arrest at G0/G1 phase, which
causes a replication suspension, consequently damaged DNA was accumulated and tumor suppression mechanisms were activated. However, in
HeLa cells, cell cycle was arrested at the G2/M phase, which triggered an
increase in the expression of tumor suppressor genes such as p53, p21,
p27, p16, and p18 and cyclin-dependent kinases inhibitors genes (CDKIs
genes). In the two cancer cell models used in the study by Chang et al.,[24]
the antiproliferative properties of linalool are particularly related to its
ability to cause cell cycle arrest.

Miyashita et al.[93] evaluated the combinatory effect between linalool
and DOX. DOX is an anthracycline anticarcinogenic agent.[101] The molecular mechanism of DOX involves the activation of the ubiquitin-protea-
some system that, in turn, regulates the NF-κB transcription factor which
promotes cell growth. In some several types of cancer DOX induces an
over-activation of NF-κB leading to the promotion of cancer cells survival.
They[93] investigated the activity of four *Humulus lupulus* (hop) components (myrcene, α-humulen, β-caryophyllene, and linalool) on DOX
permeability in P388 leukemia cells. The combination between linalool
(0.01, 0.1, and 1 µM) and DOX (9.0 µM) increased DOX concentration
in P388 leukemia cells leading to an increase in DOX cytotoxicity. The
oral administration of DOX combined with linalool (DOX—2.0 mg/kg
and linalool—1.0 mg/kg) in P388 leukemia tumor-bearing mice significantly decreased the tumor weight in comparison with DOX treated group.
Linalool enhanced the anticancer activity of DOX due to its action on Na^+-
dependent nucleoside transporter 3, which increased the transport of DOX
leading to an increased in DOX concentration in P388 leukemia cells. The
combination of linalool enables the use of lower concentrations of DOX
in cancer cells which in turn decreases the adverse effects of this chemotherapeutic agent in normal tissue. Thus, linalool can be applied in cancer
chemotherapy as DOX adjuvant in order to improve the quality of life of
leukemia cancer patients.

4.5 CLINICAL ADVANCES ON THE USE OF PLANT-DERIVED MONOTERPENES

In traditional medicine, natural compounds derived mainly from plants, such as essential oils, are commonly used as therapeutic agents. For example, chamomile essential oil is known to relief muscular pain as well as eucalyptus oil.[138] Moreover, chamomile and lavender essential oils exhibit relaxing properties which help patients with insomnia. Recent scientific studies emphasize the therapeutic value of essential oils.

The medicinal properties of essential oils are associated with the large percentage of monoterpenes present in their composition. As mentioned earlier, monoterpenes exhibit anticancer properties. Additionally, monoterpenes also possessed anti-inflammatory, analgesic, antimicrobial, and anxiolytic properties which make plant-derived monoterpenes suitable pharmacological candidates for the development of new medicines.[110]

4.5.1 ANTI-INFLAMMATORY APPLICATIONS

Frankincense oil from *Boswellia carterii* is used in traditional Chinese medicine to relief inflammation and pain.[76] α-pinene, linalool, and 1-octanol are the main ingredients in frankincense oil composition and contribute to the topical anti-inflammatory and analgesic properties of the oil.

In the inflammatory response, macrophages have the ability to produce pro-inflammatory cytokines (such as IL-1β, IL-6, and TNF-α) in order to recruit immune cells.[67,111] Therapeutic agents that inhibit the expression of these inflammatory cytokines have potential to treat inflammatory diseases.

Kim et al.[67] confirmed the anti-inflammatory efficacy of α-pinene. Their study demonstrated that α-pinene treatment reduced the production of IL-6, TNF-α, and NO in mouse peritoneal macrophages stimulated with the endotoxin, lipopolysaccharide. The inhibited expression of inflammatory mediators occurred because α-pinene blocks the activation of MAPK and NF-κB pathways.

Acute pancreatitis is a multifactorial inflammatory disease characterized by an increase in cytokine production and the release of digestive enzymes to systemic circulation which causes adverse effects not only in pancreas but also in other systemic organs.[68] The results of Bae et al.[11] suggested that α-pinene possessed anti-inflammatory effect on cerulein-induced acute pancreatitis. The authors administered intraperitoneally α-pinene after inducing acute pancreatitis with cerulein in mice models. α-pinene reduced myeloperoxidase activity and the production of pro-inflammatory cytokines. Furthermore, the administration of α-pinene inhibited cerulein-induced cell death in pancreatic cells.

Osteoarthritis is a painful degenerative joint disease that causes articular cartilage damages, which lead to the progressive reduction of joint function. Local inflammation in joint characterizes osteoarthritis.[115] The essential oil extracted from *Juniperus oxycedrus* L. subsp. *Oxycedrus* leaves demonstrated the capacity to reduce joint inflammation in osteoarthritis disease models. In fact, Rufino et al.[115] confirmed that the beneficial effects of the essential oil of *J. oxycedrus* L. subsp. *Oxycedrus* is due to the α-pinene, the main component of this plant species. In human chondrocytes, the enantiomer (+)-α-pinene inhibited the activation of NF-κB and JNK pathways, production of IL-1β and the expression of inflammatory genes. Therefore, (+)-α-pinene maybe a potential antiosteoarthritic drug.

According to recent studies, oral supplementation of limonene demonstrated intestinal anti-inflammatory effects.[29–30] D'alessio et al.[30] confirmed that limonene reduced serum TNF-α levels, an effect comparable to ibuprofen in mice 5,6-trinitrobenzene sulfonic acid-induced colitis. In the human trials, the diet supplementation with D-limonene-containing orange peel extract also decreased peripheral IL-6 levels in the healthy elderly patients. Thus, limonene decreases the level of inflammatory cytokines and according to the in vitro study performed by D'alessio et al.[30] it has a protective effect in human colon carcinoma cells (HT-29/B6).

Naproxen, is a nonsteroidal anti-inflammatory drug (NSAID) to reduce pain due to its anti-inflammatory and antinociceptive properties.[102] Furthermore, naproxen, as well as the majority of NSAIDs, has adverse effects in the gastrointestinal tract and in the cardiovascular system. In the study by Ortiz et al.[102] the oral administration of naproxen–citral combination did not induce gastric side effects thus suggesting that the synergy between naproxen and citral may be beneficial in the treatment of inflammatory diseases.

Similarly, D-limonene is a suitable candidate to be used as an adjuvant of DOX in cancer therapy.[111] As mentioned earlier in this chapter, DOX is a chemotherapeutic agent that can cause oxidative stress and local inflammation due to the increase in TNF-α levels. The oxidative stress can lead to organ injury. In Wistar rat models, pretreatment with D-limonene has a protective effect against DOX-induced renal-toxicity since it reduces oxidative stress and the production of inflammatory cytokines.[111]

4.5.2 ANALGESIC APPLICATIONS

According to several studies, linalool is the more effective phytochemical used in pain relief among the selected monoterpenes. Batista et al.[14] reported that linalool enantiomer, (−)-linalool, inhibited the nociceptive response caused by intraplantar injection of glutamate in mice. The enantiomer, (−)-linalool, was administered intraperitoneally (10–200 mg/kg), orally (5–100 mg/kg), through intrathecal (0.1–3 μg/site) and intraplantar (10–300 ng/paw) injection. In all the administration routes studied, (−)-linalool interacted with N-methyl-D-aspartate receptors which in turn interact with ionotropic glutamatergic-dependent mechanisms reducing pain stimulus. Their findings,[13,14] revealed that (−)-linalool was able to control the response to a pain stimulus over a long period in two different chronic pain in vivo mice models. The two reports draw positive conclusions about the potential of (−)-linalool in the relief of chronic pain.

The chemotherapeutic treatment with paclitaxel, an antimitotic agent, can have adverse effects and can induce peripheral neuropathy. The intraplantar administration of linalool reduces paclitaxel-induced acute pain symptoms (allodynia and hyperalgesia).[66]

Carpal tunnel syndrome (CTS) is a peripheral idiopathic neuropathy caused by the compression of the median nerve. The symptoms of this medical condition are pain, numbness, and tingling in the wrist and hands. The available nonsurgical treatments have limitations and the progression of the condition may restrain the patient daily life. Seol et al.[118] demonstrated the inhalation of linalool can increase antioxidant activity and reduce blood pressure in CTS and non-CTS patients. Linalool inhalation may constitute an additional therapeutic component in CTS treatment.

4.5.3 ANTIMICROBIAL APPLICATIONS

Nowadays, the uncontrolled access to antibiotics has led to a worldwide increased incidence of antibiotic resistant bacterial pathogens. Ciprofloxacin is an antibiotic commonly used in the treatment of gastroenteritis caused by *Campylobacter jejuni*. Since *C. jejuni* is highly resistant to ciprofloxacin action, Kovač et al.[71] tested, through molecular genetics approaches, the antimicrobial activity of (−)-α-pinene. The results concluded that (−)-α-pinene inhibited microbial efflux, decreased membrane integrity and caused metabolic disruption in *C. jejuni*. Kovač et al.[71] suggested that (−)-α-pinene can control *Campylobacter* antibiotic resistance.

Raut et al.[110] inferred that terpenes can be used to prevent and eradicate *Candida albicans* biofilms. *C. albicans* is a yeast and opportunistic pathogens to form colonies in the host tissues as well as in implanted prostheses.[23] These *C. albicans* colonies, also designated as biofilms, can induce drug resistant infections primarily in immunosuppressed patients.[107] Citral, geraniol, thymol, and carvacrol inhibited the adhesion of mature biofilms.[110] On the other hand, thujone, carvone, menthol, isopulegol, nerol, and linalool prevented the formation of yeast colonies.

The antiviral action of monoterpenes, namely α-pinene, against infectious bronchitis virus (IBV), a member of *Coronaviridae* was confirmed by Yang et al.[144] Their study indicated that (−)-α-pinene and (−)-β-pinene inhibited IBV virus replication by suppressing RNA and IBV nucleocapsid (N) protein binding. The two monoterpenes, (−)-α-pinene and (−)-β-pinene, demonstrated the anti-IBV activity and are suitable candidates for novel antiviral drug pharmaceutical development.

4.5.4 ANXIOLYTIC APPLICATIONS AND EFFECTS ON MEMORY

Aromatherapy is an alternative complementary therapy that uses aromatic essential oils extracted from plant materials. Aromatherapy is beneficial in the treatment of chronic pain, depression, cognitive maladies, anxiety, and insomnia since it creates a sense of general well-being in the individuals.[106] The vast majority of the essential oils applied in aromatherapy have in their composition linalool.[77] Moreover, linalool is also present in the composition of several analgesic, anxiolytic, and sedative traditional medicines.

The inhalation of linalool induces a sedative effect since it decreases mobility without side effects on the coordination and reduces body temperature in mice models.[78] Inhalation of linalool in mice models exhibited similar anxiolytic effects of diazepam (psychotropic drug).[77] In mice, inhaled linalool enhanced social interaction since it reduced the aggressive behavior. Only higher concentrations of inhaled linalool affected mice memory. According to the aforementioned studies, linalool is an efficient anxiolytic and relaxation agent that requires further development and research by the pharmaceutical industry.

As previously mentioned, higher doses of linalool can negatively affect mice memory. Nevertheless, small doses (0.1 mg/kg) of citral can improve rat spatial learning and memory due to an increase in the concentration of retinoic acid in hippocampus.[143] A high dose of citral (1.0 mg/kg) induces a reduction of retinoic acid in hippocampus causing limitations in the spatial learning capacity and memory. Therefore, citral has a dual effect in rats' spatial learning and memory.

4.6   ENCAPSULATION OF MONOTERPENES IN LIPID-BASED DRUG DELIVERY SYSTEMS

Essential oils exhibit a wide range of therapeutic applications albeit the limitations. These essential oils constrains include mainly low water solubility and high volatility.[17] The encapsulation of essential oils in drug delivery systems reduces constrains and makes feasible the biomedical application of these substances. Essential oil encapsulation in nanodelivery systems is advantageous since it increases essential oil physical stability, protects the essential oil from degradation, easily penetrates tissues, and enables a controlled release of substance locally.[17]

Polymer-based nanoparticles and lipid-base nanoparticles are the two main nanodelivery systems. Lipid-based nanocarriers include microemulsions, nanoemulsions, liposomes, micelles, NLCs, and SLNs.[17] Liposomes are colloidal dispersions formed spontaneously. SLNs are dispersions composed by physiological solid lipid, surfactants and aqueous phase.[119,120,126] NLCs are second generation lipid nanoparticles and their composition has four main ingredients solid lipid (at room and human body temperatures), a liquid lipid, surfactants, and an aqueous phase.[31,95]

Stratum corneum (SC) is the most superficial layer of the skin and exerts a protective barrier function.[104] The SC impose skin penetration restrictions based on molecular size and lipophilicity of the topical administered active compounds.[19] Terpenes can act as skin-penetration enhancers. In fact, the addition of terpenes in ultradeformable liposomes enhanced the skin penetration of hydrophilic drugs.[128] Hoppel et al.[56] indicated that R-(+)-limonene increased the skin permeation of diclofenac sodium. Terpenes enhanced skin permeation and can be used as ingredients in microemulsion and nanoemulsions production in order to create less toxic skin delivery systems.[56]

In addition, monoterpenes are prone to be encapsulated in nanodelivery systems. In fact, studies reported that D-limonene containing nanoemulsions with a nanometric droplet size (54 nm) demonstrated to be interesting transdermal delivery nanocarriers.[75,80]

The advanced stages of cancer, as well as the resistance of cancer cells to the conventionally used chemotherapeutic agents incite the production of new anticancer therapies. As previously mentioned, linalool exhibits antiproliferative and anticancer proprieties, thus is a candidate for the pharmaceutical development of new anticancer therapies. Furthermore, the nanoencapsulation of linalool surpasses its limitations (poor waters olubility and volatility) since it increases permeability and drug retention limiting the cytotoxic effects in normal cell and consequently increasing linalool concentrations in the target tissue.[53,121] In Han et al.'s[53] study, linalool-loaded nanoparticles (LIN-NPs) were spontaneously produced by constant stirring using polyoxyethylene sorbitan monooleate 20 or 80 (Tween 20 or Tween 80). LIN-NPs induced cytotoxic and pro-apoptotic effects in epithelial ovarian cancer cells. Notably in epithelial ovarian cancer cells, linalool increased ROS production which decreases mitochondrial membrane potential and increases caspase-3 levels inducing apoptosis and consequently cancer cells death. In in vivo models, the combination of LIN-NPs with the conventionally chemotherapeutic agent, paclitaxel, led to a reduction in tumor weight.

The recent study[121] reported the linalool encapsulation in NLCs. The authors produced a formulation using a high-pressure homogenizer with 2.5 wt.% of glycerin monostearate (solid lipid), 2.5 wt.% of decanoyl/octanoyl-glycerides (liquid lipid), 2.0 wt.% of sorbitane monooleate (Span 80), and 4.0 wt.% of Tween 80. The encapsulation efficiency and the drug-loading capacity of linalool-loaded nanostructured lipid carriers (LL-NLCs) were 79.563 and 7.555 wt.%, respectively. Shi et al.[121] suggested that linalool encapsulation in NLC's enhances the monoterpene bioavailability. Thus, LL-NLCs are suitable drug delivery systems candidates to be used in cancer therapy.

4.7 SUMMARY

Phytochemicals, in particular, monoterpenes have demonstrated their beneficial effects in the treatment of a wide spectrum of diseases. Monoterpenes are significant pharmaceutical active ingredients, which can be used in the development of new medicines due to their extensive therapeutic properties. Recent reports have shown antiproliferative and anticancer properties of monoterpenes in several different types of cancer models. Currently, the search for new anticancer agents is a necessity due to an increase in the cases of cancer cells resistance to the conventional chemotherapeutic agents, which lead to a decreasing in cancer patient's survival. Furthermore, conventional chemotherapeutic agents have also shown adverse side-effects which have repercussions in the cancer patient's quality of life.

Monoterpenes are a valuable alternative to the conventional chemotherapeutic agents since they are less toxic to normal cells regardless of the administration route. In addition, monoterpenes can be administered in combination with conventional anticancer agents as adjuvants reducing side effects of chemotherapy.

The encapsulation of monoterpenes in drug delivery systems is possible and advantageous since it reduces their limitations (poor water-solubility and volatility) and increases their therapeutic efficacy. In fact, the monoterpene-loaded drug delivery systems have the ability to selectively target tissue reducing the toxic effects on normal cells and increasing the concentration of drug in cancer cells. Therefore, in an effort to develop delivery systems loaded with monoterpenes further research is essential.

KEYWORDS

- 2-C-methyl-D-erythritol 4-phosphate pathway
- acceptable daily intake
- acute pancreatitis
- allodynia
- α-pinene
- α-pinene isomerization
- analgesic applications
- antibiotic resistance
- anticancer activity
- anticancer agents
- anti-inflammatory applications

REFERENCES

1. Abedini, S.; Sahebkar, A.; Hassanzadeh-Khayyat, M. Chemical Composition of the Essential Oil of *Thymus vulgaris* L. Grown in Iran. *J. Essent. Oil Bear. Plants* **2014,** *17,* 538–543.

2. Adams, T. B.; Gavin, L. C.; Mcgowen, M. M.; Waddell, W. J.; Cohen, S. M.; Feron, V. J.; Marnett, L. J.; Munro, I. C.; Portoghese, P. S.; Rietjens, I. M. C. M.; Smith, R. L. The FEMA GRAS Assessment of Aliphatic and Aromatic Terpene Hydrocarbons Used as Flavor Ingredients. *Food Chem. Toxicol.* **2011,** *49,* 2471–2494.

3. Amiri, P.; Shahpiri, A.; Asadollahi, Ma.; Momenbeik, F.; Partow, S. Metabolic Engineering of *Saccharomyces cerevisiae* for Linalool Production. *Biotechnol. Lett.* **2016,** *38,* 503–508.

4. Amorim, J. L.; Simas, D. L. R.; Pinheiro, M. M. G.; Moreno, D. S. A.; Alviano, C. S.; Silva, A. J. R.; Fernandes, P. D. Anti-Inflammatory Properties and Chemical Characterization of the Essential Oils of Four *Citrus* Species. *Public Library Sci. One* **2016,** *11,* 1–18.

5. Andersch, Y. B.; Hagvall, L.; Siwmark, C.; Niklasson, B.; Karlberg, A. T.; Bråred, J. C. Air-Oxidized Linalool Elicits Eczema in Allergic Patients—A Repeated Open Application Test Study. *Contact Dermatitis* **2014,** *70,* 129–138.

6. Api, A. M.; Belsito, D.; Bhatia, S.; Bruze, M.; Calow, P.; Dagli, M. L.; Dekant, W.; Fryer, A. D.; Kromidas, L.; La Cava, S.; Lalko, J.F.; Lapczynski, A.; Liebler, D. C.; Miyachi, Y.; Politano, V. T.; Ritacco, G.; Salvito, D.; Shen, J.; Schultz, T. W.; Sipes, I. G.; Wall, B.; Wilcox, D. K. RIFM Fragrance Ingredient Safety Assessment, Linalool, CAS Registry Number 78–70–6. *Food Chem. Toxicol.* **2015,** *82,* S29–S38.

7. Aprotosoaie, A. C.; Hăncianu, M.; Costache, I. I.; Miron, A. Linalool: A Review on a Key Odorant Molecule with Valuable Biological Properties. *Flavour Fragrance J.* **2014,** *29,* 193–219.

8. Asbahani, A. E.; Miladi, K.; Badri, W.; Sala, M.; Addi, E. H. A.; Casabianca, H.; Mousadik, A. E.; Hartmann, D.; Jilale, A.; Renaud, F. N. R.; Elaissari, A. Essential Oils: From Extraction to Encapsulation. *Int. J. Pharm.* **2015,** *483,* 220–243.

9. Astani, A.; Schnitzler, P. Antiviral Activity of Monoterpenes Beta-Pinene and Limonene against *Herpes simplex* Virus In Vitro. *Iran. J. Microbiol.* **2014,** *6,* 149–155.

10. Austin, C. A.; Shephard, E. A.; Pike, S. F.; Rabin, B. R.; Phillips, I. R. The Effect of Terpenoid Compounds on Cytochrome P-450 Levels in Rat Liver. *Biochem. Pharmacol.* **1988,** *37,* 2223–2229.

11. Bae, G. S.; Park, K. C.; Choi, S. B.; Jo, I. J.; Choi, M. O.; Hong, S. H.; Song, K.; Song, H. J.; Park, S. J. Protective Effects of Alpha-Pinene in Mice with Cerulein-Induced Acute Pancreatitis. *Life Sci.* **2012,** *91,* 866–871.

12. Banerjee, A.; Sharkey, T. D. Methylerythritol 4-Phosphate (MEP) Pathway Metabolic Regulation. *Nat. Prod. Rep.* **2014,** *31,* 1043–1055.

13. Batista, P. A.; Werner, M. F. P.; Oliveira, E. C.; Burgos, L.; Pereira, P.; Brum, L. F. S.; Story, G. M.; Santos, A. R. S. The Antinociceptive Effect of (−)-Linalool in Models of Chronic Inflammatory and Neuropathic Hypersensitivity in Mice. *J. Pain* **2010,** *11,* 1222–1229.

14. Batista, P. A.; Werner, M. F. P.; Oliveira, E. C.; Burgos, L.; Pereira, P.; Brum, L. F. S.; Santos, A. R. S. Evidence for the Involvement of Ionotropic Glutamatergic Receptors on the Antinociceptive Effect of (−)-Linalool in Mice. *Neurosci. Lett.* **2008,** *440,* 299–303.

15. Belsito, D.; Bickers, D.; Bruze, M.; Calow, P.; Greim, H.; Hanifin, J. M.; Rogers, A. E.; Saurat, J. H.; Sipes, I. G.; Tagami, H. A Toxicologic and Dermatologic Assessment of Cyclic Acetates when Used as Fragrance Ingredients.*Food Chem. Toxicol.* **2008,** *46,* S1–S27.

16. Bhalla, Y.; Gupta, V. K.; Jaitak, V. Anticancer Activity of Essential Oils: A Review. *J. Sci. Food Agric.* **2013,** *93,* 3643–3653.

17. Bilia, A. R.; Guccione, C.; Isacchi, B.; Righeschi, C.; Firenzuoli, F.; Bergonzi, M. C. Essential Oils Loaded in Nanosystems: A Developing Strategy for a Successful Therapeutic Approach. *Evidence-Based Complementary Altern. Med.* **2014,** *2014,*1–14.

18. Bouzabata, A.; Cabral, C.; Goncalves, M. J.; Cruz, M. T.; Bighelli, A.; Cavaleiro, C.; Casanova, J.; Tomi, F.; Salgueiro, L. *Myrtus communis* L. as Source of a Bioactive and Safe Essential Oil. *Food Chem. Toxicol.* **2015,** *75,* 166–172.

19. Bruno, B. J.; Miller, G. D.; Lim, C. S. Basics and Recent Advances in Peptide and Protein Drug Delivery. *Ther. Delivery* **2013,** *4,* 1443–1467.

20. Burg, W.; Bouma, K.; Schakel, D. J.; Wijnhoven, S. W. P.; Engelen, J.; Loveren, H.; Ezendam, J. Assessment of the Risk of Respiratory Sensitization from Fragrance Allergens Released by Air Fresheners. *Inhalation Toxicol.* **2014,** *26,* 310–318.

21. Carrasco, A.; Martinez-Gutierrez, R.; Tomas, V.; Tudela, J. *Lavandula angustifolia* and *Lavandula latifolia* Essential Oils from Spain: Aromatic Profile and Bioactivities. *Planta Med.* **2016,** *82,* 163–170.

22. Cerchiara, T.; Straface, S. V.; Brunelli, E.; Tripepi, S.; Gallucci, M. C.; Chidichimo, G. Antiproliferative Effect of Linalool on RPMI 7932 Human Melanoma Cell Line: Ultrastructural Studies. *Nat. Prod. Commun.* **2015,** *10,* 547–549.

23. Chandra, J.; Kuhn, D. M.; Mukherjee, P. K.; Hoyer, L. L.; Mccormick, T.; Ghannoum, M. A. Biofilm Formation by the Fungal Pathogen *Candida albicans*: Development, Architecture, and Drug Resistance. *J. Bacteriol.* **2001,** *183,* 5385–5394.

24. Chang, M. Y.; Shieh, D. E.; Chen, C. C.; Yeh, C. S.; Dong, H. P. Linalool Induces Cell Cycle Arrest and Apoptosis in Leukemia Cells and Cervical Cancer Cells Through CDKIs. *Int. J. Mol. Sci.* **2015,** *16,* 28169–28179.

25. Chen, W.; Liu, Y.; Li, M.; Mao, J.; Zhang, L.; Huang, R.; Jin, X.; Ye, L. Anti-Tumor Effect of α-Pinene on Human Hepatoma Cell Lines Through Inducing G2/M Cell Cycle Arrest. *J. Pharmacol. Sci.* **2015,** *127,* 332–338.

26. Chen, W. Q.; Xu, B.; Mao, J. W.; Wei, F. X.; Li, M.; Liu, T.; Jin, X. B.; Zhang, L. R. Inhibitory Effects of α-Pinene on Hepatoma Carcinoma Cell Proliferation. *Asian Pac. J. Cancer Prev.: APJCP* **2014,** *15,* 3293–3297.

27. Corma, A.; Iborra, S.; Velty, A. Chemical Routes for the Transformation of Biomass into Chemicals. *Chem. Rev.* **2007,** *107,* 2411–2502.

28. Costa, R.; Bisignano, C.; Filocamo, A.; Grasso, E.; Occhiuto, F.; Spadaro, F. Antimicrobial Activity and Chemical Composition of *Citrus aurantifolia* (Christm.) Swingle Essential Oil from Italian Organic Crops. *J. Essent. Oil Res.* **2014,** *26,* 400–408.

29. D'alessio, P.; Bennaceur-Griscelli, A.; Ostan, R.; Franceschi, C. New Targets for the Identification of an Anti-inflammatory Anti-senescence Activity.Chapter 27, In

Senescence; Nagata, T., Ed.; In Tech Open Access Publisher, 2012; pp 647–666.DOI: 10.5772/34303.

30. D'alessio, P. A.; Ostan, R. O; Bisson, J. F.; Schulzke, J. D.; Ursini, M. V.; Béné, M. C. Oral Administration of d-Limonene Controls Inflammation in Rat Colitis and Displays Anti-Inflammatory Properties as Diet Supplementation in Humans. *Life Sci.* **2013,** *92,* 1151–1156.

31. Das, S.; Ng, W. K.; Tan, R. B. H. Are Nanostructured Lipid Carriers (NLCs) Better Than Solid Lipid Nanoparticles (SLNs): Development, Characterizations and Comparative Evaluations of Clotrimazole-Loaded SLNs and NLCs? *Eur. J. Pharm. Sci.* **2012,** *47,* 139–151.

32. Dhar, P.; Chan, P.; Cohen, D. T.; Khawam, F.; Gibbons, S.; Snyder-Leiby, T.; Dickstein, E.; Rai, P. K.; Watal, G. Synthesis, Antimicrobial Evaluation, and Structure—Activity Relationship of α-Pinene Derivatives. *J. Agric. Food Chem.* **2014,** *62,* 3548–3552.

33. Družić, J.; Jerković, I.; Marijanović, Z.; Roje, M. Chemical Biodiversity of the Leaf and Flower Essential Oils of *Citrus aurantium* L. from Dubrovnik Area (Croatia) in Comparison with *Citrus sinensis* L. Osbeck cv. Washington Navel, *Citrus sinensis* L. Osbeck cv. Tarocco and *Citrus sinensis* L. Osbeck cv. Doppio Sanguigno. *J. Essent. Oil Res.* **2016,** *28*(4), 283–291.

34. Edris, A. E. Pharmaceutical and Therapeutic Potentials of Essential Oils and Their Individual Volatile Constituents: A Review. *Phytother. Res.* **2007,** *21,* 308–323.

35. Elsherbiny, E. A.; El-Khateeb, A. Y.; Azzaz, N. A. Chemical Composition and Fungicidal Effects of *Ocimum basilicum* Essential Oil on *Bipolaris* and *Cochliobolus* Species. *J. Agric. Sci. Technol.* **2016,** *18,* 1143–1152.

36. Falk-Filipsson, A.; Löf, A.; Hagberg, M.; Hjelm, E. W.; Wang, Z. d-Limonene Exposure to Humans by Inhalation: Uptake, Distribution, Elimination, and Effects on the Pulmonary Function. *J. Toxicol. Environ. Health* **1993,** *38,* 77–88.

37. Faraldos, J. A.; Kariuki, B. M.; Coates, R. M. 2-Azapinanes: Aza Analogues of the Enantiomeric Pinyl Carbocation Intermediates in Pinene Biosynthesis. *Organ.Lett.* **2011,** *13,* 836–839.

38. Fernandes, J. Antitumor Monoterpenes. Chapter 8, In*Bioactive Essential Oils and Cancer;* Sousa, D. P., Ed.; Springer, 2015; pp 175–200.

39. Filip, S.; Vidović, S.; Vladić, J.; Pavlić, B.; Adamović, D.; Zeković, Z. Chemical Composition and Antioxidant Properties of *Ocimum basilicum* L. Extracts Obtained by Supercritical Carbon Dioxide Extraction: Drug Exhausting Method. *J. Supercrit. Fluids* **2016,** *109,* 20–25.

40. Friedrich, K.; Delgado, I. F.; Santos, L. M. F.; Paumgartten, F. J. R. Assessment of Sensitization Potential of Monoterpenes Using the Rat Popliteal Lymph Node Assay. *Food Chem. Toxicol.* **2007,** *45,* 1516–1522.

41. Fukumoto, S.; Morishita, A.; Furutachi, K.; Terashima, T.; Nakayama, T.; Yokogoshi, H. Effect of Flavour Components in Lemon Essential Oil on Physical or Psychological Stress. *Stress Health* **2008,** *24,* 3–12.

42. Galata, M.; Sarker, L. S.; Mahmoud, S. S. Transcriptome Profiling, and Cloning and Characterization of the Main Monoterpene Synthases of *Coriandrum sativum* L. *Phytochemistry* **2014,** *102,* 64–73.

43. George, K. W.; Alonso-Gutierrez, J.; Keasling, J. D.; Lee, T. S. Isoprenoid Drugs, Biofuels, and Chemicals—Artemisinin, Farnesene, and Beyond. *Adv. Biochem. Eng. Biotechnol.* **2015,** *148,* 355–389.

44. Geron, C.; Rasmussen, R.; Arnts, R. R.; Guenther, A. A Review and Synthesis of Monoterpene Speciation from Forests in the United States. *Atmos. Environ.* **2000,** *34,* 1761–1781.

45. Gialeli, C.; Theocharis, A. D.; Karamanos, N. K. Roles of Matrix Metalloproteinases in Cancer Progression and Their Pharmacological Targeting. *Fed. Eur. Biochem. Soc. J.* **2011,** *278,* 16–27.

46. Gminski, R.; Tang, T.; Mersch-Sundermann, V. Cytotoxicity and Genotoxicity in Human Lung Epithelial A549 Cells Caused by Airborne Volatile Organic Compounds Emitted from Pine Wood and Oriented Strand Boards. *Toxicol. Lett.* **2010,** *196,* 33–41.

47. Gomes-Carneiro, M. R.; Viana, M. E. S.; Felzenszwalb, I.; Paumgartten, F. J. R. Evaluation of β-Myrcene, α-Terpinene and (+)- and (−)-α-Pinene in the *Salmonella/Microsome Assay. Food Chem. Toxicol.* **2005,** *43,* 247–252.

48. Gu, Y.; Ting, Z.; Qiu, X.; Zhang, X.; Gan, X.; Fang, Y.; Xu, X.; Xu, R. Linalool Preferentially Induces Robust Apoptosis of a Variety of Leukemia Cells via Upregulating p53 and Cyclin-Dependent Kinase Inhibitors. *Toxicology* **2010,** *268,* 19–24.

49. Gunaseelan, S.; Balupillai, A.; Govindasamy, K.; Muthusamy, G.; Ramasamy, K.; Shanmugam, M.; Prasad, N. R. The Preventive Effect of Linalool on Acute and Chronic UVB-Mediated Skin Carcinogenesis in Swiss Albino Mice. *Photochem. Photobiol. Sci.* **2016,** *15,* 851–860.

50. Haag, J. D.; Lindstrom, M. J.; Gould, M. N. Limonene-Induced Regression of Mammary Carcinomas. *Cancer Res.* **1992,** *52,* 4021–4026.

51. Hădărugă, D. I.; Hădărugă, N. G.; Costescu, C. I.; David, I.; Gruia, A. T. Thermal and Oxidative Stability of the *Ocimum basilicum* L. Essential Oil/β-Cyclodextrin Supramolecular System. *Beilstein J. Organ. Chem.* **2014,** *10,* 2809–2820.

52. Hagvall, L.; Bråred Christensson, J. Cross-Reactivity Between Citral and Geraniol—Can It Be Attributed to Oxidized Geraniol? *Contact Dermatitis* **2014,** *71,* 280–288.

53. Han, H. D.; Cho, Y. J.; Cho, S. K.; Byeon, Y.; Jeon, H. N.; Kim, H. S.; Kim, B. G.; Bae, D. S.; Lopez-Berestein, G.; Sood, A. K.; Shin, B. C.; Park, Y. M.; Lee, J. W. Linalool-Incorporated Nanoparticles as a Novel Anticancer Agent for Epithelial Ovarian Carcinoma. *Mol. Cancer Ther.* **2016,** *15,* 618–627.

54. Herman, A.; Herman, A. P. Essential Oils and Their Constituents as Skin Penetration Enhancer for Transdermal Drug Delivery: A Review. *J. Pharm. Pharmacol.* **2015,** *67,* 473–485.

55. Heydorn, S.; Menné, T.; Andersen, K. E.; Bruze, M.; Svedman, C.; White, I. R.; Basketter, D. A. Citral a Fragrance Allergen and Irritant. *Contact Dermatitis* **2003,** *49,* 32–36.

56. Hoppel, M.; Caneri, M.; Glatter, O.; Valenta, C. Self-Assembled Nanostructured Aqueous Dispersions as Dermal Delivery Systems. *Int. J. Pharm.* **2015,** *495,* 459–462.

57. Ilc, T.; Parage, C.; Boachon, B.; Navrot, N.; Werck-Reichhart, D. Monoterpenol Oxidative Metabolism: Role in Plant Adaptation and Potential Applications. *Front. Plant Sci.* **2016,** *7,* 110–115.

58. Jana, S.; Patra, K.; Sarkar, S.; Jana, J.; Mukherjee, G.; Bhattacharjee, S.; Mandal, D. P. Antitumorigenic Potential of Linalool is Accompanied by Modulation of Oxidative

Stress: An In Vivo Study in Sarcoma-180 Solid Tumor Model. *Nutr. Cancer* **2014,** *66,* 835–848.

59. Jiang, H.; Wang, J.; Song, L.; Cao, X.; Yao, X.; Tang, F.; Yue, Y. GC× GC-TOFMS Analysis of Essential Oils Composition from Leaves, Twigs and Seeds of *Cinnamomum camphora* L. Presl and Their Insecticidal and Repellent Activities. *Molecules* **2016,** *21,* 423.

60. Jongedijk, E.; Cankar, K.; Buchhaupt, M.; Schrader, J.; Bouwmeester, H.; Beekwilder, J. Biotechnological Production of Limonene in Microorganisms. *Appl. Microbiol. Biotechnol.* **2016,** *100,* 2927–2938.

61. Kaimoto, T.; Hatakeyama, Y.; Takahashi, K.; Imagawa, T.; Tominaga, M.; Ohta, T. Involvement of Transient Receptor Potential A1 Channel in Algesic and Analgesic Actions of the Organic Compound Limonene. *Eur. J. Pain* **2016,** *20,* 1155–1165.

62. Kanavy, H. E.; Gerstenblith, M. R. Ultraviolet Radiation and Melanoma. In *Seminars in Cutaneous Medicine and Surgery;* Frontline Medical Communications, 2011, pp 222–228.

63. Kang, E.; Lee, H. D.; Jung, Y. J.; Shin, S. Y.; Koh, D.; Lee, Y. H. α-Pinene Inhibits Tumor Invasion Through Downregulation of Nuclear Factor (NF)-κB-Regulated Matrix Metalloproteinase-9 Gene Expression in MDA-MB-231 Human Breast Cancer Cells. *Appl. Biol. Chem.* **2016,** *59,* 1–6.

64. Kang, M.; Eom, J.; Kim, Y.; Um, Y.; Woo, H. M. Biosynthesis of Pinene from Glucose Using Metabolically-Engineered *Corynebacterium glutamicum. Biotechnol. Lett.* **2014,** *36,* 2069–2077.

65. Karin, M. How NF-κB is Activated: The Role of the IκB Kinase (IKK) Complex. *Oncogene* **1999,** *18,* 6867–6874.

66. Katsuyama, S.; Kuwahata, H.; Yagi, T.; Kishikawa, Y.; Komatsu, T.; Sakurada, T.; Nakamura, H. Intraplantar Injection of Linalool Reduces Paclitaxel-Induced Acute Pain in Mice. *Biomed. Res.* **2012,** *33,* 175–181.

67. Kim, D.; Lee, H.; Jeon, Y.; Han, Y.; Kee, J.; Kim, H.; Shin, H; Kang, J.; Lee, B. Su.; Kim, S.; Kim, S.; Park, S.; Choi, B.; Park, S.; Um, J.; Hong, S. H. Alpha-Pinene Exhibits Anti-Inflammatory Activity Through the Suppression of MAPKs and the NF-κB Pathway in Mouse Peritoneal Macrophages. *Am. J. Chin. Med.* **2015,** *43,* 731–742.

68. Kim, H. Cerulein Pancreatitis: Oxidative Stress, Inflammation, and Apoptosis. *Gut Liver* **2008,** *2,* 74–80.

69. Kim, Y. W.; Kim, M. J.; Chung, B. Y.; Bang Du, Y.; Lim, S. K.; Choi, S. M.; Lim, D. S.; Cho, M. C.; Yoon, K.; Kim, H. S.; Kim, K. B.; Kim, Y. S.; Kwack, S. J.; Lee, B. Safety Evaluation and Risk Assessment of d-Limonene. *J. Toxicol. Environ. Health Part B* **2013,** *16,* 17–38.

70. Kordali, S.; Usanmaz, A.; Cakir, A.; Komaki, A.; Ercisli, S. Antifungal and Herbicidal Effects of Fruit Essential Oils of Four *Myrtus Communis*Genotypes. *Chem. Biodiversity* **2016,** *13,* 77–84.

71. Kovač, J.; Šimunović, K.; Wu, Z.; Klančnik, A.; Bucar, F.; Zhang, Q.; Možina, S. S. Antibiotic Resistance Modulation and Modes of Action of (−)-α-Pinene in *Campylobacter jejuni. Public Library Sci. One* **2015,** *10,* 1–14.

72. Kusuhara, M.; Urakami, K.; Masuda, Y.; Zangiacomi, V.; Ishii, H.; Tai, S.; Maruyama, K.; Yamaguchi, K. Fragrant Environment with Alpha-Pinene Decreases Tumor Growth in Mice. *Biomed. Res.* **2012,** *33,* 57–61.

73. Lapczynski, A.; Letizia, C. S.; Api, A. M. Addendum to Fragrance Material Review on Linalool. *Food Chem. Toxicol.* **2008**, *46*, S190–S192.

74. Lehman-Mckeeman, L. D.; Caudill, D. d-Limonene Induced Hyaline Droplet Nephropathy in α2u-Globulin Transgenic Mice. *Toxicol. Sci.* **1994**, *23*, 562–568.

75. Li, P. H.; Chiang, B. H. Process Optimization and Stability of D-Limonene-in-Water Nanoemulsions Prepared by Ultrasonic Emulsification Using Response Surface Methodology. *Ultrason. Sonochem.* **2012**, *19*, 192–197.

76. Li, X.; Yang, Y.; Li, Y.; Zhang, W. K.; Tang, H. α-Pinene, Linalool, and 1-Octanol Contribute to the Topical Anti-inflammatory and Analgesic Activities of Frankincense by Inhibiting COX-2. *J. Ethnopharmacol.* **2016**, *179*, 22–26.

77. Linck, V. M.; Silva, A. L.; Figueiró, M.; Caramão, E. B.; Moreno, P. R. H.; Elisabetsky, E. Effects of Inhaled Linalool in Anxiety, Social Interaction and Aggressive Behavior in Mice. *Phytomedicine* **2010**, *17*, 679–683.

78. Linck, V. M.; Silva, A. L.; Figueiró, M.; Piato, A. L.; Herrmann, A. P.; Birck, F. D.; Caramão, E. B.; Nunes, D. S.; Moreno, P. R. H.; Elisabetsky, E. Inhaled Linalool-Induced Sedation in Mice. *Phytomedicine* **2009**, *16*, 303–307.

79. Liou, G. Y.; Storz, P. Reactive Oxygen Species in Cancer. *Free Radical Res.* **2010**, *44*, 479–486. DOI:10.3109/10715761003667554.

80. Lu, W.; Chiang, B.; Huang, D.; Li, P. Skin Permeation of D-Limonene-Based Nanoemulsions as a Transdermal Carrier Prepared by Ultrasonic Emulsification. *Ultrason. Sonochem.* **2014**, *21*, 826–832.

81. Lücker, J.; Tamer, M. K.; Schwab, W.; Verstappen, F. W. A.; Plas, L. H. W.; Bouwmeester, H. J.; Verhoeven, H. A. Monoterpene Biosynthesis in Lemon (*Citrus limon*). *Eur. J. Biochem.* **2002**, *269*, 3160–3171.

82. Maeda, H.; Yamazaki, M.; Katagata, Y. Kuromoji (*Lindera umbellata*) Essential Oil-Induced Apoptosis and Differentiation in Human Leukemia HL-60 Cells. *Exp. Ther. Med.* **2012**, *3*, 49–52.

83. Maksymiuk, C. S.; Gayahtri, C.; Gil, R. R.; Donahue, N. M. Secondary Organic Aerosol Formation from Multiphase Oxidation of Limonene by Ozone: Mechanistic Constraints via Two-Dimensional Heteronuclear NMR Spectroscopy. *Phys. Chem. Chem. Phys.* **2009**, *11*, 7810–7818.

84. Marcus, J.; Klossek, M. L.; Touraud, D.; Kunz, W. Nano-droplet Formation in Fragrance Tinctures. *Flavour Fragrance J.* **2013**, *28*, 294–299.

85. Marnett, L. J.; Cohen, S. M.; Fukushima, S.; Gooderham, N. J.; Hecht, S. S.; Rietjens, I. M. C. M.; Smith, R. L.; Adams, T. B.; Bastaki, M.; Harman, C. L.; Mcgowen, M. M.; Taylor, S. V. GRASr2 Evaluation of Aliphatic Acyclic and Alicyclic Terpenoid Tertiary Alcohols and Structurally Related Substances Used as Flavoring Ingredients. *J. Food Sci.* **2014**, *79*, R428–R441.

86. Matsuo, A. L.; Figueiredo, C. R.; Arruda, D. C.; Pereira, F. V.; Scutti, J. A. B.; Massaoka, M. H.; Travassos, L. R.; Sartorelli, P.; Lago, J. H. G. α-Pinene Isolated from *Schinus terebinthifolius* Raddi (*Anacardiaceae*) Induces Apoptosis and Confers Antimetastatic Protection in a Melanoma Model. *Biochem. Biophys. Res. Commun.* **2011**, *411*, 449–454.

87. Matura, M.; Goossens, A.; Bordalo, O.; Garcia-Bravo, B.; Magnusson, K.; Wrangsjö, K.; Karlberg, A. T. Patch Testing with Oxidized R-(+)-Limonene and Its Hydroperoxide Fraction. *Contact Dermatitis* **2003**, *49*, 15–21.

88. Mercier, B.; Prost, J.; Prost, M. The Essential Oil of Turpentine and Its Major Volatile Fraction (α-and β-Pinenes): A Review. *Int. J. Occup. Med. Environ. Health* **2009**, *22*, 331–342.

89. Miller, J. A.; Thompson, P. A.; Hakim, I. A.; Lopez, A. M.; Thomson, C. A.; Chew, W.; Hsu, C. H.; Chow, H. H. S. Safety and Feasibility of Topical Application of Limonene as a Massage Oil to the Breast. *J. Cancer Ther.* **2012**, *3*, 5A.DOI: 10.4236/jct.2012.325094.

90. Miller, J. A.; Lang, J. E.; Ley, M.; Nagle, R.; Hsu, C. H.; Thompson, P. A.; Cordova, C.; Waer, A.; Chow, H. H. S. Human Breast Tissue Disposition and Bioactivity of Limonene in Women with Early-Stage Breast Cancer. *Cancer Prev. Res.* **2013**, *6*, 577–584.

91. Miller, J. A.; Pappan, K.; Thompson, P. A.; Want, E. J.; Siskos, A. P.; Keun, H. C.; Wulff, J.; Hu, C.; Lang, J. E.; Chow, H. H. S. Plasma Metabolomic Profiles of Breast Cancer Patients After Short-Term Limonene Intervention. *Cancer Prev. Res.* **2015**, *8*, 86–93.

92. Mitić-Ćulafić, D.; Žegura, B.; Nikolić, B.; Vuković-Gačić, B.; Knežević-Vukčević, J.; Filipič, M. Protective Effect of Linalool, Myrcene and Eucalyptol Against t-Butyl Hydroperoxide Induced Genotoxicity in Bacteria and Cultured Human Cells. *Food Chem. Toxicol.* **2009**, *47*, 260–266.

93. Miyashita, M.; Sadzuka, Y. Effect of Linalool as a Component of Humulus Lupulus on Doxorubicin-Induced Antitumor Activity. *Food Chem. Toxicol.* **2013**, *53*, 174–179.

94. Müller, M.; Buchbauer, G. Essential Oil Components as Pheromones. A Review. *Flavour Fragrance J.* **2011**, *26*, 357–377.

95. Müller, R. H.; Radtke, M.; Wissing, S. A. Solid Lipid Nanoparticles (SLN) and Nanostructured Lipid Carriers (NLC) in Cosmetic and Dermatological Preparations. *Adv. Drug Delivery Rev.* **2002**, *54*, S131–S155.

96. Neuenschwander, U.; Guignard, F.; Hermans, I. Mechanism of the Aerobic Oxidation of α-Pinene. *Chem. Sus. Chem.* **2010**, *3*, 75–84.

97. Nikolić, M.; Jovanović, K. K.; Marković, T.; Marković, D.; Gligorijević, N.; Radulović, S.; Soković, M. Chemical Composition, Antimicrobial, and Cytotoxic Properties of Five *Lamiaceae* Essential Oils. *Ind. Crops Prod.* **2014**, *61*, 225–232.

98. O'neill, M. J. (Ed.). *The Merck Index: An Encyclopedia of Chemicals, Drugs, and Biologicals;* Cambridge, UK: Royal Society of Chemistry, England; 2006; p 2708.

99. Ohtsubo, S.; Fujita, T.; Matsushita, A.; Kumamoto, E. Inhibition of the Compound Action Potentials of Frog Sciatic Nerves by Aroma Oil Compounds Having Various Chemical Structures. *Pharmacol. Res. Perspect.* **2015**, *3*, 1–13.

100. Opdyke, D. L. J. Monographs on Fragrance Raw Materials. *Food Cosmet. Toxicol.* 1979, *17,* 509–511, 513, 515–519, 521–523, 525, 527, 529, 531–533.

101. Ortiz-Lazareno, P. C.; Bravo-Cuellar, A.; Lerma-Díaz, J. M.; Jave-Suárez, L. F.; Aguilar-Lemarroy, A.; Domínguez-Rodríguez, J. R.; González-Ramella, O.; Célis, R.; Gómez-Lomelí, P.; Hernández-Flores, G. Sensitization of U937 Leukemia Cells to Doxorubicin by the MG132 Proteasome Inhibitor Induces an Increase in Apoptosis by Suppressing NF-Kappa B and Mitochondrial Membrane Potential Loss. *Cancer Cell Int.* **2014**, *14*, 1.

102. Ortiz, M. I.; González-García, M. P.; Ponce-Monter, H. A.; Castañeda-Hernández, G.; Aguilar-Robles, P. Synergistic Effect of the Interaction Between Naproxen and Citral on Inflammation in Rats. *Phytomedicine* **2010**, *18*, 74–79.

103. Pal, M.; Mishra, T.; Kumar, A.; Baleshwar; Upreti, D. K.; Rana, T. S. Chemical Constituents and Antimicrobial Potential of Essential Oil from *Betula utilis* Growing in High Altitude of Himalaya (India). *J. Essent. Oil Bear. Plants* **2015**, *18*, 1078–1082.

104. Pappas, A. Epidermal Surface Lipids. *Dermatoendocrinol* **2009**, *1*, 72–76.

105. Pereira, S. L.; Marques, A. M.; Sudo, R. T.; Kaplan, M. A. C.; Zapata-Sudo, G. Vasodilator Activity of the Essential Oil from Aerial Parts of *Pectis Brevipedunculata* and its Main Constituent Citral in Rat Aorta. *Molecules* **2013**, *18*, 3072–3085.

106. Perry, N.; Perry, E. Aromatherapy in the Management of Psychiatric Disorders: Clinical and Neuropharmacological Perspectives. *CNS Drugs* **2006**, *20*, 257–280.

107. Privett, B. J.; Nutz, S. T.; Schoenfisch, M. H. Efficacy of Surface-Generated Nitric Oxide Against *Candida albicans* Adhesion and Biofilm Formation. *Biofouling J. Bioadhes. Biofilm Res.* **2010**, *26*, 973–983.

108. Program, National Toxicology. NTP Toxicology and Carcinogenesis Studies of d-Limonene (CAS No. 5989–27–5) in F344/N Rats and B6C3F1 Mice (Gavage Studies). *Nat. Toxicol. Prog. Tech. Rep. Ser.* **1990**, *347*, 1–165.

109. Raguso, R. A. More Lessons from Linalool: Insights Gained from A Ubiquitous Floral Volatile. *Curr. Opin. Plant Biol.* **2016**, *32*, 31–36.

110. Raut, J. S.; Shinde, R. B.; Chauhan, N. M.; Karuppayil, S. M. Terpenoids of Plant Origin Inhibit Morphogenesis, Adhesion, and Biofilm Formation by *Candida albicans*. *Biofouling J. Bioadhes. Biofilm Res.* **2013**, *29*, 87–96.

111. Rehman, M. U.; Tahir, M.; Khan, A. Q.; Khan, R.; Oday-O-Hamiza; Lateef, A.; Hassan, S. K.; Rashid, S.; Ali, N.; Zeeshan, M.; Sultana, S. D-Limonene Suppresses Doxorubicin-Induced Oxidative Stress and Inflammation via Repression of COX-2, iNOS, and NFκB in Kidneys of Wistar Rats. *Exp. Biol. Med.* **2014**, *239*, 465–476.

112. Richa, S. R. P.; Hader, D. P. Physiological Aspects of UV-Excitation of DNA. *Top. Curr. Chem.* **2015**, *356*, 203–248.

113. Rottava, L.; Cortina, P. F.; Zanella, C. A.; Cansian, R. L.; Toniazzo, G.; Treichel, H.; Antunes, O. A. C.; Oestreicher, E. G.; Oliveira, D. Microbial Oxidation of (−)-α-Pinene to Verbenol Production by Newly Isolated Strains. *Appl. Biochem. Biotechnol.* **2010**, *162*, 2221–2231.

114. Rouanet, P.; Linares-Cruz, G.; Dravet, F.; Poujol, S.; Gourgou, S.; Simony-Lafontaine, J.; Grenier, J.; Kramar, A.; Girault, J.; Le Nestour, E.; Maudelonde, T. Neoadjuvant Percutaneous 4-Hydroxytamoxifen Decreases Breast Tumoral Cell Proliferation: A Prospective Controlled Randomized Study Comparing Three Doses of 4-Hydroxytamoxifen Gel to Oral Tamoxifen. *J. Clin. Oncol.* **2005**, *23*, 2980–2987.

115. Rufino, A. T.; Ribeiro, M.; Judas, F.; Salgueiro, L.; Lopes, M. C.; Cavaleiro, C.; Mendes, A. F. Anti-inflammatory and Chondroprotective Activity of (+)-α-Pinene: Structural and Enantiomeric Selectivity. *J. Nat. Prod.* **2014**, *77*, 264–269.

116. Sarigiannis, D. A.; Karakitsios, S. P.; Gotti, A.; Liakos, I. L.; Katsoyiannis, A. Exposure to Major Volatile Organic Compounds and Carbonyls in European Indoor Environments and Associated Health Risk. *Environ. Int.* **2011**, *37*, 743–765.

117. Sarria, S.; Wong, B.; Martín, H. G.; Keasling, J. D.; Peralta-Yahya, P. Microbial Synthesis of Pinene. *Am. Chem. Soc. Synth. Biol.* **2014**, *3*, 466–475.

118. Seol, G. H.; Kang, P.; Lee, H. S.; Seol, G. H. Antioxidant Activity of Linalool in Patients with Carpal Tunnel Syndrome. *BioMed Central Neurol.* **2016**, *16*, 1–6.

119. Severino, P.; Santana, M. H. A.; Souto, E. B. Optimizing SLN and NLC by 2 2 Full Factorial Design: Effect of Homogenization Technique. *Mater. Sci. Eng.: C* **2012,** *32,* 1375–1379.

120. Severino, P.; Andreani, T.; Jäger, A.; Chaud, M. V.; Santana, M. H. A.; Silva, A. M.; Souto, E. B. Solid Lipid Nanoparticles for Hydrophilic Biotech Drugs: Optimization and Cell Viability Studies (Caco-2 and HEPG-2 cell lines). *Eur. J. Med. Chem.* **2014,** *81,* 28–34.

121. Shi, F.; Zhao, Y.; Firempong, C. K.; Xu, X. Preparation, Characterization and Pharmacokinetic Studies of Linalool-Loaded Nanostructured Lipid Carriers. *Pharm. Biol.* **2016,** *54,* 2320–2328.

122. Silva, A. C. R.; Lopes, P. M.; Azevedo, M. M. B.; Costa, D. C. M.; Alviano, C. S.; Alviano, D. S. Biological Activities of α-Pinene and β-Pinene Enantiomers. *Molecules* **2012,** *17,* 6305–6316.

123. Sobral, M. V.; Xavier, A. L.; Lima, T. C.; Sousa, D. P. Antitumor Activity of Monoterpenes Found in Essential Oils. *Sci. World J.* **2014,** *2014,* 1–35.

124. Soković, M.; Glamočlija, J.; Marin, P. D.; Brkić, D.; Griensven, L. J. L. D. Antibacterial Effects of the Essential Oils of Commonly Consumed Medicinal Herbs Using an In Vitro Model. *Molecules* **2010,** *15,* 7532–7546.

125. Sourmaghi, M. H. S.; Kiaee, G.; Golfakhrabadi, F.; Jamalifar, H.; Khanavi, M. Comparison of Essential Oil Composition and Antimicrobial Activity of *Coriandrum sativum* L. Extracted by Hydrodistillation and Microwave-Assisted Hydrodistillation. *J. Food Sci. Technol.* **2015,** *52,* 2452–2457.

126. Souto, E. B.; Severino, P.; Santana, M. H. A.; Pinho, S. C. Solid Lipid Nanoparticles: Classical Methods of Lab Production. *Quím. Nova* **2011,** *34,* 1762–1769.

127. Spracklen, D. V.; Bonn, B.; Carslaw, K. S. Boreal Forests, Aerosols and the Impacts on Clouds and Climate. *Philos. Trans. R. Soc. A: Math. Phys. Eng. Sci.* **2008,** *366,* 4613–4626.

128. Subongkot, T.; Duangjit, S.; Rojanarata, T.; Opanasopit, P.; Ngawhirunpat, T. Ultradeformable Liposomes with Terpenes for Delivery of Hydrophilic Compound. *J. Liposome Res.* **2012,** *22,* 254–262.

129. Sun, X. B.; Wang, S. M.; Li, T.; Yang, Y. Q. Anticancer Activity of Linalool Terpenoid: Apoptosis Induction and Cell Cycle Arrest in Prostate Cancer Cells. *Trop. J. Pharm. Res.* **2015,** *14,* 619–625.

130. Sundaravel, B.; Babu, C. M.; Vinodh, R.; Cha, W. S.; Jang, H. T. Synthesis of Campholenic Aldehyde from α-Pinene Using Bi-Functional PrAlPO-5 Molecular Sieves. *J. Taiwan Inst. Chem. Eng.* **2016,** *63,* 157–165.

131. Tomazoni, E. Z.; Pansera, M. R.; Pauletti, G. F.; Moura, S.; Ribeiro, R. T. S.; Schwambach, J. In Vitro Antifungal Activity of Four Chemotypes of *Lippia alba* (*Verbenaceae*) Essential Oils Against *Alternaria solani* (*Pleosporeaceae*) Isolates. *Ann. Braz. Acad. Sci.* **2016,** *88,* 999–1010.

132. Uter, W.; Yazar, K.; Kratz, E. M.; Mildau, G.; Lidén, C. Coupled Exposure to Ingredients of Cosmetic Products: I. Fragrances. *Contact Dermatitis* **2013,** *69,* 335–341.

133. Wang, E. C.; Wang, A. Z. Nanoparticles and Their Applications in Cell and Molecular Biology. *Integr. Biol.* **2014,** *6,* 9–26.

134. Wang, J.; Hua, W.; Yue, Y.; Gao, Z. MSU-S Mesoporous Materials: An Efficient Catalyst for Isomerization of α-Pinene. *Bioresour. Technol.* **2010,** *101,* 7224–7230.

135. Wang, W.; Li, N.; Luo, M.; Zu, Y.; Efferth, T. Antibacterial Activity and Anticancer Activity of *Rosmarinus officinalis* L. Essential Oil Compared to that of its Main Components. *Molecules* **2012,** *17,* 2704–2713.

136. Webb, D. R.; Ridder, G. M.; Alden, C. L. Acute and Subchronic Nephrotoxicity of d-Limonene in Fischer 344 Rats. *Food Chem. Toxicol.* **1989,** *27,* 639–649.

137. Webb, D. R.; Kanerva, R. L.; Hysell, D. K.; Alden, C. L.; Lehman-Mckeeman, L. D. Assessment of the Subchronic Oral Toxicity of d-Limonene in Dogs. *Food Chem. Toxicol.* **1990,** *28,* 669–675.

138. Wei, A.; Shibamoto, T. Antioxidant Activities and Volatile Constituents of Various Essential Oils. *J. Agric. Food Chem.* **2007,** *55,* 1737–1742.

139. Wörz, N.; Claus, P.; Lang, S.; Hampe, M. J. Thermodynamics and Transport Properties of Citral. *Am. Inst. Chem. Eng.* **2012,** *58,* 2557–2562.

140. Xie, Q.; Liu, Z.; Li, Z. Chemical Composition and Antioxidant Activity of Essential Oil of Six *Pinus* Taxa Native to China. *Molecules* **2015,** *20,* 9380–9392.

141. Yang, J.; Nie, Q.; Ren, M.; Feng, H.; Jiang, X.; Zheng, Y.; Liu, M.; Zhang, H.; Xian, M. Metabolic Engineering of *Escherichia coli* for the Biosynthesis of Alpha-Pinene. *Biotechnol. Biofuels* **2013,** *6,* 1.

142. Yang, X.; Lu, T.; Chen, C.; Zhou, L.; Wang, F.; Su, Y.; Xu, J. Synthesis of Hierarchical AlPO-*n* Molecular Sieves Templated by Saccharides. *Microporous Mesoporous Mater.* **2011,** *144,* 176–182.

143. Yang, Z.; Xi, J.; Li, J.; Qu, W. Biphasic Effect of Citral, a Flavoring and Scenting Agent, on Spatial Learning and Memory in Rats. *Pharmacol. Biochem. Behav.* **2009,** *93,* 391–396.

144. Yang, Z.; Wu, N.; Zu, Y.; Fu, Y. Comparative Anti-Infectious Bronchitis Virus (IBV) Activity of (−)-Pinene: Effect on Nucleocapsid (N) Protein. *Molecules* **2011,** *16,* 1044–1054.

145. Zhao, K.; Singh, J. Mechanism(s) of in Vitro Percutaneous Absorption Enhancement of Tamoxifen by Enhancers. *J. Pharm. Sci.* **2000,** *89,* 771–780.

146. Zhao, Y.; Chen, R.; Wang, Y.; Qing, C.; Wang, W.; Yang, Y. In Vitro and in Vivo Efficacy Studies of *Lavender angustifolia* Essential Oil and its Active Constituents on the Proliferation of Human Prostate Cancer. *Integr. Cancer Ther.* **2016.** DOI:10.1177/1534735416645408.

CHAPTER 5

ROLE OF NUTRACEUTICALS IN PREVENTION OF NONALCOHOLIC FATTY LIVER

SAHAR Y. AL-OKBI

CONTENTS

ABSTRACT

The nonalcoholic fatty liver disease (NAFLD) is a component of metabolic syndrome and the mortality among NAFLD is higher than general population. NAFLD spectrum includes simple steatosis, steatohepatitis, fibrosis, cirrhosis, hepatocellular carcinoma, and liver failure. It is important to recognize that underactivity, insulin resistance, lipotoxicity, adipokines, oxidative stress, overnutrition, genetic factors, and other inflammatory mediators have strong link to the development of NAFLD/NASH (Nonalcoholic Steatohepatitis). Treatment of fatty liver is crucial not only due to the possible progression to more aggressive liver diseases but also because it may cause cardiovascular diseases. Dietary regimen and exercise play an important role in the management of NAFLD. Ideal therapy or nutraceuticals for NAFLD must possess insulin-sensitizing, antioxidants, lipid-lowering, hepatic protective, and fat mass reducing effect.

5.1   INTRODUCTION

Fatty liver represents a major health problem in both developed and developing countries. The problem is more prominent in obese and diabetics. The risk does not reside in the deposition of fat in liver but the progression may occur beyond this. Severe accumulation of fat can lead to initiation of high oxidative stress that consequently results in the development of inflammatory conditions leading to nonalcoholic steatohepatitis (NASH). Nonalcoholic fatty liver diseases (NAFLD) are among the most prevalent and dangerous chronic liver diseases that affect a high proportion of the world's population. It maybe an initiator of metabolic syndrome and is accompanied by all the changes seen in metabolic syndrome. It is regarded as the hepatic component of the metabolic syndrome.[26]

It is estimated that NAFLD prevalence is about 20% of total population in developed countries. Its prevalence is presumably much higher in obese and diabetic persons. Mortality among patients having NAFLD is more than other populations and is higher in elderly, cirrhotic patients, and those with impaired fasting glucose.[2] NAFLD refers to a wide spectrum of liver disease, which occurs in subjects that do not consume alcohol.

Simple fatty liver (steatosis) is a mild form, which involves nonphysiological accumulation of fat in the liver cells, especially triglycerides. Steatosis by itself is not harmful to the liver. However, the progression of nonalcoholic steatosis to NASH, which is a state of steatosis accompanied by inflammation along with hepatocyte damage and necrosis, is a real serious condition. NASH can ultimately lead to fibrosis and cirrhosis. Cirrhosis can progress to hepatocellular carcinoma or liver failure. Generally, the simple fatty liver prognosis is good; meanwhile, the progression to hepatic cirrhosis and cancer can occur in 10–15% in patients with NASH. Obese, dyslipidemic, diabetics, and individuals with insulin resistance are more exposed to NAFLD than other populations.[2,15,44,65,70,74]

Food has a major role in both incidence and prevention of NAFLD. Therefore, the specific dietary regimen may participate greatly in the management and prevention of NASH. Nutraceuticals, a concentrated form of bioactive constituents prepared from food, might also possess protective action and may prevent the progression of fatty liver to more advanced liver diseases as well as guarding against the incidence of cardiovascular diseases.

This chapter reviewed different studied nutraceuticals that may have therapeutic efficiency and/or protective effect toward NAFL.

5.2 ROLE OF DIETS IN THE PATHOGENESIS OF NONALCOHOLIC FATTY LIVER DISEASE (NAFLD)

The type of carbohydrate is important for NAFLD development. Food with high glycemic index increased deposition of fat in the liver. Fructose, which is a simple sugar, enhances hepatic lipogenesis through the de novo pathway and reduces beta-fatty acid oxidation, increases the deposition of lipid in the hepatocytes. Fat intake broadly leads to hepatic fat deposition. Although main hepatic triacylglycerols (TGs) come from fatty acids derived from adipose tissue hydrolysis, yet the composition of dietary fat could participate in the induction of fatty liver.[28,83] Inflammation-related factors could be stimulated by fat and carbohydrates that can cause NAFLD.[47] It was reported that deficiency of omega-3 fatty acids, high dietary cholesterol, and female gender are more accused in the development of NAFLD than fat saturation index.[25]

5.3 CAUSES OF NONALCOHOLIC STEATOHEPATITIS (NASH) AND NAFLD

The mechanism underlying the cause of NASH is not well documented; however, insulin resistance is evidenced to be involved in all stages of NAFLD. Other factors have been reported to participate in the incidence of fatty liver. High dietary fats, sugars, and carbohydrates enhance the induction of insulin resistance. There is evidence that the increase in insulin resistance is correlated with the elevation of abdominal fat and body mass index. There is also a strong relation between insulin resistance and elevated low-density lipoprotein cholesterol (LDL-ch) and TGs. Insulin resistance has effects on hepatic, muscles, and adipocytes metabolism of fat and sugar. These effects lead to enhanced uptake, infiltration, and deposition of TGs in hepatic cells. TGs are delivered from the diet and transported from abdominal fat and peripheral muscles. These large quantities of TGs are stored in tiny sacs inside the hepatocytes.

Inflammatory process and death of hepatic tissue that occurs in NASH remain to be clearly explained. Different theories including insulin resistance have been proposed. It is suggested that fat deposition in the liver by itself might induce NASH. This theory claims that the large quantity of lipid in the liver could be a source of peroxidation and free radicals that damage proteins and organelles in the hepatocytes. This process could result in death of cells and/or an inflammatory cell event. Deregulation of variable cytokines such as interleukin-6 (Int-6), tumor necrosis factor-α (TNF-α), and transforming growth factor-β that induce inflammation, cell death, and even increase insulin resistance may participate in the events. The cell obtains its energy requirement from mitochondria; mitochondria may malfunction during the fatty liver resulting in decrease of cell energy with subsequent reduction in mitochondrial fatty acid beta-oxidation and cell death. Lipid peroxidation together with elevated reactive oxygen species could be induced by cytochrome enzymes that are involved in different metabolic pathways. Peroxisome proliferator activating receptors (PPAR) in the cell nucleus that are essential for insulin to elucidate its effect may stop working, leading to insulin resistance associated by liver cell inflammation and scarring. Reduced transportation of triglycerides from the liver as very low-density lipoprotein (VLDL) and induction of apoptosis maybe involved in this process.[22,64,66,71]

Leptin resistance may contribute to the development of NASH.[48] In response to eating, leptin is secreted from stomach, fat, and brain cells to curb the appetite. Surprisingly, NASH patients though have high leptin level yet their appetite was not suppressed which may point to leptin resistance. One of the important functions of leptin is to prevent hepatocytes inflammation and scarring. It is also interesting that leptin is insulin sensitizer. Since NASH patients are insulin resistant, this could support the theory of malfunctioning of leptin receptors leading to leptin resistance as well.

The level of adiponectin was reported to be low in obese and NAFLD patients as well as type two diabetic patients compared with healthy subjects. Reduction of adiponectin may lead to inflammation of hepatocyte.[14] The role of adipokines (leptin and adiponectin) in NAFLD progression is probably pivotal, mediated by oxidative stress.

Dysbiosis, a perturbation of gut microflora favoring the pathogenic flora at the expense of the beneficial, has been proposed as one of the recent causes of NASH due to the exporting inflammation to the liver. The suggested underlying mechanism is the systemic translocation of bacteria and intestinal and hepatic inflammation due to the deregulation of the epithelial barrier function. Moreover, liver pathology is further modulated by microbiome metabolite including short-chain fatty acids, lipopolysaccharides, bile acids, and ethanol.[17]

5.4 PROGRESSION OF LIVER FIBROSIS

Most of the changes that lead to liver fibrogenesis are mediated by hepatic stellate cells. On liver injury, stellate cells are stimulated and converted to profibrogenic myofibroblastic phenotype that leads to collagen and matrix protein deposition in the liver. Platelet-derived growth factor is the main component that stimulates hepatic stellate cells proliferation.[16] Weight gain and prominent insulin resistance usually accompany the progression to liver fibrosis.[29] As a result of continual chronic inflammation, liver cirrhosis may develop over time.

5.5 METABOLIC SYNDROME AND NAFLD INTERRELATION

Metabolic syndrome is a real worldwide health problem accounting for 22% of general population. The components of metabolic syndrome are central adiposity (and/or obesity), dyslipidemia, glucose intolerance, insulin resistance, hypertension, and NAFLD. These clinical manifestations may lead to cardiovascular diseases and type 2 diabetes that could result in increased mortality rates.[1,4,31,33,39,40,84] Metabolic syndrome is considered as one of the most causative factors of NAFLD that puts a very large population at high risk of liver failure.[52,58]

It maybe speculated that insulin resistance is a common factor between hypertension, diabetes, obesity, dyslipidemia, steatosis, and the progression to more advanced liver diseases such as NASH, cirrhosis, and liver cancer.[53] The regulation of glucose metabolism and postprandial lipoprotein is greatly affected by central obesity and insulin resistance. In insulin resistance state, there is an impairment of the suppression of lipolysis mediated by insulin which could induce increased release of fatty acids from adipose tissue followed by deposition in liver cells leading to enhanced steatosis with increased risk of atherothrombotic diseases. In insulin-resistant subjects, a fatty liver could increase both the VLDL production and the glucose level.[82] It has been reported that insulin resistance could develop atherogenic dyslipidemia due to the presence of small dense atherogenic LDL particles. Int-6 and TNF-α released from visceral adipose tissue are accompanied by reduction in insulin sensitivity,[43] that could further reduce hepatic cell integrity and increase fatty infiltration.

5.6 THERAPEUTIC INTERVENTION

Treatment of fatty liver is a crucial not only due to the possible progression to more aggressive liver diseases but also because it is a provider of cardiovascular diseases. There is no established treatment for NAFLD except modifying risk factors such as the different metabolic syndrome components such as diabetes, dyslipidemia, and obesity as well as reduction of body weight and experiencing exercise.[13,56,79] The therapy of NAFLD based on the modification of lifestyle including dietary intervention and weight loss. Dietary intervention represented by intake of hypocaloric diet low in saturated fatty acids and high in monounsaturated fatty acids and omega-3 fatty acids are of beneficial effect due to their antihypertensive, insulin-sensitizing, and hypotriglyceridemic effect.[1] Reduction in the consumption of high glycemic index carbohydrates may also be important since these carbohydrates appear deleterious, as they favor hyperglycemia and hyperinsulinemia and stimulates de novo lipogenesis.[49] Decreasing total fat intake is necessary.[57]

A moderate weight loss is preferred in patients with fatty infiltration (0.5 kg/week for a child and 1.6 kg/week for the adult), since rapid weight loss could enhance necroinflammation, bile stasis, and portal fibrosis.[13,32,49,57]

Therapeutic agents of promising results include insulin-sensitizing drugs (such as biguanides such as metformin) that increase the insulin sensitivity in the liver by reversing insulin resistance induced by TNF-α. Also, glitazones improve insulin sensitivity through the activation of PPAR gamma. Other therapeutic agents of beneficial effect were reported such as antioxidants, lipid-lowering, hepatic protective agents, ursodeoxycholic acid, pentoxifylline (pentoxil, trental), inhibitors of angiotensin-converting enzyme, and glucagon-like peptide-1 receptor agonist.[18,20,27,51,60,75,77,78] However, the efficacy of these agents remains questionable.[64]

5.7   POTENTIAL BENEFITS OF NUTRACEUTICALS TOWARD NAFLD

Bioactive food components in form of nutraceuticals together with an energy-restricted diet might be a promising approach to manage the metabolic syndrome manifestations[1] including NAFLD. Strategy for the choice of nutraceuticals may rely on providing a hypolipidemic, antioxidant, anti-inflammatory, insulin sensitizer, hepatoprotective, and reducer of fat mass. Nutraceuticals must be evaluated carefully controlled clinical trials.

Vitamin E reduced the amount of hepatic steatosis[75] and NASH.[3] Vitamin E and C therapy reduce NASH.[36] Vitamin E and C combination is an efficient therapy for fatty liver, which could be comparable to ursodeoxycholic acid.[30] Vitamin E can stabilize mitochondrial function, improves biochemical markers of liver inflammation, and possess antioxidant activity. Moreover, n-acetylcysteine and s-adenosylmethionine showed promising results for the treatment of NAFLD. Vitamin E combined with ursodeoxycholic acid significantly improved liver function tests in long-term study.[67]

Investigation of the therapeutic effect of caffeic acid and chlorogenic acid toward fatty liver in mice fed high-fat diet showed promising reduction in body weight, visceral fat mass and liver, heart, adipose tissue, and plasma triglycerides. Both phenolic acids reduced cholesterol content in plasma, adipose tissue, and heart. Chlorogenic acid supplementation produced an increase in plasma adiponectin, which could indicate its anti-inflammatory effect. The mechanism underlying the effect of caffeic and chlorogenic acids was ascribed to their significant inhibition of fatty acid synthase, enhancement of fatty acids beta-oxidation, and increased PPAR alpha expression in liver.[23] Caffeic acid is present in tea, pear, basil, thyme, and apple, whereas chlorogenic acid is present in fruits, green tea, and green coffee bean.

Ferulic acid, which is a phenolic compound, effectively reduced oxidative stress and provides some protection against lipid peroxidation[73] suggesting its beneficial use toward NAFLD. Ferulic acid is present in oats, rice, orange, pineapple, apple, and peanut.

Syringic acid, which is O-methylated trihydroxybenzoic acid, is present in the açaí palm (*Euterpe oleracea*) and in *Ardisia elliptica*, and grapes. Oral administration of syringic acid demonstrated significant hepatoprotective effect in rats, similar to silymarin, represented by significant improvement in the activity of alkaline phosphatase, gamma-glutamyl transpeptidase, aspartate aminotransferase, and alanine aminotransferase. Syringic acid possesses antihyperlipidemic activity by reducing total cholesterol (T. Ch), TGs, VLDL-ch, and LDL-ch as well as producing significant increase in high-density lipoprotein cholesterol. It significantly reduced TGs, T. Ch in liver and kidney.[68] Therefore, the syringic acid may possess protective effect toward fatty liver.

Chronic administration of quercetin reduces hepatic deposition of fat that followed the intake of high-fat, cholesterol, and sucrose diets in mice. It also produced improvements in the different metabolic syndrome-related parameters since it reduces oxidative stress and the expression of hepatic genes that have a relation to steatosis.[45] The activity of mRNA involved in liver fatty acids synthesis was reduced after treatment of an animal with quercetin,[63] thereby inhibits the synthesis of both fatty acids and triglycerides in hepatic cells.[35] Quercetin is present in onions, apples, berries, tea, and grapes.

Fenugreek seed polyphenols have a promising insulin-sensitizing effect in a rat model of insulin resistance, which was ascribed to improving insulin signaling.[41] This action as well as the previously reported anti-inflammatory effect of the saponin present in fenugreek[42] could point to the importance of fenugreek as a source of bioactive compounds of potential therapy toward fatty liver.

Resveratrol, which is present in high level in red grapes, especially the pomace, inhibited NAFLD in rat model.[21] This effect is partially due to the inhibition of TNF-α and oxidative stress.

Genistein, an isoflavone present in soybeans, improved NASH that was induced in rats by feeding high-fat diet. This effect was related to the reduction of fat accumulation and inflammation in the liver with a lower incidence of apoptotic cells in the liver. The reduction in fat deposition was mediated by reduction of fat synthesis and enhancement of fatty acid turnover due to increasing the beta fatty acid oxidation.[38]

L-Carnitine is a quaternary ammonium compound. It is synthesized from methionine and lysine. L-Carnitine is essential for the transfer of long-chain acyl groups from fatty acids to mitochondria to be broken by beta-oxidation to acetyl-CoA to produce energy through the citric acid cycle. L-carnitine improved liver histology in a randomized controlled trial in NASH.[59] A significant effect on normalization of alanine transaminase was reported in NAFLD patients treated with 3 g carnitine/day.

Lecithin or dietary phosphatidylcholine enhances beta-oxidation of fatty acids and reduces triglycerides synthesis in liver thereby reduces the accumulation of fat in the liver which leads to improvement of liver steatosis and hepatomegaly.[19,61]

Xylitol, a food and medicinal sweetener is a sugar alcohol composed of five-carbon and provides three calories. It is present in vegetables and fruits such as plums, cauliflower, raspberries, and strawberries.[34] Xylitol is commonly consumed as an alternative to high energy supplements in diabetics. Xylitol was reported to significantly reduce plasma T. Ch, triglycerides, and total lipids together with lowering visceral fat mass significantly in high-fat diet fed rats. The genes related to liver fatty acids beta-oxidation were significantly high with consumption of xylitol in high-fat diet fed rats compared to those fed on high-fat diet only. Therefore, xylitol intake may prevent the incidence of obesity and development of metabolic syndrome components[12] including fatty liver.

Betaine, which is *N, N, N*-trimethylglycine, is a methyl donor. It is present in sugar beet, broccoli, grains, shellfish, and spinach. The methyl donation is very important for proper liver function and reproduction of cells. Betaine is also important for the synthesis of carnitine in the body. Feeding high-fat diet that may lead to NAFLD and disturb S-containing substances metabolism significantly pointed to the possible inhibition of hepatic transsulfuration in NAFLD. Supplementation of betaine proved successful in protection from nonalcoholic steatosis in addition to reducing oxidative stress through an effect on transsulfuration reactions.[46]

Health implications of whey protein isolate could be seen in improving the different biomarkers of fatty liver and type 2 diabetes. Mice treated with whey protein isolate showed significant reduction in liver fats. Glucose tolerance and insulin sensitivity were further improved by whey protein isolate; thereby, it maybe effective in slowing the progression of fatty liver as well as type 2 diabetes.[76] It has also been reported that liver function was improved in NASH patients on treatment with soy protein through reducing blood and liver lipids, improving insulin sensitivity and elevating antioxidant capacity.[85]

Hypertriglyceridemic patients showed improvement in transaminase activity and ultrasonographic evidence of fatty liver on administration of omega-3 fatty acids.[37] Omega-3 fatty acids are capable of improving the pathophysiological changes that occur in NASH represented by liver damage, inflammation, and oxidative stress as well as reducing hepatic fat, TNF-α, and Int.[54] Also, it was noticed that supplementation of docosahexaenoic acid in two dose levels (250 and 500 mg/day) for 6 months reduced hepatic fat in children with NAFLD.[62]

Some compounds have shown benefit in preventing hepatic fibrosis through blocking proliferation of liver stellate cells. Hepatic stellate cells were shown to be inhibited by more than 75% on administration of apigenin, genistein, quercetin, biochanin, and daidzin. Also, gamma-linolenic, eicosapenteanoic, and alpha-linolenic acids were reported to inhibit liver stellate cells when used in a concentration of 50 nmol/l.[16]

Curcumin limits the fibrogenic evolution of experimental steatohepatitis in mice and reduces tissue inhibitor of metalloprotease secretion and oxidative stress in cultured stellate cells.[81] Curcumin has been reported to alter serum and liver lipid peroxidation and possess anti-inflammatory effect in rats.[72]

The edible shoots or sprouts of bamboo from different bamboo species such as *Bambusa vulgaris* and *Phyllostachys edulis* are consumed in many Asian broths and dishes are present in the market in fresh, canned, and dried forms. Bamboo shoot oil has the ability to reduce blood T. Ch, triglycerides, LDL-ch, hepatic lipase, and serum atherogenic index, in addition to increase the excretion of cholesterol in feces significantly. The oil also decreases relative liver weight and hepatic fat content significantly. The hypolipidemic activity of the oil might be ascribed to its ability to reduce the absorption of cholesterol and increasing cholesterol excretion.[50]

Although buckwheat contains the word wheat, it is not a cereal, grain, or grass, yet it has no relation to wheat. Germinated buckwheat possesses remedial effect toward fatty liver that partially could be through suppression of gene expression of adipogenic transcription factors. This effect may be attributed to the presence of quercetin, rutin, and other flavonoids.[24]

Oral administration of *Nigella sativa* crude oil, prepared by pressing seeds on cold, produced significant improvement of inflammatory fatty liver model in rats which was induced by feeding high fructose diet. The effect was reflected in reduction in liver total lipids, triglycerides, and cholesterol. Treatment with the crude oil also produced significant improvement of dyslipidemia, plasma TNF-α, malondialdehyde, and liver dysfunction. Thereby, *N. sativa* crude oil reduced oxidative stress and inflammation together with hypolipidemic effect. This study showed that *N. sativa* crude oil not only improved steatohepatitis but also reduced the risk of developing cardiovascular diseases. The bioactivity of *N. sativa* crude oil could be attributed to the presence of the monounsaturated fatty acids (oleic and eicosenoic) and the volatile constituents, p-cymene and thymoquinone.[6] Tocopherols and sterols reported to present in the fixed oil[5,69] could participate in the bioactivity of *Nigella* crude oil.

Pumpkin seed oil and stabilized rice bran oil afford hepatoprotection against NASH in rat model. The mechanism of action involved antioxidant and anti-inflammatory activity along with the reduction of liver fat and improvement of hypercholesterolemia. Both oils improved liver histopathological changes but pumpkin seed oil was superior in this respect where it reversed all histopathological abnormalities in liver tissue which became comparable to normal.[9] The bioactivity of both oils could be related to the presence of bioactive constituents. It has been reported that rice bran oil contains stigmasterol, campesterol, beta-sitosterol, and triterpenoid compounds (alpha and beta-amyrin). Beta-carotene and alpha, gamma, and delta tocopherols and tocotrienols were also identified in rice bran oil. Gamma-oryzanol, policosanol (long-chain primary fatty alcohol), and oleic and linoleic fatty acids are among the important phytochemicals and nutrients, which are present in rice bran oil, and possess health benefits.[7] Pumpkin seed oil has been reported to contain alpha and delta tocopherols, phytosterols, beta-carotene, phenolic compounds, oleic acid as monounsaturated fatty acid, and linoleic acid which is omega-6 fatty acid.[11,55,80,85] Nanoemulsion prepared from pumpkin seed oil showed superior effect in preventing the progression of fatty liver compared to the native form of the original oil. The effect of the nanoemulsion was manifested in a very low dose with a sustainable action due to the increased absorption and bioavailability.[10]

Daily administration of clove essential oil conventional emulsion, clove essential oil microemulsion, or eugenol microemulsion has beneficial healthy effect toward fatty liver in rats accompanied by improvement in dyslipidemia. The different emulsion forms helped to reduce complications of fatty liver, especially CVDs. Eugenol is the major bioactive constituent of clove essential oil. The biological effect is mediated through antioxidant and anti-inflammatory mechanism together with the reduction of liver total fat, triglycerides, and cholesterol.[8]

5.8 CONCLUSION

The NAFLD is a component of metabolic syndrome and the mortality among NAFLD is higher than general population. NAFLD spectrum includes simple steatosis, steatohepatitis, fibrosis, cirrhosis, hepatocellular carcinoma, and liver failure. It is important to recognize that under activity, insulin resistance, lipotoxicity, adipokines, oxidative stress, overnutrition, genetic factors, and other inflammatory mediators have strong link to the development of NAFLD/NASH. Treatment of fatty liver is crucial not only due to the possible progression to more aggressive liver diseases but also because it is a provider of cardiovascular diseases. Dietary regimen and exercise play an important role in the management of NAFLD. Ideal therapy or nutraceuticals for NAFLD must possess insulin-sensitizing, antioxidants, lipid-lowering, hepatic protective, and fat mass reducing effect.

5.9 SUMMARY

NAFLD represents the hepatic component of metabolic syndrome. Accumulation of fat itself in the liver is not dangerous, but the hazards reside in its progression to steatohepatitis, which is inflammation with fat deposition. Different factors of such progression have been proposed, insulin resistance remains to be the most important factor. Steatohepatitis may lead to hepatic cirrhosis and carcinoma and to cardiovascular diseases. Diets and body weight play a major role in both the induction and management of NAFLD. So far, no established treatment for NAFLD is present except for modifying risk factors such as components of the metabolic syndrome. Nutraceuticals together with low-calorie foods might be an effective approach to control metabolic syndrome aspects including fatty liver. This chapter reviewed different studied nutraceuticals that could have therapeutic efficiency and/or protective effect toward nonalcoholic fatty liver.

KEYWORDS

- apigenin
- bamboo shoot
- betaine
- biochanin
- buckwheat
- caffeic acid
- carnitine
- chlorogenic acid
- clove
- curcumin
- daidzin
- diets
- fenugreek
- ferulic acid
- genistein
- insulin resistance
- lecithin
- metabolic syndrome
- *Nigella sativa*
- nonalcoholic fatty liver
- nutraceuticals
- polyunsaturated fatty acids
- pumpkin seed oil
- quercetin
- resveratrol
- stabilized rice bran oil
- syringic acid
- vitamin E
- whey protein
- xylitol

REFERENCES

1. Abete, I.; Goyenechea, E.; Zulet, M. A.; Martínez, J. A. Obesity and Metabolic Syndrome: Potential Benefit from Specific Nutritional Components. *Nutr. Metab. Cardiovasc. Dis.* **2011,** *20,* 1–15.

2. Adams, L. A.; Lymp, J. F.; Sauver, J. S. T.; Sanderson, S. O.; Lindor, K. D.; Feldstein, A.; Angulo, P. The Natural History of Nonalcoholic Fatty Liver Disease: A Population-Based Cohort Study. *Gastroenterology* **2005**, *129*, 113–121.

3. Adinolfi, L. E.; Restivo, L. Does Vitamin E Cure Nonalcoholic Steatohepatitis? *Expert Rev. Gastroenterol. Hepatol.* **2011**, *5*(2), 147–150.

4. Alberti, K. G.; Zimmet, P. Z. Definition, Diagnosis and Classification of Diabetes Mellitus and its Complications, Part 1: Diagnosis and Classification of Diabetes Mellitus Provisional Report of a WHO Consultation. *Diabetes Med.* **1998**, *15*(7), 539–553.

5. Al-Okbi, S.; Ammar, N.; Abd El-Kader, M. Studies of Some Biochemical, Nutritional and Anti-inflammatory Effects of *Nigella sativa* Seeds. *Egypt J. Pharm. Sci.* **1997**, *38*, 451–469.

6. Al-Okbi, S. Y.; Mohamed, D. A.; Hamed, T. E.; Edris, A. E. Potential Protective Effect of *Nigella sativa* Crude Oils Towards Fatty Liver in Rats. *Eur. J. Lipid Sci. Technol.* **2013**, *115*, 774–782.

7. Al-Okbi, S. Y.; Ammar, N. M.; Mohamed, D. A.; Hamed, I. M.; Desoky, A. H.; El-Bakry, H. F.; Helal, A. M. Egyptian Rice Bran Oil: Chemical Analysis of the Main Phytochemicals. *La Rivista Italiana delle Sostanze Grasse* (The Italian Magazine of Grasse Substances) **2014**, *91*(1), 47–58.

8. Al-Okbi, S. Y.; Mohamed, D. A.;Hamed, T. E.; Edris, A. E. Protective Effect of Clove Oil and Eugenol Microemulsions on Fatty Liver and Dyslipidemia as Components of Metabolic Syndrome. *J. Med. Food* **2014**, *17*(7), 764–771.

9. Al-Okbi, S. Y.; Mohamed, D. A.; Hamed, T. E.; Esmail, R. S. H. Rice Bran Oil and Pumpkin Seed Oil Alleviate Oxidative Injury and Fatty Liver in Rats Fed High Fructose Diet. *Pol. J. Food Nutr. Sci.* **2014**, *64*(2), 127–133.

10. Al-Okbi, S. Y.; Mohamed, D. A.; Hamed, T. E.; Kassem, A. A.; Abd El-Alim, S. H.; Mostafa D. M. Enhanced Prevention of Progression of Non-alcoholic Fatty Liver to Steatohepatitis by Incorporating Pumpkin Seed Oil in Nanoemulsions. *J. Mol. Liq.* **2017**, *225*(Jan), 822–832. https://doi.org/10.1016/j.molliq.2016.10.138.

11. Al-Okbi, S. Y., Mohamed, D. A., Kandil, E., Ahmed, E. K., Mohammed, S. E. Functional Ingredients and Cardiovascular Protective Effect of Pumpkin Seed Oils. *Grasas y Aceites* (Fats & Oils) **2014**, *65*(1), e007.

12. Amo, K.; Arai, H.; Uebanso, T.; Fukaya, M., Koganei, M., Sasaki, H.; Yamamoto, H.; Taketani, Y.; Takeda, E. Effects of Xylitol on Metabolic Parameters and Visceral Fat Accumulation. *J. Clin. Biochem. Nutr.* **2011**, *49*(1),1–7.

13. Angulo, P.; Lindor, K. Non-alcoholic Fatty Liver Disease. *J. Gastroenterol. Hepatol.* **2002**, *17*(Suppl.), S186–S190.

14. Aygun, C.; Senturk, O.; Hulagu, S.; Uraz, S.; Celebi, A.; Konduk, T.; Mutlu, B.; Canturk, Z. Serum Levels of Hepatoprotective Peptide Adiponectin in Non-alcoholic Fatty Liver Disease. *Eur. J. Gastroenterol. Hepatol.* **2006**, *18*, 175–180.

15. Bacon, B. R.; Farahvash, M. J.; Janney, C. G.; Neuschwander-Tetri, B. A. Non-alcoholic Steatohepatitis: An Expanded Clinical Entity. *Gastroenterology* **1994**, *107*, 1103–1109.

16. Badria, F. A.; Dawidar, A. A.; Houssen, W. E.; Shier, W. T. In vitro Study of Flavonoids, Fatty Acids, and Steroids on Proliferation of Rat Hepatic Stellate Cells. *Z. Naturforsch. C.* (J. Biosci.) **2005**, *60*(1–2), 139–142.

17. Bashiardes, S.; Shapiro, H.; Rozin, S.; Shibolet, O.; Elinav, E. Non-Alcoholic Fatty Liver and the Gut Microbiota. *Mol. Metab.* **2016,** *5*(9), 782–794.

18. Belfort, R.; Harrison, S. A.; Brown, K.; Darland, C.; Finch, J.; Hardies, J.; *et al.* A Placebo-Controlled Trial of Pioglitazone in Subjects with Nonalcoholic Steatohepatitis. *N. Engl. J. Med.* **2006,** *355*, 2297–2307.

19. Buang, Y.; Wang, Y.; Cha, J.; Nagao, K.; Yanagita, T. Dietary Phosphatidylcholine Alleviates Fatty Liver Induced by Orotic Acid. *Nutrition* **2005,** *21*(7–8), 867–873.

20. Bugianesi, E.; Gentilcore, E.; Manini, R.; Natale, S.; Vanni, E.; Villanova, N.; David, E.; Rizzetto, M.; Marchesini, G. A Randomized Controlled Trial of Metformin Versus Vitamin E or Prescriptive Diet in Nonalcoholic Fatty Liver Disease. *Am. J. Gastroenterol.* **2005,** *100*, 1082–1090.

21. Bujanda, L.; Hijona, E.; Larzabal, M.; Beraza, M.; Aldazabal, P.; García-Urkia, N.; Sarasqueta, C.; Cosme, A.; Irastorza, B.; González, A.; Arenas, J. I. Jr. Resveratrol Inhibits Nonalcoholic Fatty Liver Disease in Rats. *BMC Gastroenterol.* **2008,** *8*, 40.

22. Cave, M.; Deaciuc, I.; Mendez, C., Song, Z.; Joshi-Barve, S.; Barve, S.; McClain, C. Nonalcoholic Fatty Liver Disease: Predisposing Factors and the Role of Nutrition. *J. Nutr. Biochem.* **2007,** *18*(3), 184–195.

23. Cho, A. S.; Jeon, S. M.; Kim, M. J.; Yeo, J.; Seo, K. I.; Choi, M. S.; Lee, M. K. Chlorogenic Acid Exhibits Anti-obesity Property and Improves Lipid Metabolism in High-Fat Diet-Induced-Obese Mice. *Food Chem. Toxicol.* **2010,** *48*(3), 937–943.

24. Choi, K. Y.; Kim, S. M.; Kim, Y. E.; Choi, D. S.; Baik, S. Y.; Lee, J. Prevalence and Cardiovascular Disease Risk of the Metabolic Syndrome Using National Cholesterol Education Program and International Diabetes Federation definitions in the Korean Population. *Metab. Clin. Exp.* **2007,** *56*, 552–558.

25. Comhair, T. M.; Caraballo, S. C. G.; Dejong, C. H. C.; Lamers, W. H.; Köhler, S. E. Dietary Cholesterol, Female Gender and n-3 Fatty Acid Deficiency are More Important Factors in the Development of Non-alcoholic Fatty Liver Disease than the Saturation Index of the Fat. *Nutr. Metab.* **2011,** *8*(4), 1–13.

26. Cortez-Pinto, H.; Camilo, M. E.; Baptista, A.; De Oliveira, A. G.; De Moura, M. C. Non-alcoholic Fatty Liver: Another Feature of the Metabolic Syndrome? *Clin. Nutr.* **1999,** *18*, 353–358.

27. Ding, X.; Saxena, N. K.; Lin, S. Exendin-4, a Glucagon-Like Protein-1 (GLP-1) Receptor Agonist, Reverses Hepatic Steatosis in ob/ob Mice. *Hepatology* **2006,** *43*, 173–181.

28. Donnelly, K. L.; Smith, C. I.; Schwarzenberg, S. J.; Jessurun, J.; Boldt, M. D.; Parks, E. J. Sources of Fatty Acids Stored in Liver and Secreted via Lipoproteins in Patients with Nonalcoholic Fatty Liver Disease. *J. Clin. Invest.* **2005,** *115*, 1343–1351.

29. Ekstedt, M.; Franzén, L. E.; Mathiesen, U. L.; Thorelius, L.; Holmqvist, M.; Bodemar, G.; Kechagias, S. Long-term Follow-up of Patients with NAFLD and Elevated Liver Enzymes. *Hepatology* **2006,** *44*(4), 865–873.

30. Ersöz, G.; Günşar, F.; Karasu, Z.; Akay, S.; Batur, Y.; Akarca, U. S. Management of Fatty Liver Disease with Vitamin E and C Compared to Ursodeoxycholic Acid Treatment. *Turk. J. Gastroenterol.* **2005,** *16*(3), 124–128.

31. Expert Panel on Detection, Evaluation, and Treatment of High Blood Cholesterol in Adults (Adult Treatment Panel III). Executive Summary of the Third Report of the National Cholesterol Education Program (NCEP). *JAMA* **2001,** *285*, 2486–2497.

32. Farrell, G. C.; Larter, C. Z. Nonalcoholic Fatty Liver Disease: From Steatosis to Cirrhosis. *Hepatology* **2006,** *43*(2, Suppl. 1), S99–S112.

33. Ford, E. S.; Giles, W. H.; Dietz, W. H. Prevalence of the Metabolic Syndrome Among US Adults: Findings from the Third National Health and Nutrition Examination Survey. *JAMA* **2002,** *287,* 356–359.

34. Georgieff, M.; Moldawer, L. L.; Bistrian, B. R.; Blackburn, G. L. Xylitol, an Energy Source for Intravenous Nutrition after Trauma. *JPEN J. Parenter. Enteral. Nutr.* **1985,** *9,* 199–209.

35. Gnoni, G. V.; Paglialonga, G.; Siculella, L. Quercetin Inhibits Fatty Acid and Triacylglycerol Synthesis in Rat-Liver Cells. *Eur. J. Clin. Invest.* **2009,** *39*(9), 761–768.

36. Harrison, S. A.; Fincke, C.; Helinski, D.; Torgerson, S.; Hayashi, P. A Pilot Study of Orlistat Treatment in Obese, Non-alcoholic Steatohepatitis Patients. *Ailment Pharmacol. Ther.* **2004,** *20,* 623–628.

37. Hatzitolios, A.; Savopoulos, C.; Lazaraki, G.; Sidiropoulos, I.; Haritanti, P.; Lefkopoulos, A. Efficacy of Omega-3 Fatty Acids, Atorvastatin and Orlistat in Non-alcoholic Fatty Liver Disease with Dyslipidemia. *Indian J. Gastroenterol.* **2004,** *23,* 131–134.

38. Huang, C.; Qiao, X.; Dong, B. Neonatal Exposure to Genistein Ameliorates High-Fat Diet-Induced Non-alcoholic Steatohepatitis in Rats. *Br. J. Nutr.* **2011,** *106*(1), 105–113.

39. Isomaa, B.; Almgren, P.; Tuomi, T.; Forsen, B.; Lahti, K.; Nissen, M.; Taskinen, M. R.; Groop, L. Cardiovascular Morbidity and Mortality Associated with the Metabolic Syndrome. *Diabetes Care* **2001,** *24,* 683–689.

40. Isomaa, B.; Henricsson, M.; Almgren, P., Tuomi, T.; Taskinen, M. R.; Groop, L. The Metabolic Syndrome Influences the Risk of Chronic Complications in Patients with Type II Diabetes. *Diabetologia* **2001,** *44,* 1148–1154.

41. Kannappan, S.; Anuradha, C. V. Insulin Sensitizing Actions of Fenugreek Seed Polyphenols, Quercetin & Metformin in a Rat Model. *Indian J. Med. Res.* **2009,** *129*(4), 401–408.

42. Kawabata, T.; Cui, M. Y.; Hasegawa, T.; Takano, F.; Ohta, T. Anti-inflammatory and Anti-melanogenic Steroidal Saponin Glycosides from Fenugreek (*Trigonella foenum-graecum* L.) Seeds. *Planta Med.* **2011,** *77*(7), 705–710.

43. Kern, P. A.; Ranganathan, S.; Li, C.; Wood, L.; Ranganathan, G. Adipose Tissue Tumor Necrosis Factor and Interleukin-6 Expression in Human Obesity and Insulin Resistance. *Am. J. Physiol. Endocrinol. Metab.* **2001,** *280,* E745–751.

44. Kim, C. H.; Younossi, Z. M. Nonalcoholic Fatty Liver Disease: A Manifestation of the Metabolic Syndrome. *Cleve. Clin. J. Med.* **2008,** *75*(10), 721–728.

45. Kobori, M.; Masumoto, S.; Akimoto, Y.; Oike, H. Chronic Dietary Intake of Quercetin Alleviates Hepatic Fat Accumulation Associated with Consumption of a Western-Style Diet in C57/BL6J Mice. *Mol. Nutr. Food Res.* **2011,** *55*(4), 530–540.

46. Kwon, do Y.; Jung, Y. S.; Kim, S. J.; Park, H. K., Park, J. H., Kim, Y. C. Impaired Sulfur-Amino Acid Metabolism and Oxidative Stress in Nonalcoholic Fatty Liver are Alleviated by Betaine Supplementation in Rats. *J. Nutr.* **2009,** *139*(1), 63–68.

47. Lê, K. A.; Bortolotti, M. Role of Dietary Carbohydrates and Macronutrients in the Pathogenesis of Nonalcoholic Fatty Liver Disease. *Curr. Opin. Clin. Nutr. Metab. Care.* **2008,** *11*(4), 477–482.

48. Lee, J. H.; Lee, J. J.; Cho, W. K.; Yim, N. H., Kim, H. K.; Yun, B.; Ma, J. Y. KBH-1, an Herbal Composition, Improves Hepatic Steatosis and Leptin Resistance in High-Fat Diet-Induced Obese Rats. *BMC Complement Altern. Med.* **2016**, *16*, 355.

49. Leclercq, I. A.; Horsmans, Y. Nonalcoholic Fatty Liver Disease: The Potential Role of Nutritional Management. *Curr. Opin. Clin. Nutr. Metab. Care.* **2008**, *11*(6), 766–773.

50. Lu, B.; Xia, D.; Huang, W.; Wu, X.; Zhang, Y.; Yao, Y. Hypolipidemic Effect of Bamboo Shoot Oil (*P. pubescens*) in Sprague-Dawley Rats. *J. Food Sci.* **2010**, *75*(6), H205–H211.

51. Marchesini, G.; Brizi, M.; Bianchi, G.; Tomassetti, S.; Zoli, M.; Melchionda, N. Metformin in Non-alcoholic Steatohepatitis. *Lancet* **2001**, *358*(9285), 893–894.

52. Marchesini, G.; Bugianesi, E.; Forlani, G.; Cerrelli, F.; Lenzi, M.; Manini, R.; Natale, S.; Vanni, E.; Villanova, N.; Melchionda, N.; Rizzetto, M. Nonalcoholic Fatty Liver, Steatohepatitis, and the Metabolic Syndrome. *Hepatology* **2003**, *37*(4), 917–923.

53. Marchesini, G.; Marzocchi, R.; Agostini, F.; Bugianesi, E. Nonalcoholic Fatty Liver Disease and the Metabolic Syndrome. *Curr. Opin. Lipidol.* **2005**, *16*(4), 421–427.

54. Marsman, H. A.; Heger, M.; Kloek, J. J.; Nienhuis, S. L.; ten Kate, F. J.; van Gulik, T. M. Omega-3 Fatty Acids Reduce Hepatic Steatosis and Consequently Attenuate Ischemia-Reperfusion Injury Following Partial Hepatectomy in Rats. *Dig. Liver Dis.* **2011**, *43*(12), 984–990.

55. Matus, Z.; Molnár, P.; Szabó, L. G. Main Carotenoids in Pressed Seeds (*Cucurbitae semen*) of Oil Pumpkin (*Cucurbita pepo convar. pepo var. styriaca*). *Acta Pharm. Hung.* **1993**, *63*(5), 247–256.

56. Mehta, K.; Van Thiel, D. H.; Shah, N.; Mobarhan, S. Nonalcoholic Fatty Liver Disease: Pathogenesis and the Role of Antioxidants. *Nutr. Rev.* **2002**, *60*(9), 289–293.

57. Mensink, R. P.; Plat, J.; Schrauwen, P. Diet and Nonalcoholic Fatty Liver Disease. *Curr. Opin. Lipidol.* **2008**, *19*(1), 25–29.

58. Mokdad, A. H.; Ford, E. S.; Bowman, B. A.; Dietz, W. H.; Vinicor, F., Bales, V. S.; Marks, J. S. Prevalence of Obesity, Diabetes, and Obesity-Related Health Risk Factors. *JAMA* **2003**, *289*, 76–79.

59. Musso, G.; Gambino, R.; Cassader, M.; Pagano, G. A Meta-Analysis of Randomized Trials for the Treatment of Nonalcoholic Fatty Liver Disease. *Hepatology* **2010**, *52*(1), 79–87.

60. Neuschwander-Tetri, B. A.; Brunt, E. M.; Wehmeier, K. R.; Wehmeier, K. R.; Oliver, D.; Bacon, B. R. Improved Nonalcoholic Steatohepatitis after 48 Weeks of Treatment with the PPAR-Gamma Ligand Rosiglitazone. *Hepatology* **2003**, *38*, 1008–1017.

61. Niebergall, L.; Jacobs, R.; Chaba, T.; Vance D. Phosphatidyl Choline Protects Against Steatosis in Mice but not Non-alcoholic Steatohepatitis. *Biochim. et Biophys. Acta (BBA): Mol. Cell Biol. Lipids* **2011**, *1811*(12), 1177–1185.

62. Nobili, V.; Bedogni, G.; Alisi, A.; Pietrobattista, A.; Risé, P.; Galli, C.; Agostoni, C. Docosahexaenoic Acid Supplementation Decreases Liver Fat Content in Children with Non-alcoholic Fatty Liver Disease: Double-Blind Randomized Controlled Clinical Trial. *Arch. Dis. Child.* **2011**, *96*(4), 350–353.

63. Odbayar, T. O.; Badamhand, D.; Kimura, T.; Takashi, Y.; Tsushida, T.; Ide T. Comparative Studies of Some Phenolic Compounds (Quercetin, Rutin, and Ferulic Acid)

Affecting Hepatic Fatty Acid Synthesis in Mice. *J. Agric. Food Chem.* **2006,** *54*(21), 8261–8265.

64. Park, S. H. Nonalcoholic Steatohepatitis: Pathogenesis and Treatment. *Korean J. Hepatol.* **2008,** *14*(1), 12–27.

65. Park, S. H.; Jeon, W. K.; Kim, S. H.; Kim, H. J.; Park, D. I.; Cho, Y. K.; Sung, I. K.; Sohn, C. I.; Keum, D. K.; Kim, B. I. Prevalence and Risk Factors of Non-alcoholic Fatty Liver Disease Among Korean Adults. *J. Gastroenterol. Hepatol.* **2006,** *21*, 138–143.

66. Park, H. S.; Park, J. Y.; Yu, R. Relationship of Obesity and Visceral Adiposity with Serumconcentrations of CRP, TNF-Alpha and IL-6. *Diabetes Res. Clin. Pract.* **2005,** *69*, 29–35.

67. Pietu, F.; Guillaud, O.; Walter, T.; Vallin, M.; Hervieu, V.; Scoazec, J.; Dumortier, J. Ursodeoxycholic Acid with Vitamin E in Patients with Nonalcoholic Steatohepatitis: Long-term Results. *Clin. Res. Hepatol. Gastroenterol.* **2012,** *36*(2), 146–155.

68. Ramachandran, V. Preventive Effect of Syringic Acid on Hepatic Marker Enzymes and Lipid Profile Against Acetaminophen-Induced Hepatotoxicity in Rats. *Int. J. Pharm. Biol. Arch.* **2010,** *1*(4), 393–398.

69. Ramadan, F.; Wahdan, K. Blending of Corn Oil with Black Cumin (*Nigella sativa*) and Coriander (*Coriandrum sativum*) Seed Oils: Impact on Functionality, Stability and Radical Scavenging Activity. *Food Chem.* **2012,** *132*, 873–879.

70. Ratziu, V.; Giral, P.; Charlotte, F.; Bruckert, E.; Thibault, V.; Theodorou, I.; Khalil, L.; Turpin, G.; Opolon, P.; Poynard, T. Liver Fibrosis in Overweight Patients. *Gastroenterology* **2000,** *118*(6), 1117–1123.

71. Ratziu, V; Saboury, M.; Poynard, T. Worsening of Steatosis and Fibrosis Progression. *Gut* **2003,** *52*, 1386–1387.

72. Reddy, A. C. P.; Lokesh, B. R. Alterations in Lipid Peroxidation in Rat Liver by Dietary n-3 Fatty Acids: Modulation of Anti-oxidant Enzymes by Curcumin, Eugenol and Vitamin E. *J. Nutr. Biochem.* **1994,** *5*, 181–188.

73. Rukkumani, R.; Aruna, K.; Varma, P. S.; Menon, V. P.; Influence of Ferulic Acid on Circular Prooxidant—Antioxidant Status During Alcohol and PUFA Inducedtoxicity. *J. Physiol. Pharmacol.* **2004,** *55*(3), 551–561.

74. Sanyal, A. J.; Campbell-Sargent, C.; Mirshahi, F.; Rizzo, W. B.; Contos, M. J.; Sterling, R. K.; Luketic, V. A.; Shiffman, M. L.; Clore, J. N. Nonalcoholic Steatohepatitis Association of Insulin Resistance and Mitochondrial Abnormalities. *Gastroenterology* **2001,** *120*, 1183–1192.

75. Sanyal, A. J.; Mofrad, P. S.; Contos, M. J.; Sargeant, C.; Luketic, V. A.; Sterling, R. K.; Stravitz, R. T.; Shiffman, M. L.; Clore, J.; Mills, A. S. A Pilot Study of Vitamin E Versus Vitamin E and Pioglitazone for the Treatment of Nonalcoholic Steatohepatitis. *Clin. Gastroenterol. Hepatol.* **2004,** *2*, 1107–1115.

76. Shertzer, H. G.; Woods, S. E.; Krishan, M.; Genter, M. B., Pearson, K. J. Dietary Whey Protein Lowers the Risk for Metabolic Disease in Mice Fed a High-Fat Diet. *J. Nutr.* **2011,** *141*(4), 582–587.

77. Socha, P.; Horvath, A.; Vajro, P.; Dziechciarz, P.; Dhawan, A.; Szajewska, H. Pharmacological Interventions for Nonalcoholic Fatty Liver Disease in Adults and in Children: A Systematic Review. *J. Pediatr. Gastroenterol. Nutr.* **2009,** *48*(5), 587–596.

78. Tiikkainen, M.; Hakkinen, A. M.; Korsheninnikova, E.; Nyman, T.; Mäkimattila, S.; Yki-Järvinen, H. Effects of Rosiglitazone and Metformin on Liver Fat Content, Hepatic Insulin Resistance, Insulin Clearance, and Gene Expression in Adipose Tissue in Patients with Type 2 Diabetes. *Diabetes* **2004**, *53*, 2169–2176.
79. Ueno, T.; Sugawara, H.; Sujaku, K.; Hashimoto, O.; Tsuji, R.; Tamaki, S.; Torimura T.; Inuzuka, S.; Sata, M.; Tanikawa, K. Therapeutic Effects of Restricted Diet and Exercise in Obese Patients with Fatty Liver. *J. Hepatol.* **1997**, *27*, 103–107.
80. Van Hoed, V.; Felkner, B.; Bavec, F.; Grobelnik, S.; Bavec, M.; Verhe, R. Influence of Processing on Antioxidants Content of Pumpkin Seed Oil. *Book of abstracts, 7th Euro Fed Lipid Congress on Lipids, Fats and Oils from Knowledge to Application*; Graz, Austria; European Federation for the Science and Technology of Lipid, Germany. October 18–21, 2009.
81. Vizzutti, F.; Provenzano, A.; Galastri, S.; Milani, S.; Delogu, W.; Novo, E.; Caligiuri, A.; Zamara, E.; Arena, U.; Laffi, G.; Parola, M.; Pinzani, M.; Marra, F. Curcumin Limits the Fibrogenic Evolution of Experimental Steatohepatitis. *Lab. Invest.—J. Tech. Methods Pathol.* **2010**, *90*(1), 104–115.
82. Vozarova, B.; Stefan, N.; Lindsay, R. S.; Saremi, A.; Pratley, R. E.; Bogardus, C.; Tataranni, P. A. High Alanine Aminotransferase is Associated with Decreased Hepatic Insulin Sensitivity and Predicts the Development of Type 2 Diabetes. *Diabetes* **2002**, *51*, 1889–1895.
83. Westerbacka, J.; Lammi, K.; Häkkinen, A. M.; Rissanen, A.; Salminen, I.; Aro, A.; Yki-Järvinen, H. Dietary Fat Content Modifies Liver Fat in Overweight Nondiabetic Subjects. *JCEM* **2005**, *90*(5), 2804–2809.
84. World Health Organization—International Society of Hypertension Guidelines for the Management of Hypertension. Guidelines Subcommittee. *J. Hypertens.* **1999**, *17*, 151–183.
85. Yang, H.; Tzeng, Y.; Chai, C.; Hsieh, A.; Chen, J.; Chang, L.; Yang, S. Soy Protein Retards the Progression of Non-alcoholic Steatohepatitis via Improvement of Insulin Resistance and Steatosis. *Nutrition* **2011**, *27*(9), 943–948.

CHAPTER 6

BLACK CUMIN: A REVIEW OF PHYTOCHEMISTRY, ANTIOXIDANT POTENTIAL, EXTRACTION TECHNIQUES, AND THERAPEUTIC PERSPECTIVES

MUHAMMAD JAWAD IQBAL, MASOOD SADIQ BUTT, and HAFIZ ANSAR RASUL SULERIA

CONTENTS

ABSTRACT

Black cumin is a rich source of basic as well as bioactive components having considerable health-promoting potential. Among various bioactive moieties, thymoquinone (TQ) is the major component. Based on literature, it is suggested that dietary components such as black cumin bioactive ingredients are being exploited as a therapeutic agent because they are economical, effective, and practical to reduce the risk associated with life-threatening disorders. Further studies are still required to explore the effects of black cumin extracts on the overall health of humans. Bioevaluation trial is a useful tool to investigate the effect of functional ingredients in mitigating various disorders owing to health-promoting potential such as antioxidant, hypoglycemic, hypocholesterolemic, anti-inflammatory, and anticancer perspectives. Furthermore, its role in boosting immune system needs further investigations.

6.1 INTRODUCTION

Black cumin is a commonly known therapeutic herb belonging to the botanical family *Ranunculaceae*.[10] The scientific name of black cumin is *Nigella sativa* L., while local names of black cumin in subcontinent are "*Kalvanji*" or "*Kalonji*" in Urdu and Hindi languages, respectively. It is an annual flowering plant cultivated in Asian as well as in Mediterranean region. The size of plant is about 20–90 cm having finely divided linear leaves. Its flowers are delicate and color varies from blue to white having 5–10 petals.[3] Its fruit is composed of several follicles united to form a capsule and each contains numerous seeds. Seeds are rich in oil, having pungent odor and bitter taste.[101,130] From ancient time, seeds, shoots, and leaves of black cumin are being used for medicinal purposes.[130,131]

For thousands of years, *N. sativa* seeds are commonly used as a spice in Indo-Pak subcontinent and Middle East region. In Islamic teachings, black cumin is known as one of the utmost forms of therapeutic medicine. In *Tibb-e-Nabwi* (Prophetic Medicine), it is recommended to be used on regular basis. Traditionally, these seeds were used in the treatment of various ailments (i.e., hyperglycemia, hypercholesterolemia, and inflammation) and have been an important part of Ayurvedic medicines.[131] The seeds of black cumin have intense hot peppery taste and commonly used in the preparation of coffee, salads, casseroles, as well as tea. It has also been used as preservative, seasoning, and flavoring agent in many food items such as bread, pickles, and other products.[119] Black cumin seeds have been recommended safe to consume orally in moderate quantities.[118]

This chapter focuses on the phytochemistry of black cumin (*N. sativa*) with special reference to thymoquinone (TQ) along with various extraction techniques that are used for the isolating bioactive moieties from black cumin and in vitro antioxidant potential of black cumin. Furthermore, therapeutic perspectives of black cumin have also been discussed to enlighten the hypoglycemic, hypocholesterolemic, anti-inflammatory, and anticancer potential. The objectives of the study were:

- Assessment of in vitro antioxidant potential of various black cumin extracts.
- Elaborate various conventional and modern extraction techniques used for the isolation of black cumin extracts.
- Explore the phytochemical significance of black cumin (*N. sativa*).
- Probing the role of black cumin in mitigating various lifestyle-related disorders.

6.2 NUTRITIONAL PROFILE AND PHYTOCHEMISTRY OF BLACK CUMIN

Black cumin seeds are rich in oil, proteins, total carbohydrates, and ash that range from 22.0–40.35, 20.85–31.20, 23.90–40.01, and 3.70–4.70%, respectively.[16,130] In a study, Tunisian and Iranian varieties were compared for their quality parameters and it has been documented that moisture, oil, proteins, ash, and carbohydrates were 8.65, 28.48, 26.7, 4.86, and 40.0%, respectively, in Tunisian variety; while in the Iranian variety, these were 4.08, 40.35, 22.6, 4.41, and 32.7%, correspondingly. The composition of black cumin seeds varies with geographical location, harvesting time, and agricultural practices.[32]

Furthermore, mineral profiling of black cumin showed potassium, calcium, magnesium, sodium, zinc, copper, and iron contents as 4842–7275, 2158–2894, 2118–2452, 98.4–178.7, 40.32–47.60, 9.45–13.43, and 8.61–30.04 mg/kg, respectively.[53] Black cumin contains both essential and fixed oil.[9] Fixed oil of *N. sativa* contains considerable quantities of polyunsaturated fatty acids, which is about 48–70% of oil, while saturated (12–25%) and monounsaturated fatty acids (18–29%) are in lesser proportions.[32,147]

Black cumin showed rich phytochemistry and nutraceutical potential owing to the presence of TQ, dihydrothymoquinone, carvacrol, β-pinene, α-pinene, α-thujene, ρ-cymene, thymol, and t-anethole.[135] Black cumin contains two types of alkaloids: one is the isoquinoline represented by nigellimin and nigellimin-N-oxide, while second the pyrazole alkaloids which include nigellidin and nigellicin.[101,105] Subsequently, two aliphatic compounds are isolated from black cumin by conventional solvent extraction method using hexane as solvent. These compounds are characterized as 6-nonadecanone-2 and 16-triecosen-7-ol-1.[137] Furthermore, saponins have also been identified in ethanolic extracts of black cumin.[97]

6.3   ANTIOXIDANT PERSPECTIVE OF BLACK CUMIN

The antioxidant potential is the ability to scavenge free radicals, chelate metal ions, and delay action of enzymes involved in oxidation. Previous evidences have revealed that foods rich in polyphenols exhibit strong in vivo and in vitro antioxidant activity.[1] Nutraceutical ingredients present in black cumin seeds, especially antioxidants, fat-soluble vitamins, phytosterols, and some pyrazanol-containing compounds, have hypocholesterolemic and hypoglycemic effects.[94,137] Antioxidant and therapeutic effects of different constituents of black cumin seed oil, such as thymoquinone (TQ) and nigellone, have been studied.[39,118] Most of the antioxidant and therapeutic characteristics of black cumin seeds are due to TQ, which is mainly present in essential oil as well as in the fixed oil but in lesser quantity.[9] TQ has been tested for its in vitro and in vivo antioxidant efficacy against several physiological threats including diabetes, hypercholesterolemia, cardiovascular diseases (CVDs), inflammation, and cancer.[58] In a scientific study, it was observed that the oral administration of black cumin oil (100 µL/kg) and TQ (100 mg/kg) for 1 week was able to uphill the serum antioxidant status of experimental rats.[42]

Prior et al.[117] narrated that no single assay precisely reflects all the characteristics of antioxidants in the mixed or complex system and there are no ultimate universal methods by which antioxidant activity can be quantified accurately. Mostly, natural antioxidants have multifunctional attributes in complex heterogeneous food items; their potential cannot be assessed by only one method.[48,92] As the interaction of antioxidant component with chemicals in each method is different under various situations, a variety of antioxidant assays have been employed for comprehensive analyses of extract such as β-carotene bleaching test, 1,1-diphenyl-2-picrylhydrazyl (DPPH), 2,2-azino-bis-3-ethylbenzothiazoline-6-sulphonic acid (ABTS), and ferric-reducing antioxidant potential (FRAP) assays.[124]

Each in vitro antioxidant assay has its own specification for different bioactive components. DPPH is a stable and highly colored oxidizing radical that results in the formation of a yellow-colored hydrazine (DPPH-H) associated with the abstraction of free hydrogen atoms from phenolic antioxidants.[142] Previous studies have depicted that the free radical scavenging activity of black cumin extracts ranged from 76.43 ± 3.59–83.52 ± 0.61 $\mu mol/100$ μmol.[82] Previously, the free radical scavenging power of various black cumin extracts was explored for different fractions, that is, water, ethyl acetate, hexane, and methanolic fraction; and values were reported as 2.65, 2.26, 2.17, and 1.89 in terms of IC_{50}, respectively.[91]

Among numerous methods of exploring the antioxidant potential of black cumin, β-carotene bleaching test is one of the most common methods that is based on coupled oxidation of β-carotene as well as linoleic acid. In absence of an antioxidant, β-carotene rapidly undergoes discoloration that reduces absorbance of resultant extract, which is detected spectrophotometrically. The main reason for this reduction is the coupled oxidation that engenders free radicals and as a result, bleaches out orange color of β-carotene. However, in the presence of an antioxidant, bleaching effect was hindered by neutralizing free radical. Mariod et al.[91] determined the antioxidant properties of black cumin by linoleic acid and β-carotene assay assessment. They observed that the discoloration effect enhances, showing the reduction in antioxidant activity, with an increase in temperature. It was noticed that the effect on the couple oxidation of linoleic acid and β-carotene was maximum for hexane and water fractions (WFs) as compared to methanolic and ethyl acetate fractions. Ramadan and Mörsel[119] interpreted that one gram of black cumin oil contains 593 $\mu g/g$ of β-carotene. In another study, Erkan et al.[43] evaluated the essential and fixed oil of black cumin for their antioxidant potential. It was reported that lipid peroxidation was inhibited about 25.62 and 92.56% by the administration of fixed and essential oils, respectively.

In the same way, FRAP and ABTS assays are also employed to analyze reducing power of black cumin extracted through solvents at various time and temperature. One of the research groups has reported in vitro antioxidant power of black cumin using FRAP assay. They found a linear relationship between total phenolic content and antioxidant activity of many plants.[152] The bioactive moieties of *N. sativa* can act as electron donor and act as free radical scavengers, converting them into stable products and ultimately breaking down the radical chain reaction. The antioxidant activity of *N. sativa* varied significantly ($p < 0.05$) among the extracts. In the study, alkaline hydrolyzed extract of black cumin showed the highest FRAP expressed as mg GAE/100 g black cumin powder (28 ± 0.159) followed by acid hydrolyzed extract (26 ± 0.096) and the control group (24 ± 0.262).[112] Moreover, time-dependent analyses showed highest values for FRAP at 50 min (24.73 ± 0.85 mg GAE/100 g) while lowest (18.20 ± 0.38 mg GAE/100 g) at 35 min in the whole experiment. Similarly, FRAP assay of black cumin was 15.88 mg/mL.[30] Moreover, it was noticed that FRAP of methanolic extracts of black cumin ranged from 24 to 28 mg GAE/100 g.[152] Likewise, Martos et al.[93] depicted that ferric-reducing power of black cumin extracts was 3.33–3.39 mmol/L Trolox. Furthermore, Erkan et al.[43] showed that the values for ABTS assay seeds after 1, 4, and 6 min of reaction were 2.0 ± 0.7, 2.4 ± 0.3, and 2.5 ± 0.6 mM Trolox, respectively.

6.4 QUANTIFICATION AND CHARACTERIZATION OF BIOACTIVE MOIETIES: THYMOQUINONE

Chemical analyses of black cumin seeds revealed that the oil is the major constituent, which comprises essential and fixed oils. TQ is one of the main active compounds of oil possessing protective activity against many ailments owing to high antioxidant activity.[103] Many experimental modules highlighted the therapeutic potential of TQ against numerous physiological disparities such as inflammation, hyperglycemia, hypercholesterolemia, hepatotoxicity, renal dysfunctions, and cancer.[25,103]

Nowadays, numerous conventional (mustard, rapeseed, groundnut, sesame) and nonconventional (flaxseed, black cumin, sunflower, soybean, safflower) oil sources are gaining importance among the masses due to their health-promoting potential.[5] Among these, black cumin is of particular importance because it has enriched phytochemistry and can be utilized for the production and formulations of various functional foods. Functional importance of essential oil extracted from black cumin is mainly due to its volatile components (0.40–1.50%), TQ (18.4–24%), and monoterpenes

(46%).[14] Sultan et al.[147] indicated that oil content of tocopherols and carotenoids collectively was 450.66 ± 16.21 mg/kg in black cumin oil, while the content of TQ was 201.31 ± 13.17 mg/kg in black cumin oil. Organic extracts of *N. sativa* seeds reported to have considerable content of TQ. In one such study, TQ content was 11.8% in acetone extract seed.[137] Nonetheless, black seed oils contain higher level of high-density lipoprotein (HDL) cholesterol while it contains lesser amounts of low-density lipoprotein (LDL) cholesterol. Furthermore, the utilization of black cumin in formulating various diets enhances the content fat-soluble vitamins and phytosterols in the particular diet.[119,120] Extraction rates of black cumin phytochemicals by different extraction techniques have been summarized in Table 6.1.

During the last decade, Burits and Bucar[26] analyzed and characterized the essential oil of black cumin using gas chromatography coupled with mass spectrometer and quantified TQ, p-cymene, carvacrol, t-anethole, 4-terpineol, and longifoline as 27.8–57.0, 7.1–15.5, 5.8–11.6, 0.25–2.3, 2.0–6.6, and 1.0–8.0%, respectively. In another study, it was noticed that the major compounds of essential oil are monoterpenes and their oxygenated derivatives (i.e., 87.7 and 9.9%, respectively). Additionally, the main constituents of oil are p-cymene, γ-terpinene, α-thujene, carvacrol, and β-pinene, which are 60.2, 12.9, 7.2, 3.0, and 2.1%, respectively.[160]

In a study, TQ in methanolic extracts of black cumin was quantified as 5.19 µg/mL by using high-performance liquid chromatography (HPLC) with higher sensitivity and short elution time.[60] Earlier, the concentration of TQ in ethanolic extracts of black cumin was estimated by using high-performance thin-layer chromatography as 0.067% w/w.[104] In a comprehensive investigation, the concentration of TQ was determined in supercritical fluid extracts at various pressure and temperature conditions. It was noticed that concentration of TQ ranged from 3.04–4.09 mg/g of oil.[139] Previously, it was observed that TQ content of methanolic, hexane, and supercritical extracts of black cumin was about 2.7, 1.2, and 12.3%, respectively. Furthermore, it was also documented that supercritical fluid extraction (SFE) at 120 bar pressure and 40°C temperature gives better recovery of TQ. The SFE-extracted TQ also experiences less degradation as compared to conventional solvent extracts.[73] In another investigation, it was noticed that both solvent and supercritical fluid extracts have rich antioxidant profile and protective effect of black cumin against various maladies due to its TQ content as well as other antioxidants.[115]

TABLE 6.1 Extraction Rates of Black Cumin Phytochemicals by Different Extraction Methods.

Phytonutrient	Extraction Method	Solvent	Concentration	Reference
Carotenoids	Solvent extraction	Hexane	88.95 ± 3.91 mg/kg oil	[136]
Hydroxybenzoic acid	Solvent extraction	Water extract	0.989 mg/100 g dry weight (DW)	[87]
Hydroxybenzoic acid	Solvent extraction	Methanol	0.188 mg/100 g DW	[87]
p-Cumaric acid	Solvent extraction	Methanol	0.631 mg/100 g DW	[87]
p-Cumaric acid	Solvent extraction	Water extract	3.83 mg/100 g DW	[87]
Syringic acids	Solvent extraction	Water extract	0.496 mg/100 g DW	[87]
Syringic acids	Solvent extraction	Methanol	0.125 mg/100 g DW	[87]
Thymoquinone (TQ)	Solvent extraction	Hexane	18.4–24% of volatile oil	[12]
TQ	Solvent extraction	Hexane	201.31 ± 13.17 mg/kg oil	[136]
TQ	Soxhlet extraction	Hexane	27.8–57.0% of essential oil	[24]
TQ	Soxhlet extraction	Acetone	11.8% of seed extract	[131]
TQ	Solvent extraction	Methanol	5.19 μg/mL	[56]
TQ	Solvent extraction	Ethanol	0.067% w/w	[100]
TQ	Supercritical fluid extraction (SFE)	Carbon dioxide	3.04–4.09 mg/g oil	[133]
TQ	Solvent extraction	Methanol	2.7%	[69]
TQ	Solvent extraction	Hexane	1.2%	[69]
TQ	SFE	Carbon dioxide	12.3%	[69]
TQ	SFE	Carbon dioxide	20.8 mg/g oil	[123]
TQ	High-pressure Soxhlet	Liquid CO_2	8.8 mg/g oil	[123]
TQ	Soxhlet extraction	Hexane	6.3 mg/g oil	123

TABLE 6.1 *(Continued)*

Phytonutrient	Extraction Method	Solvent	Concentration	Reference
TQ	Soxhlet extraction	Ethanol	5.0 mg/g oil	[123]
TQ	Soxhlet extraction	Methanol	82.96 mg/kg seeds	[63]
TQ	SFE	Carbon dioxide	$2.00 \pm 0.17\%$ w/w	[57]
TQ	Soxhlet extraction	Hexane	$0.57 \pm 0.01\%$ w/w	[57]
TQ	Solid/liquid extraction	Methanol	286.50 mg/kg seeds	[63]
TQ	Microwave-assisted extraction (MAE)	Methanol	628 mg/kg black seeds	[63]
TQ	MAE	Water	38.23%	[77]
TQ	MAE	Methanol	353.04–571.90 mg/kg seeds	[63]
Total tocopherols	Solvent extraction	Hexane	361.71 ± 10.23 mg/kg oil	[136]
p-cymene	Hydrodistillation	Et₂O:pentane	60.2% of essential oil	[148]
p-cymene	MAE	Water	28.61%	[77]
p-cymene	Soxhlet extraction	Hexane	7.1–15.5% of essential oil	[24]

Literature confirmed differences in the levels of the isolated tocopherols, that is, α and γ tocopherols that are active components of black cumin and niger seed oil. It was estimated that α-tocopherol is about 48% and γ-tocopherol is about 53% of total vitamin E content in *N. sativa* oil. Furthermore, phylloquinone was also present in considerable quantity, that is, 0.1% of black cumin seed oil.[119]

The concentration of total sterols in black cumin seeds ranged from 1993.07–2887.28 mg/kg. Moreover, sitosterol, campesterol, and stigmasterol were estimated as 59.10, 10.0, and 16.50% of total sterol, respectively.[38] In another study, both types of seed oils (fixed and essential oils) were examined: β-stesterol (32.2–34.1% of total phytosterol content) was observed as the main component of phytosterols followed by Δ5-avenasterol (27.8–27.9% of total phytosterol content) and Δ7-venasterol (18.5–22.0% of total phytosterol content).[119] From these values, it can be seen that both seed oils extracted are remarkably similar in their sterol compositions. Ramadan and Morsel[120] reported the significant presence of stigmasterol (0.314±0.02 g/kg), β-sitosterol (1.182±0.05 g/kg), Δ5-avenasterol (1.025±0.04 g/kg), and Δ7-avenasterol (0.809±0.02 g/kg) in crude seed oil of black cumin.

Moreover, in a scientific study, three phenolic components were present in black cumin extract as syringic, *p*-cumaric, and hydroxybenzoic acids. All these active moieties were identified in the crude methanolic extracts (CME) and WF of black cumin. The content of *p*-cumaric acid was comparatively higher in both fractions, contributing about 66.8 and 72.1% to the total phenolics and exhibiting the levels of 0.631 and 3.83 mg/100 g dry weight (DW) in CME and WF, correspondingly. Syringic and hydroxybenzoic acids were also predominant, but somewhat higher in WF (0.496 and 0.989 mg/100 g DW) than in CME (0.125 and 0.188 mg/100 g DW), respectively.[91]

The amount of the detected *p*-cumaric acid in both CME and WFs of seeds is much higher (0.36 mg/100 g dry sample) than that of CME of roots, while a higher amount of hydroxybenzoic acid (1.73 mg/100 g) was reported in CME of black cumin roots. Surprisingly, syringic acid was not identified in black cumin shoot or root extracts.[25] The levels of total phenolic compounds in black cumin CME and WF determined by HPLC were 0.94 and 5.3 mg/100 g DW, respectively.[91] Furthermore, crude fixed oil of *N. sativa* seeds proved to be a valuable source of phytosterols, essential fatty acids, phospholipids, and glycolipids.

6.5   EXTRACTION OF BIOACTIVE MOLECULES

Black cumin seeds are conventionally used as a cooking ingredient in the form of spice in Asian countries and are also used as medicine for treatment of various ailments. Chemical analyses of seed revealed that oil is its major constituent, which comprises both essential and fixed oil. TQ is among the major active constituents of oil and reported to have protective activity against many disorders owing to its high antioxidant activity.[103] Numerous studies have been conducted on black cumin seeds, shoots, and roots, which confirmed the presence of phenolic components.[25] These phenolics are investigated for their bioactivity and it is documented that TQ along with other phenolic acids such as *p*-cumaric, etc. is responsible for the antioxidant activity of seeds. There are several methods for the extraction of active components from black cumin: Soxhlet extraction using conventional organic solvents such as ethanol, methanol, or hexane, and hydrodistillation along with some modern techniques such as microwave-assisted extraction (MAE), ultrasonic extraction, and SFE technique using carbon dioxide as supercritical fluid.

6.5.1   SOXHLET EXTRACTION TECHNIQUE

Traditionally, different solvents and extraction techniques are used for extraction of active components. One of these techniques includes Soxhlet extraction followed by rotary evaporator for removal of chemical solvents. The technology of Soxhlet extraction and Soxhlet extraction apparatus are detailed in: [http://nptel.ac.in/courses/102103016/module4/lec33/images/1.jpg].

Different solvents are being utilized for the extraction, fractionation, and characterization of black cumin, as various solvents have unique capabilities to extract bioactive substances from seeds. Generally, it is observed that extraction ratio enhances with increase in polarity of solvents. A study on black cumin seedcake showed that it contains a noticeable amount of extractable components and extraction of these substances with polar solvents was found most effective.[91] Another study on black cumin seedcake extracts was carried out to optimize the extraction efficiency of aqueous solvents and results showed that highest extraction of antioxidant compounds was obtained through aqueous methanolic extraction.[95]

Earlier, Khattak et al.[70] used conventional solvent extraction by employing different solvents such as water, methanol, acetone, and hexane for the extraction of bioactive moieties from *N. sativa*. Afterward, the resultant extracts were subjected to total phenolic content determination and free radical scavenging activity estimation. In this context, it was noticed that acetone extracts give the highest yield. One of their peers conducted a similar research in which black cumin seeds were homogenized with hexane for 4 h at 180 U/min and afterward centrifugation was done and supernatant was taken. Resultant extract was concentrated using rotary evaporator and further subjected to antioxidant assay estimation.[32]

Recently, a study was conducted to compare the two methods (Soxhlet and cold press) to extract black cumin oil. Oil extraction yield in solvent extraction and cold press method was 37 and 27%, respectively.[50] Salea et al.[129] compared various extraction methods such as SFE technique, high-pressure Soxhlet, n-hexane Soxhlet extraction, and percolation with ethanol. It was observed that n-hexane Soxhlet technique gave maximum yield followed by SFE, percolation with ethanol, and high-pressure Soxhlet as 19.1, 14.0, 12.4, and 5.8%, respectively.

6.5.2   SUPERCRITICAL FLUID EXTRACTION TECHNIQUE

SFE technique is an effective and convenient alternative to conventional methods of extracting active components due to numerous benefits, for example, the use of environment-friendly and compatible fluids, such as CO_2, lesser use of solvent compared to conventional methods, extraction environment is oxygen free so reduces the chance of oxidation, and the removal of the solute from solvent is easy by simple expansion and short extraction time.[138] A fluid is called a supercritical fluid when its pressure and temperature reach thermodynamically above its critical point. Supercritical fluids have distinct properties of both gases and liquids as these diffuse and effuse like gases, while it possesses solubilization characteristic of liquids.[161] Furthermore, the density of supercritical fluids can be modified significantly by making minute changes in pressure and temperature. Supercritical carbon dioxide has been used to extract essential oils, fatty acids, and bioactive compounds from various fruits, vegetables, and spices.[54]

A study was conducted to explore the antimicrobial potential of *N. sativa* extract using SFE technique.[7] Later on, in a study, black cumin oil was extracted through SFE technique to explore the best extraction condition (temperature, extraction time, and pressure) in order to obtain high yield, antioxidant activity, and TQ quantity. The supercritical extraction system and flowchart are explained in detail in: [https://commons.wikimedia.org/wiki/File: SFEschematic.jpg].

Maximum TQ content (4.09 mg/mL) was obtained from the extract taken at 40°C temperature, 150 bar pressure, and an extraction time of 120 min, while, highest extraction yield (23.20%) was obtained at 350 bar pressure and 60°C temperature in 120 min. In case of antioxidant activity, maximum was shown by the extracts taken at 350 bar pressure, 50°C temperature, and 60 min time as depicted by IC_{50} value, that is, 2.59 mg/mL using DPPH radical scavenging activity method.[139] In another study, Salea et al.[129] compared various extraction methods, that is, SFE, high-pressure Soxhlet with liquid CO_2, n-hexane Soxhlet extraction, and percolation with ethanol; and maximum TQ content was obtained through SFE (20.8 mg/g oil) followed by high-pressure Soxhlet (8.8 mg/g oil), n-hexane Soxhlet extraction (6.3 mg/g oil), and percolation with ethanol (5.0 mg/g oil).

In addition, SFE allows the extraction of oil at a lower temperature, without residues.[113] SFE techniques have few more advantages over conventional solvent extraction techniques, such as by adjusting the temperature and pressure of supercritical fluid, its density can be modified which ultimately results in adjusting the solubilizing power of a supercritical fluid. This supercritical fluid has a more promising mass transfer in SFE owing to lower surface tension and viscosity, while, higher diffusion coefficient as compared to liquid solvent. These characteristics make it the most appropriate alternate of conventional organic solvents.[161]

In another study, the TQ-rich fraction was extracted using supercritical fluid extractor at 600 bars pressure and 40°C temperature. The extraction process lasted for 3 h and solvent (CO_2) flow rate was 30 g/min. The experiment showed that the extracted TQ was 2.00±0.17% w/w while TQ content obtained from conventional solvent Soxhlet extraction was 0.57±0.01% w/w.[61]

Recently, a research has explored the better extraction conditions of black cumin oil using SFE. Various extraction conditions were applied such as temperature, pressure, and extraction time. A study was conducted using two conventional solvents: hexane and methanol in contrast to SFE obtained at 60 MPa pressure and 40°C temperature to check the efficiency of two methods.[114] Similarly, it was demonstrated that maximum extraction of essential oil from black cumin was obtained at 40 MPa pressure and maximum TQ content was also extracted at the same pressure.[85]

6.5.3 MICROWAVE-ASSISTED EXTRACTION (MAE)

MAE is a comparatively modern technique of extracting bioactive moieties from solid samples. MAE is based on the principle of absorption of electromagnetic radiation by the components of the extractant. The microwaves transfer heat majorly in two ways: one is regarding the polarization dipole, while the second one is related to the mobility of ions. In this method, heat is produced by the friction of silver ions and solution molecules. Among numerous organic solvents, the best solvents for carrying out MAE are water, ethanol, and acetone.[140]

The significance of TQ as an additive/ingredient in the preparation of foods, cosmetic industries, and pharmaceutical industries has been well known and praised in previous studies throughout the globe. Although the extraction techniques of TQ from natural sources such as black cumin has not been considered thoroughly, especially in the case of MAE, yet very limited data is available. For this purpose, recently a study was carried out for comparing the MAE method with conventional methods, that is, solid/liquid extraction and Soxhlet extraction methods. It was reported that the enriched extracts obtained through MAE contained 2 and 7 times higher TQ content as compared to solid/liquid extraction and Soxhlet extraction. The optimal conditions for obtaining the highest yield of TQ were: 30 mL solvent per gram of seeds, 30°C temperature, and 10 min of extraction time. The measured yield of TQ was about 628 mg/kg of black cumin seeds.[67]

Likewise, a study was conducted to evaluate the influence of various extraction methods including MAE on physicochemical characteristics and stability of black cumin oil. *N. sativa* oil was obtained using various extraction techniques such as cold pressing, Soxhlet extraction, and MAE. Afterward, the oils were analyzed for bioactive components and fatty acids. It was reported that black cumin oil extracted through MAE technique contained more concentration of TQ as compared to other two methods, that is, cold press and Soxhlet extraction methods.[71]

Liu and his colleagues[81] extracted volatile components from *N. sativa* seeds by employing MAE technology. The major bioactive component isolated was TQ followed by p-cymene, 4-isopropyl-9-methoxy-1-methyl-1-cyclohexene, longifolene, α-thujene, and carvacrol as 38.23, 28.61, 5.74, 5.33, 3.88, and 2.31%, respectivly.[81] Furthermore, studies were conducted to compare the two methods, that is, MAE and hydrodistillation extraction, for the extraction of volatile components and essential oil of black cumin. Results showed that MAE provided higher/better yield of volatile components in lesser time and energy consumption as compared to hydrodistillation[22,23].

Recently, it was shown that the concentration of TQ obtained from Soxhlet and conventional solid/liquid techniques was 82.96 and 286.50 mg/kg black cumin seeds, respectively, while it was in the range of 353.04–571.90 mg/kg seeds with MAE.[67] MAE technique is considered as one of the superior techniques, having several benefits (simple, less process time, and less usage of solvent) over conventional extraction methods. Proper control of temperature in MAE can produce more reliable and reproducible outcomes in terms of extraction efficiency.[18] The most important factor that should be monitored while dealing with MAE is the temperature, while other factors also have significance such as process time,[162] microwave power level,[83] solvent system,[128] and surface area of contact.[72]

6.6 THERAPEUTIC POTENTIAL OF BLACK CUMIN AGAINST VARIOUS METABOLIC SYNDROMES

Metabolic syndromes are defined as a permutation of several medical ailments occurring simultaneously that enhance the risk of causing inflammation, CVD, diabetes, and renal and hepatic stress along with cancer insurgence.[116] Black cumin seed oils contain a significant amount of polyunsaturated fatty acids, which have a crucial role in preventing CVDs as well as improving the immune system.[63,126] The poor dietary habits are major causes of developing numerous complications and ailments.[144,145] Various diseases such as inflammation, cardiovascular complications, diabetes, cancer, and immune dysfunctions are the prominent cause of high mortality rates around the globe.[20,151]

Black cumin possesses medicinal properties and is rich in phytochemicals, which have free radical scavenging activity to prevent ailments.[133,165] Phytochemicals, antioxidants, polyphenols, and flavonoids have shown anti-inflammatory, hypocholesterolemic, and hypoglycemic potential.[29,34,92,122,125] The choice of diet and a healthy lifestyle can control

30–40% of diseases.[20,35,44,106] Bioactive molecules of black cumin make it a crucial choice for a healthy lifestyle. Black cumin can play an important role in the formulation and development of diet-based therapies against many physiological threats.[13,65,120]

Black cumin contains both fixed and essential oils. Fixed oil comprises fatty acids, fat-soluble vitamins, and volatile constituents in considerable amounts; however, essential oil comprise only volatiles. In addition, its antioxidant and hypolipidemic potential was attributed to the presence of dihomolinolenic acid, which was almost 1.9–2.3% of the oil.[32,105]. Oily fraction also contains considerable amounts of tocopherols and related bioactive compounds along with balanced fatty acid profile. Literature has shown that the yield of black cumin fixed oil ranges from 22.00–40.35% of total seed content.[9,32] In Indo-Pak subcontinent, black cumin is well known for treatment of indigestion and stomach disorders. Scientific research shows that people consuming functional and nutraceutical components enriched diet have less chances of developing chronic ailments along with lower mortality rates than those individuals not taking bioactive components in their diet.[62,157]

6.6.1 *HYPOGLYCEMIC PERSPECTIVES*

In the recent era, *N. sativa* is in limelight as an antidiabetic drug owing to its ability to maintain the integrity of β-cells of liver. Diabetes mellitus is one of the leading causes of mortality all over the globe and, if uncontrolled, can target multi-organ systems.[169] It was observed that most diabetic cases are type II, while type I diabetes mostly occurs in childhood. According to the estimates in 2030, about 376 million peoples will be affected with diabetes worldwide.[163] Generally, the diabetes is regarded as a complete or partial deficiency in the secretion of insulin which is linked with chronic hyperglycemia as well as disturbed metabolism of lipids, carbohydrates, and proteins. As a result of the metabolic abnormalities during diabetes, cardiovascular complications, oxidative stress, and immune dysfunction may also develop.[108,121]

Diet diversification is an important segment of diet-based therapies. A slight modification in the daily diet can possibly prevent the onset of diabetes mellitus.[28] The use of organic natural food components is attaining wide popularity. These natural compounds are generally known as micronutrients and in the recent era, much emphasis has been paid to explain their therapeutic mechanisms.[57] In a research project, it was concluded that occurrence of myocardial necrosis is more prevalent in diabetes mellitus.[147] It was observed that myocardial membrane damage was due to the production of free radical and occurrence of LDL oxidation. At the molecular level, loss of pancreatic β-cells plays an important role in the development of insulin deficiency and progression of diabetes mellitus.[74,78] Drug therapies often lose their effectiveness and consensus is being sought to explore the natural compounds for the treatment of diabetes mellitus and its complications.[74,164] Many researchers designed studies to evaluate the role of black cumin for management of diabetes.[89] Extracts of black cumin possessed blood glucose-lowering effects, but the exact antidiabetic mechanism is not yet established.[40,66]

Using immunohistochemical staining, Fararh et al.[46] demonstrated that black cumin oil has insulinotropic properties in type 2 diabetes. Similar results were also reported by other research groups.[6,66,96] Similarly, it was demonstrated that petroleum ether extract of black cumin exerted an insulin-sensitizing action in vivo.[76] Earlier, lower level of serum glucose (194.41 mg/dL) was observed in diabetic rats treated with black cumin as compared to control group (340.43 mg/dL).[99] Glucose-lowering effects of *N. sativa* oil may be attributed to non-pancreatic action instead of being associated with the stimulated release of insulin or by fractional proliferation/regeneration of β-cells in pancreas.[66,90] It was also reported that insulinotropic characteristics of black cumin may also positively influence the gluconeogenesis in liver.[40]

Black cumin lowers the hepatic glucose synthesis lowering the precursors (lactate, glycerol, and alanine) required for gluconeogenic reactions in liver, and ultimately decreases the gluconeogenesis in treated diabetic hamsters.[45] Black cumin, owing to reactive oxygen species (ROS) scavenging potential, may be beneficial in regulating the diabetic complications in experimental diabetic rats.[64] Furthermore, black cumin also protects biological membranes from lipid peroxidation preventing liver damage by inhibiting lipid peroxidation in diabetic rabbits.[99]

Diabetic complications create major problems, which affect health and ultimately cause the death of an individual. The antioxidants present in black cumin are helpful to control the free radical production and can also check the elevated glucose as well as cholesterol levels, so diabetic complications can be avoided.[64,89,90] In people suffering from diabetes, hyperglycemia is the most common metabolic disorder in addition to the heterogeneity of this disease.[141] In this case, pancreatic β-cells are damaged due to which level of insulin decreases, which is the basic cause of initiating diabetes mellitus.[74,78]

The onset of diabetes mellitus may be prevented by slight changes in the daily diet.[28] For the treatment of these life-threating disorders, plant-based compounds are widely used. Previously, lower glucose level (194.41 mg/dL) in diabetic rats was observed when they were fed black cumin in contrast to control diet (340.43 mg/dL) without the addition of black cumin.[99] Hypoglycemic effect of black cumin oil may be beneficial for extra-pancreatic action rather than enhancing insulin release and partially increasing of pancreatic β-cells.[40,66,90] Insulinotropic properties of black cumin resulted in high release of insulin that may also affect the production of adenosine triphosphate (ATP) in liver by different cycles.[109] Hypoglycemic effect of black cumin oil may also be beneficial to lower level of hepatic glucose.[45] The raised concentration of free radical could be one of the most important causes for the onset of diabetes mellitus.[33] The increase in free radicals during diabetes mellitus lowers the enzymes activities and antioxidant status of the body.[156]

Hepatic enzymes play a major role in maintaining proper blood glucose metabolism and the enzymes include glutathione peroxidase (GPx), catalase (CAT), glucose-6-phosphate dehydrogenase (G6PDH), and glutathione S-transferase (GST). The long-term hyperglycemia resulted in decreased activities of GPx (26%), CAT (34%), GST (38%), and G6PDH (27%) in diabetic patients.[74,155,156] The various researchers [38,164] have shown a marked rise in lipid peroxides and lowered antioxidant enzymes in diabetes mellitus. Scientists narrated that nonenzymatic glycosylations of enzymes and increased glucose concentration give rise to an increase in the production of lipid peroxides, related to hyperglycemia.[50]

In a study, lower level of lipid peroxidation was observed in diabetic rabbits treated with N. sativa and hence protected from liver damage.[99] This effect might be due to a high level of antioxidant production that protects tissues from the dangers of free radicals.[65] The occurrence of oxidative decomposition of liver is said to be due to the raised level of lipid peroxidation.[96,154]

The black cumin seed oil causes the stimulation of insulin, which indicates the extra-pancreatic action and becomes the ultimate reason for the hypoglycemic effect of black cumin oil. Modifications in insulin sensitivity also need detailed study as enhanced insulin sensitivity results in a low risk of cardiovascular problem associated with diabetes mellitus.[148] The beneficial plant oils for the control of diabetes are getting popular among researchers in the last few decades. Therefore, the phytochemistry of black cumin enhances its worth as a natural antidiabetic agent.[39,118]

6.6.2   HYPOCHOLESTEROLEMIC POTENTIAL

Bioactive ingredients of black cumin seed, especially antioxidants, fat-soluble vitamins, phytosterols, and some pyrazanols, have functional properties that are important in lowering serum cholesterol level.[4,94,136] The active ingredients such as TQ and its derivatives work as a safeguard against the raised level of cholesterol, triglycerides, and LDLs.[43] Studies showed that dyslipidemia is a major cause of CVDs, ultimately resulting in high rates of morbidity and mortality among people all over the world.[2] It has been explored in various research studies that maintained a level of plasma cholesterol is important for protection from CVD, as hyper-cholesterolemia plays a vital role in the occurrence of atherosclerosis.[136] Protection from CVDs is also important in a community when people are already suffering from other chronic diseases such as inflammation, diabetes mellitus, and cancer.[37]

It has been reflected in literature that consumption of natural plant bioactive compounds can improve serum lipid profile.[24] Accordingly, communities consuming diets enriched with functional components are at lower risk of developing metabolic syndromes.[27,29,35,51,111] Atherosclerosis is a chronic stage linked with deposition of lipids along the blood vessels leading to the difficulty in blood flow.[75]

Administration of herbal extract and black cumin seed oil can decrease cholesterol and total lipids levels in experimental rats. Supplementation of black cumin seeds in diets of experimental rats also decreased their serum triglyceride level.[170] Similarly, Badary et al.[17] observed similar trends while studying the effect of TQ on serum triglycerides in experimental rats. The dyslipidemic effect can be minimized either by reducing the synthesis of cholesterol by hepatocytes or decreasing its fractional reabsorption from the small intestine, thus lowering serum cholesterol level.

Black cumin fixed oil could have a positive effect on serum lipid profile by lowering total cholesterol, LDL, and triglycerides while increasing the HDL.[39] Black cumin has antihypertensive effects and it lowers the incidence of CVD. Black cumin seeds antioxidants along with essential oil might be beneficial to lowering the peroxidative damage in living cells.[153] A significant improvement in lipid profile was observed as a result of six different doses of TQ after 4 days.[19] It was observed that *N. sativa* oil is rich in sterol and reduces the absorption of cholesterol.[16] After sometime, another component nigellamines was observed and it decreased triglyceride levels in primary cultured mouse.[100]

A rodent trial was carried out to evaluate the cholesterol-lowering potential of black cumin extracts, and it was observed that serum cholesterol, triglycerides, glucose levels, leukocytes, and platelets count were lowered significantly by 15.5, 22, 16.5, 35, and 32%, respectively, in contrast to control values, while hematocrit and hemoglobin levels were raised gradually by 6.4 and 17.4%, respectively.[171] Later, it was observed that black cumin seeds caused a decline in serum triglycerides, LDL, and total cholesterol contents while increasing the serum HDL cholesterol level.[4]

In recent years, several investigations have been done related to supplement foods with antioxidants isolated from natural herbs/spices such as *N. sativa* to mitigate oxidative stress and hyperlipidemia.[11] The risk of CVDs can be reduced by dietary intake of antioxidants owing to their property of inhibiting oxidation of LDL and ultimately reducing the risk of developing cardiovascular disparities.[158] Likewise, Sultan et al.[147] narrated from bioefficacy study of black cumin that it has good potential to improve lipid profile of blood serum by lowering total cholesterol and LDL levels while increasing the serum HDL content. It was also expressed that black cumin powder given to rats at a dose rate of 1 and 2% reduced cholesterol level by 4.48 and 6.73% and LDL by 24.32 and 24.79%, respectively, while HDL contents improved nonsignificantly.

Formerly, it was suggested that black cumin has a great role in the activation of LDL receptor that resulted in the rapid removal of LDL cholesterol from moving lipids in blood, thus lowering the chances of potential atherogenecity.[40] According to another study, the black cumin fixed oil lowered the allied problems through regulating high blood pressure and managing respiration.[9] The high level of β-sterol in black cumin fixed oil makes it effective for decreasing blood cholesterol and avoiding coronary heart disease.[32] Similarly, it was reported that addition of 180 mg black cumin/kg/day in flour dough effects transient changes in the lipid content of experimental rats.[10]

In summary, it can be concluded that CVDs are one of the major causes of morbidity and mortality all over the world. High cholesterol and oxidation of LDL trigger the events leading to initiation of atherosclerosis.[12,94,159] A defective immune system results in the initiation of various health disparities, categorized as autoimmune disorders and immune dysfunction.[31,79,80,171] Dietary nutritional status has a significant effect on the antioxidant immune system of the body and deficiency of certain nutrients, such as vitamin E, selenium, and polyphenols, increases the onset of oxidative stress, diabetes mellitus, and atherosclerosis.[134] Functional foods or their functionally active molecules have shown therapeutic potential owing to the antioxidant, anti-inflammatory, anticancer, and immunomodulation perspectives.[28,166]

6.6.3   ANTI-INFLAMMATORY POTENTIAL

Inflammatory reactions are basically the protective biological processes occurring in the body to alleviate the harmful stimuli and these are carried out by endogenous mediators such as oxidants, eicosanoids, cytokines, lytic enzymes, and chemokines.[143] These mediators are majorly secreted by neutrophils and macrophages, though these can also be produced by the damaged tissues themselves.[88]

Although inflammation is a protective biological process, yet it can cause detrimental effects in tissues leading to tissue damage, especially in the case when the inflammatory reaction is accompanied by the synthesis of ROS. Along with ROS, nitric oxide is a highly reactive free radical and it can initiate toxic oxidative reactions, causing tissue damage and inflammation.[55]

The main bioactive ingredient of black cumin seeds "TQ" is responsible for the therapeutic effects against chronic inflammation and related conditions such as asthma, arthritis, and neurodegeneration. Inflammation has been recognized as one of the major factors in the development of tumor malignancies, while recently TQ has found to induce apoptosis and inhibit the proliferation in pancreatic ductal adenocarcinoma cells. TQ was proved to be a novel inhibitor of proinflammatory pathways and it also provides a promising strategy by combining proapoptotic and anti-inflammatory mode of action.[150]

It was observed that lipoxygenase (LO) and cyclooxygenase (COX) enzymes are key factors in formation of leukotrienes (LTs) and prostaglandins, and these are significantly involved in inflammatory reactions and responses. In this regard, Houghton et al.[55] have explored that the fixed oil of *N. sativa* expressively inhibited the 5-LO and COX pathways of arachidonate metabolism by preventing the production of LT B_4 and thromboxane B_2 metabolites in rat peritoneal leukocytes. Treatment of *N. sativa* oil inhibited the production of 5-hydroxyleicosa-tetra-enoic acid and 5-LO products at half maximal inhibitory concentration (IC_{50}) of 24 ± 01 and 25 ± 01 µg/mL in rat polymorphonuclear leukocytes.[87]

Furthermore, in the study on the effectiveness of black cumin seeds oil against hind paw edema and granuloma, it was reported that intraperitoneal injection of oil at the dose rate of 1.55 and 0.66 mL/kg significantly lowered the hind paw edema by 96.3 and 64.1%, respectively. Moreover, granuloma weight was reduced by 46.9 and 17.6% by injecting 0.66 and 0.33 mL/kg oil intraperitoneally, respectively.[102] Bioefficacy trials showed that black cumin is effective against various inflammatory conditions such as allergic encephalomyelitis, rheumatoid arthritis (RA), and ulcerative colitis induced by acetic acid.[86] Anti-inflammatory role of *N. sativa* has been investigated in allergic rhinitis patients by administrating them with placebo capsules and capsules containing *N. sativa* oil (0.5 mL) on daily basis for a period of 1 month. Treatment of *N. sativa* predominantly suppresses the nasal mucosal congestion, sneezing attacks, nasal itching, mucosal pallor, and turbinate hypertrophy. The biochemical pathways of black cumin oil in ameliorating rhinitis were not explored in the study.[107]

Furthermore, hexane extracts of *N. sativa* were reported to lower the clinical symptoms related to ovalbumin-induced allergic diarrhea when orally administrated to mice, while it has nonsignificant effect on total immunoglobulin E (IgE) levels, ovalbumin-specific IgE levels, as well as the release of interferon γ from mesenteric lymphocytes.[36] In a scientific study, the potential of black cumin was compared with synthetic medicines named as betamethasone and eucerin. The outcomes of exploration showed that *N. sativa* exhibited similar effects as of betamethasone in improving the clinical conditions and quality of life of eczema patients by decreasing the harshness of the situation.[168]

In another scientific study, preventive effects of aqueous ethanolic extracts of black cumin were on tracheal responsiveness in guinea pigs suffering from chronic obstructive pulmonary disease.[69] In a very recent study, a group of scientist examined the anti-inflammatory and antioxidant activity of black cumin oil in persons suffering from RA. It was noticed that oral administration of black cumin oil at dose rate of 1 g/day for the period of 2 months remarkably increased the levels of interleukin 10 while it has nonsignificant effect on tumor necrosis factor α, suggesting that black cumin oil can be used with RA medications.[52]

In a biological trial, it was shown that complete body irradiation of Wistar rats resulted in a noticeable reduction in hemolysin antibodies titers, globulin concentration, total protein concentration of plasma, hypersensitivity, depletion of lymphoid follicles of spleen, leukopenia, as well as thymus gland.[15] Additionally, it was observed that irradiation causes the significant elevation in the levels of malondialdehyde, CAT, as well as superoxide dismutase. Prior to irradiation, oral administration of black cumin oil at a dose rate of 1 ml/kg body weight per day for a period of 5 days per week significantly alleviated the harmful effects of irradiation.[15]

In another study, it was demonstrated that treatment of mouse splenic lymphocytes with ethanolic extracts of black cumin 1 h before irradiation prohibited the production of lipid peroxides as well as intracellular reactive species (ROS), which are the major factors that caused radiation-induced DNA damage and apoptosis.[123] Oral administration of black cumin ethanolic extracts at a dose rate of 100 mg/kg body weight per day for a period of 5 days protected the albino mice from oxidative damage to liver tissues and spleen after irradiation as estimated by the activity of natural antioxidant enzymes and lipid peroxidation. It was concluded that the radioprotective potential of black cumin is due to its radical scavenging perspectives.[123]

6.6.4 ANTICANCER PERSPECTIVES

Epidemiological studies have depicted that incidence of cancer is less prominent among people who rely on a variety of fruits and vegetables,[59] due to the bioactive compounds that occur mostly in plant foods known as polyphenols and flavonoids. Numerous scientific evidences have proved chemopreventive potential of polyphenols and flavonoids.[167] Oxidative stress in body causes damage to DNA thus leading to genetic mutation. These changes diverge cell division cycle (mitosis) and as a result, unwanted cell aggregation, known as "tumor," forms. Flavonoids protect DNA from oxidative damage by scavenging free radicals produced in the vicinity of DNA and interact with carcinoma produced from detoxification process.[21]

In a recent scenario, the role of phytochemicals for cancer prevention has been highlighted by the researchers due to their effectiveness and safety. Among different herbal sources, black cumin provides protection against various cancer insurgences. The quest to find out promising chemopreventive agent demands multidimensional perspectives to tackle carcinoma at various stages. In this context, the ability of black cumin as the anti-oncogenic agent has been revealed through various experimental models.[56,77]

There are evidences indicating the inverse association between black cumin consumption and occurrence of numerous oncogenic events. It tackles the cancer cells by soothing inflammation and inducing apoptosis. Moreover, it suppresses the expressions involved in nuclear factor kappaB in a dose-dependent manner by its active constituent TQ.[118,130]

Previously, colon cancer cell lines were treated with black cumin and noticed a marked reduction in the cell population. It was inferred that TQ in black cumin is responsible for apoptosis induction, arresting the cells in post-initiation stage, and targets different marker proteins.[49] Similarly, Rooney and Ryan[127] treated HT-29 cancer cell line with black cumin essential oil and observed a marked decline in cell proliferation. Additionally, it imparts apoptosis and arrests the cells in G_1 phase. Later, a significant reduction was observed in cell viability of SW-626 human colon cancer cells by TQ application. It was confirmed that cellular destruction, obstruction of cell metabolism, and apoptosis are the major pathways for cancer reduction.[110]

Besides, TQ treatment has proved efficient to check various cancer cells. It performs the antiproliferative action due to its ability to induce apoptosis, arrests the cells in G_1 phase, and reduces metastasis and angiogenesis. Moreover, it modulates the activity of various expressions such as p53, p73, PTEN, STAT3, PPAR-G, and activates the caspases 3, 6, and 8 for apoptosis induction.

The cell cycle arrest is one of the promising mechanisms by which black cumin curtails cancer insurgences. The G_1 arrest by black cumin is reported in various cancers such as colon, mammary, and prostate by downregulating the expression of cyclin D1 and p53 and increases the expression of p16. On contrary, lung and breast cancers are inhibited by arresting S and G_2 phases through suppressing, E2F-1, E2F-2-regulated proteins, cyclin-dependent kinases-4 (CDK-4), CDK-2, cyclin A, p21 Cip1, and p27 Kip1 expressions.[8,49,68]

6.6.4.1   COLON CANCER

Literature showed that TQ is proapoptotic and antineoplastic against colon cancer cell line HCT116.[49] In post-initiation stage, scientists concluded that volatile oil of black cumin has the ability to prevent colon carcinogenesis in rats and it imparts no adverse effect.[132] Afterward, it was recommended that TQ is an effective chemotherapeutic agent when it acted upon SW-626 colon cancer cells and its potency is analogous to 5-flurouracil. Moreover, TQ imparts favorable effects on HT-29 (colon adenocarcinoma) cell.[127]

6.6.4.2   LUNG CANCER

The α-hederin isolated from black cumin has antitumor activity against Lewis lung carcinoma (LL/2) in mice.[149] Research has demonstrated that diet supplemented with black cumin and honey has a shielding effect against oxidative stress induced by methylnitrosurea as well as carcinogenesis in colon, skin, and lungs.[84] Moreover, some reports showed that TQ and α-hederin (two important bioactive moieties of black cumin) neither enhance cytotoxicity nor apoptosis in lung carcinoma (A549) cells.

6.7   SUMMARY

Poor dietary habits, changing lifestyles, and escalating consumption of processed food have paved the way toward various physiological dysfunctions. Metabolic dysfunctions have become a great threat for sustaining healthy human life. Plant-based bioactive components have gained considerable attention in this regard to mitigate such disorders. Herbs and spices are considered as a rich source of phytochemicals with disease-modulating potential.

Black cumin (*N. sativa*) has been known for its therapeutic potential from ancient times. In the recent era, black cumin is attaining immense attention among researchers owing to its enriched composition and antioxidant potential along with auspicious therapeutic effects. Biologically active moieties of black cumin can be isolated by using various techniques such as Soxhlet extraction, SFE technique, and MAE technique. These isolated bioactive components can be used to tailor food products with health-promoting activity.

The major bioactive component of black cumin is TQ. Among various extraction techniques, SFE and MAE techniques are considered as novel techniques providing better extraction efficiency and purity of active moieties. TQ possesses antioxidant, anti-inflammatory, hypoglycemic, hypocholesterolemic, and anticarcinogenic activities. This review chapter highlights the biologically active components of black cumin (*N. sativa*) along with their extraction techniques and health-promoting potentials.

KEYWORDS

- ABTS
- anti-inflammation
- antioxidants
- Ayurveda medicines
- black cumin
- carcinogenesis
- cardiovascular disease
- diet-based therapies
- DPPH
- FRAP
- functional foods
- hypercholesterolemia
- hyperglycemia
- immune dysfunctions
- in vitro
- in vivo
- metabolic syndromes
- microwave-assisted extraction technique
- monoterpenes
- nigellone
- nutraceuticals
- oxidative stress
- phytochemicals
- proximate composition
- Soxhlet extraction technique
- spectrophotometer
- supercritical fluid
- supercritical fluid extraction technique
- thymoquinone
- tocopherol
- total flavonoids
- total phenolic content
- β-carotene bleaching assay
- ρ-cymene

REFERENCES

1. Abbas, M.; Saeed, F.; Anjum, F. M.; Afzaal, M.; Tufail, T.; Bashir, M. S.; Ishtiaq, A.; Hussain, S.; Suleria, H. A. R. Natural Polyphenols: An Overview. *Int. J. Food Prop.* **2016**, 1–11.

2. AHA (American Heart Association). *Heart Disease and Stroke Statistics-2005: Update.* American Heart Association: Dallas, TX, 2005; pp 3–4.

3. Ahmad, A.; Husain, A.; Mujeeb, M.; Khan, S. A.; Najmi, A. K.; Siddique, N. A. A Review on Therapeutic Potential of *Nigella sativa*: A Miracle Herb. *Asian Pac. J. Trop. Biomed.* **2013**, *3*, 337–352.

4. Akhtar, M. S.; Nasir, Z.; Abid, A. R. Effect of Feeding Powdered *Nigella sativa* L. Seeds on Poultry Egg Production and Their Suitability for Human Consumption. *Vet. Arh.* **2003**, *73*, 181–190.

5. Akhtar, S.; Khalid, N.; Ahmed, I.; Shahzad, A.; Suleria, H. A. R. Physicochemical Characteristics, Functional Properties, and Nutritional Benefits of Peanut Oil: A Review. *Crit. Rev. Food Sci. Nutr.* **2014**, *54*(12), 1562–1575.

6. Al-Hader, A.; Aqel, M.; Hasan, Z. Hypoglycemic Effect of Volatile Oil of *Nigella sativa* Seeds. *Int. J. Pharm.* **1993**, *2*, 96–100.

7. Alhaj, N. A.; Shamsudin, M. N.; Zamri, H. F.; Abdullah, R. Extraction of Essential Oil from *Nigella sativa* Using Supercritical Carbon Dioxide: Study of Antibacterial Activity. *Am. J. Pharmacol. Toxicol.* **2008**, *3*(4), 225–228.

8. Alhosin, M.; Abusnina, A.; Achour, M.; Sharif, T.; Muller, C.; Peluso, J. Induction of Apoptosis by Thymoquinone in Lymphoblastic Leukemia Jurkat Cells Is Mediated by a p73-Dependent Pathway Which Targets the Epigenetic Integrator UHRF1. *Biochem. Pharmacol.* **2010**, *79*, 1251–1260.

9. Ali, B.H.; Blunden, G. Pharmacological and Toxicological Properties of *Nigella sativa*. *Phytother. Res.* **2003**, *7*, 299–305.

10. Al-Jishi, S. A.; Hozaifa, B. A. Effect of *Nigella sativa* on Blood Hemostatic Function in Rats. *J. Ethnopharmacol.* **2003**, *85*, 7–14.

11. Al-Naqeep, G.; Ismail, M.; Yazan, L. S. Effects of Thymoquinone Rich Fraction and Thymoquinone on Plasma Lipoprotein Levels and Hepatic Low Density Lipoprotein Receptor and 3-Hydroxy-3-methylglutaryl Coenzyme A Reductase Genes Expression. *J. Funct. Foods* **2009**, *1*, 298–303.

12. Andican, G.; Seven, A.; Uncu, M.; Cantasdemir, M.; Numan, F.; Burcak, G. Oxidized LDL and Anti-oxLDL Antibody Levels in Peripheral Atherosclerotic Disease. *Scand. J. Clin. Lab. Invest.* **2008**, *18*, 1–7.

13. Arayne, M. S.; Sultana, N.; Mirza, A. Z.; Zuberi, M. H.; Siddiqui, F. A. In Vitro Hypoglycemic Activity of Methanolic Extract of Some Indigenous Plants. *Pak. J. Pharm. Sci.* **2007**, *20*(4), 268–273.

14. Ashraf, M.; Ali, Q.; Iqbal, Z. Effect of Nitrogen Application Rate on the Content and Composition of Oil, Essential Oil and Minerals in Black Cumin (*Nigella sativa* L.) Seeds. *J. Sci. Food Agric.* **2006**, *86*, 871–876.

15. Assayed, M. E. Radioprotective Effects of Black Seed (*Nigella sativa*) Oil Against Hemopoietic Damage and Immunosuppression in Gamma-Irradiated Rats, Immunopharmacol. *J. Immunotoxicol.* **2010**, *32*(2), 284–296.

16. Atta, M. B. Some Characteristics of Nigella (*Nigella sativa* L.) Seed Cultivated in Egypt and Its Lipid Profile. *Food Chem.* **2003**, *83*, 63–68.

17. Badary, O. A.; Taha, R. A.; Gamal-el-Din, A. M.; Abdel-Wahab, M. H. Thymoquinone Is a Potent Superoxide Anion Scavenger. *Drug Chem. Toxicol.* **2003**, *26*, 87–98.

18. Ballard, T. S.; Mallikarjunan, P.; Zhou, K.; O'Keefe, S. Microwave-Assisted Extraction of Phenolic Antioxidant Compounds from Peanut Skins. *Food Chem.* **2010**, *120*, 1185–1192.

19. Bamosa, A. O.; Ali, B. A.; Al-Hawsawi, Z. A. The Effect of Thymoquinone on Blood Lipids in Rats. *Indian J. Physiol. Pharmacol.* **2002**, *46*, 195–201.

20. Bárta, I.; Smerák, P.; Polívková, Z.; Sestáková, H.; Langová, M.; Turek, B.; Bártová, J. Current Trends and Perspectives in Nutrition and Cancer Prevention. *Neoplasma* **2006**, *53*, 19–25.

21. Benavente-garci, O.; Castillo, J.; Marin, F.; Ortun, A.; Del Ri, J. Uses and Properties of Citrus Flavonoids. *Am. Chem. Soc.* **1997**, *45*, 4505–4515.

22. Benkaci-Ali, F.; Akloul, R.; Boukenouche, A.; Pauw, E. D. Chemical Composition of the Essential Oil of *Nigella sativa* Seeds Extracted by Microwave Steam Distillation. *J. Essent. Oil Bear. Plants* **2013**, *16*, 781–94.

23. Benkaci-Ali, F.; Baaliouamer, A.; Meklati, B. Y.; Chemat, F. Chemical Composition of Seed Essential Oils from Algerian *Nigella sativa* Extracted by Microwave and Hydrodistillation. *Flavor Fragr. J.* **2007**, *22*, 148–53.

24. Borek, C. Garlic Reduces Dementia and Heart-Disease Risk. *J. Nutr.* **2006**, *136*, 810S–812S.

25. Bourgou, S.; Ksouri, R.; Bellila, A.; Skandrani, I.; Falleh, H.; Marzouk, B. Phenolic Composition and Biological Activities of Tunisian *Nigella sativa* L. Shoots and Roots. *C. R. Biol.* **2008**, *33*, 48–55.

26. Burits, M.; Bucar, F. Antioxidant Activity of *Nigella sativa* Essential Oil. *Phytother. Res.* **2000**, *14*, 323–328.

27. Butt, M. S.; Imran, A.; Sharif, M. K.; Ahmad, R. S.; Xiao, H.; Imran, M.; Rasool, H. A. Black Tea Polyphenols: A Mechanistic Treatise. *Crit. Rev. Food Sci. Nutr.* **2014**, *54*(8), 1002–1011.

28. Butt, M. S.; Sultan, M. T. Green Tea, Nature's Defense Against Malignancies. *Crit. Rev. Food Sci. Nutr.* **2009**, *49*, 463–473.

29. Butt, M. S.; Tahir-Nadeem, M.; Khan, M. K. I.; Shabir, R. Oat, Unique Among the Cereals. *Eur. J. Nutr.* **2008**, *47*(2), 68–79.

30. Chan, K. W.; Iqbal, S.; Khong, N. M. H.; Babji, A. S. Preparation of Deodorized Antioxidant Rich Extracts from 15 Selected Spices Through Optimized Aqueous Extraction. *J. Med. Plants Res.* **2011**, *5*(25), 6067–6075.

31. Chanana, V.; Majumdar, S.; Rishi, P. Involvement of Caspase-3, Lipid Peroxidation and TNF-α in Causing Apoptosis of Macrophages by Coordinately Expressed Salmonella Phenotype Under Stress Conditions. *Mol. Immunol.* **2007**, *44*(7), 1551–1558.

32. Cheikh-Rouhou, S.; Besbes, S.; Hentati, B.; Blecker, C.; Deroanne, C.; Attia, H. *Nigella sativa* L.: Chemical Composition and Physicochemical Characteristics of Lipid Fraction. *Food Chem.* **2007**, *101*, 673–681.

33. Clarkson, P. M. Antioxidants and Physical Performance. *Critical Reviews in Food Science and Nutrition*, **1995**, *35*, 131–141.

34. Colonna, M.; Danzon, A.; Delafosse, P.; Mitton, N.; Bara, S.; Bouvier, A. M.; Ganry, O.; Guizard, A. V.; Launoy, G.; Molinie, F.; Sauleau, E. A.; Schvartz, C.; Velten, M.; Grosclaude, P.; Tretarre, B. Cancer Prevalence in France, Time Trend, Situation in 2002 and Extrapolation to 2012. *Eur. J. Cancer* **2008,** *44*(1), 115–122.

35. Divisi, D.; Di Tommaso, S.; Salvemini, S.; Garramone, M.; Crisci, R. Diet and Cancer. *Actabiomedica* **2006,** *77,* 118–123.

36. Duncker, S. C.; Philippe, D.; Martin-Paschoud, C.; Moser, M.; Mercenier, A.; Nutten, S. *Nigella sativa* (Black Cumin) Seed Extract Alleviates Symptoms of Allergic Diarrhea in Mice, Involving Opioid Receptors. *PLoS One* **2012,** *7*(6), 39841.

37. Eidi, A.; Eidi, M.; Esmaeili, E. Antidiabetic Effect of Garlic (*Allium sativum* L.) in Normal and Streptozotocin-induced Diabetic Rats. *Phytomedicine* **2006,** *13,* 624–629.

38. El-Abhar, H. S.; Abdallah, D. M.; Saleh, S. Gastroprotective Activity of *Nigella sativa* Oil and its Constituent, Thymoquinone, Against Gastric Mucosal Injury Induced by Ischaemia/Reperfusion in Rats. *J. Ethnopharmacol.* **2003,** *84,* 251–258.

39. El-Dakhakhny, M.; Mady, N. I.; Halim, M. A. *Nigella sativa* L. Oil Protects Against Induced Hepatotoxicity and Improves Serum Lipid Profile in Rats. *Arzneimittel-Forschung* **2000,** *50,* 832–836.

40. El-Dakhakhny, M.; Mady, N.; Lembert, N.; Ammon, H. P. The Hypoglycemic Effect of *Nigella sativa* Oil is Mediated by Extrapancreatic Actions. *Planta Med.* **2002,** *68,* 465–466.

41. El-Missiry, M. A.; El-Gindy, A. M. Amelioration of Alloxan Induced Diabetes Mellitus and Oxidative Stress in Rats by Oil of *Eruca sativa* Seeds. *Ann. Nutr. Metab.* **2000,** *44,* 97–100.

42. El-Saleh, S. C.; Al-Sagair, O. A.; M. I. Al-Khalaf. Thymoquinone and *Nigella sativa* Oil Protection Against Methionine-induced Hyperhomocysteinemia in Rats. *Int. J. Cardiol.* **2004,** *93*(1), 19–23.

43. Erkan, N.; Ayranci, G.; Ayranci, E. Antioxidant Activities of Rosemary (*Rosmarinus officinalis* L.) Extract, Blackseed (*Nigella sativa L.*) Essential Oil, Carnosic Acid, Rosmarinic Acid and Sesamol. *Food Chem.* **2008,** *110*(1), 76–82.

44. Farah, I. O. Assessment of Cellular Responses to Oxidative Stress using MCF-7 Breast Cancer Cells, Black Seed (*N. sativa* L.) Extracts and H_2O_2. *Int. J. Environ. Res. Pub. Health* **2005,** *2,* 411–419.

45. Fararh, K. M.; Atoji, Y.; Shimizu, Y.; Shiina, T.; Nikami, H.; Takewaki, T. Mechanisms of the Hypoglycaemic and Immunopotentiating Effects of *Nigella sativa* L. Oil in Streptozotocin-induced Diabetic Hamsters. *Res. Vet. Sci.* **2004,** *77,* 123–129.

46. Fararh, K. M.; Atoji, Y.; Shimizu, Y.; Takewaki, T. Isulinotropic Properties of *Nigella sativa* Oil in Streptozotocin plus Nicotinamide Diabetic Hamster. *Res. Vet. Sci.* **2002,** *73,* 279–282.

47. Faria, A.; Oliveira, J.; Neves, P.; Gameiro, P.; Santos-Buelga, C.; de Freitas, V.; Mateus, N. Antioxidant Properties of Prepared Blueberry (*Vaccinium myrtillus*) Extracts. *J. Agric. Food Chem.* **2005,** *53,* 6896–6902.

48. Frankel, E. N.; Meyer, A. S. The Problems of Using One-dimensional Methods to Evaluate Multifunctional Food and Biological Antioxidants. *J. Agric. Food Chem.* **2000,** *80,* 1925–1941.

49. Gali-Muhtasib, H.; Diab-Assaf, M.; Boltze, C.; Al-Hmaira, J.; Hartig, R.; Roessner, A.; Schneider-Stock, R. Thymoquinone Extracted from Black Seed Triggers

Apoptotic Cell Death in Human Colorectal Cancer Cells via a p53-dependent Mechanism. *Int. J. Oncol.* **2004**, *25*, 857–866.

50. Gharby, S.; Harhar, H.; Guillaume, D.; Roudani, A.; Boulbaroud, S.; Ibrahimi, M.; Ahmad, M.; Sultana, S.; Hadda, T. B.; Chafchaouni-Moussaoui, I.; Charrouf, Z. Chemical Investigation of *Nigella sativa* L. Seed Oil Produced in Morocco. *J. Saudi Soc. Agric. Sci.* **2015**, *14*, 172–177.

51. Gonzalez, C. A. Nutrition and Cancer: The Current Epidemiological Evidence. *Br. J. Nutr.* **2006**, *96*, S42–S45.

52. Hadi, V.; Kheirouri, S.; Alizadeh, M.; Khabbazi, A.; Hosseini, H. Effects of *Nigella sativa* Oil Extract on Inflammatory Cytokine Response and Oxidative Stress Status in Patients with Rheumatoid Arthritis, a Randomized, Double-Blind, Placebo-Controlled Clinical Trial. *Avicenna J. Phytomed.* **2016**, *6*(1), 34–43.

53. Haron; Grace-Lynn, C.; Shahar, S. Comparison of Physicochemical Analysis and Antioxidant Activities of *Nigella sativa* Seeds and Oils from Yemen, Iran and Malaysia (Perbandingan Analisis Fizikokimia dan Aktiviti Antioksidan dalam Biji dan Minyak *Nigella sativa* dari Yemen, Iran dan Malaysia), *Sains Malays.* **2014**, *43*(4), 535–542.

54. Herrero, M.; Thornton, P. K.; Notenbaert, A. M.; Wood, S.; Msangi, S.; Freeman, H. A.; Bossio, D.; Dixon, J.; Sere, C.; Rosegrant, M. Smart Investments in Sustainable Food Production: Revisiting Mixed Crop-Livestock Systems. *Science* **2010**, *327*, 822–825.

55. Houghton, P. J.; Zarka, R.; de las Heras, B.; Hoult, J. R. Fixed Oil of *Nigella sativa* and Derived Thymoquinone Inhibit Eicosanoid Generation in Leukocytes and Membrane Lipid Peroxidation. *Planta Med.* **1995**, *61*, 33–36.

56. Howe, H. L.; Wu, X.; Ries, L. A.; Cokkinides, V.; Ahmed, F.; Jemal, A.; Miller, B.; Williams, M.; Ward, E.; Wingo, P. A.; Ramirez, A.; Edwards, B. K. Annual Report to the Nation on the Status of Cancer, 1975–2003, Featuring Cancer Among U.S. Hispanic/Latino Populations. *Cancer* **2006**, *107*(8), 1711–1742.

57. Hu, C. C.; Lin, J. T.; Lu, F.J.; Chou, F. P.; Yang, D. J. Determination of Carotenoids in *Dunaliella salina* Cultivated in Taiwan and Antioxidant Capacity of the Algal Carotenoid Extract. *Food Chem.* **2008**, *109*, 439–446.

58. Hussein, M. R.; Abu-Dief, E. E.; Abd El-Reheem, M. H.; Abd-Elrahman, A. Ultrastructural Evaluation of the Radioprotective Effects of Melatonin 237 Against X-ray-induced Skin Damage in Albino Rats. *Int. J. Exp. Pathol.* **2005**, *86*, 45–55.

59. Imran, M.; Saeed, F.; Nadeem, M.; Arshad, M. U.; Ullah, A.; Suleria, H. A. R. Cucurmin; Anticancer and Antitumor Perspectives: A Comprehensive Review. *Crit. Rev. Food Sci. Nutr.* **2016**, *22*(November), 1–23. http://dx.doi.org/10.1080/1040839 8.2016.1252711.

60. Iqbal, M.; Alam, P.; Anwer, M. T. High Performance Liquid Chromatographic Method with Fluorescence Detection for the Estimation of Thymoquinone in *Nigella sativa* Extracts and Marketed Formulations. *Sci. Rep.* **2013**, *2*, 655.

61. Ismail, H. I.; Chan, K. W.; Mariod, A. A.; Ismail, M. Phenolic Content and Antioxidant Activity of Cantaloupe (*Cucumis melo*) Methanolic Extracts. *Food Chem.* **2010**, *119*, 643–647.

62. Jenkins, D. J.; Kendall, C. W.; Nguyen, T. H.; Marchie, A.; Faulkner, D. A.; Ireland, C.; Josse, A. R.; Vidgen, E.; Trautwein, E. A.; Lapsley, K. G.; Holmes, C.; Josse, R.

G.; Leiter, L. A.; Connelly, P. W.; Singer, W. Effect of Plant Sterols in Combination with Other Cholesterol-Lowering Foods. *Metabolism* **2008,** *57*(1), 130–139.

63. Jones, P. J.; Varady, K. A Are Functional Foods Redefining Nutritional Requirements? *Appl. Physiol. Nutr. Metab.* **2008,** *33*, 118–123.

64. Kaleem, M.; Kirmani, D.; Asif, M.; Ahmed, Q.; Bano, B. Biochemical Effects of *Nigella sativa* L Seeds in Diabetic Rats. *Indian J. Exp. Biol.* **2006,** *44*, 745–748.

65. Kanter, M. Effects of *Nigella sativa* and its Major Constituent, Thymoquinone on Sciatic Nerves in Experimental Diabetic Neuropathy. *Neurochem. Res.* **2008,** *33*, 87–96.

66. Kanter, M.; Meral, I.; Yener, Z.; Ozbek, H.; Demir, H. Partial Regeneration/Proliferation of the Beta-Cells in the Islets of Langerhans by *Nigella sativa* L. in Streptozotocin-Induced Diabetic Rats. *Tohoku J. Exp. Med.* **2003,** *201*, 213–219.

67. Karacabey, E. Optimization of Microwave-assisted Extraction of Thymoquinone from *Nigella sativa* L. Seeds. *Maced. J. Chem. Chem. Eng.* **2016,** *35*, 2.

68. Kaseb, A. O.; Chinnakannu, K.; Chen, D.; Sivanandam, A.; Tejwani, S.; Menon, M. Androgen Receptor and E2F-1 Targeted Thymoquinone Therapy for Hormonerefractory Prostate Cancer. *Cancer Res.* **2007,** *67*, 7782–7788.

69. Keyhanmanesh, R.; Nazemiyeh, H.; Mazouchian, H.; Bagheri Asl, M. M.; Karimi Shoar, M.; Alipour, M. R.; Boskabady, M. H. *Nigella sativa* Pretreatment in Guinea Pigs Exposed to Cigarette Smoke Modulates *in vitro* Tracheal Responsiveness. *Iran. Red Crescent Med. J.* **2014,** *16*(7), e10421.

70. Khattak, K. F.; Simpson, T. J.; Ihasnullah. Effect of Gamma Irradiation on the Extraction Yield, Total Phenolic Content and Free Radical-Scavenging Activity of *Nigella sativa* Seed. *Food Chem.* **2008,** *110*, 967–972.

71. Kiralan, M.; Ulaş, M.; Özaydin, A.; Özdemır, N.; Özkan, G.; Bayrak, A.; Ramadan, M. F. Blends of Cold Pressed Black Cumin Oil and Sunflower Oil with Improved Stability: A Study Based Changes in the Levels of Volatiles, Tocopherols and Thymoquinone During Accelerated Oxidation Conditions. *J. Food Biochem.* **2017,** *41*(1), e12272.

72. Kothari, V.; Seshadri, S. Antioxidant Activity of Seed Extracts of *Annona squamosa* and *Carica papaya*. *Nutr. Food Sci.* **2010,** *40*, 403–408.

73. Kumar, T. V. S.; Negi, P.S.; Sankar, K. U. Antibacterial Activity of *Nigella sativa* L. Seed Extracts. *Br. J. Pharmacol. Toxicol.* **2010,** *1*(2), 96–100.

74. Lapshina, E. A.; Sudnikovich, E. J.; Maksimchik, J. Z.; Zabrodskaya, A.; Zavodnik, L. B.; Kubyshin, V. L.; Nocun, M.; Kazmierczak, P.; Dobaczewski, M.; Watala, C.; Zavodnik, I. B. Anti-oxidative Enzyme and Glutathione S-Transferase Activities in Diabetic Rats Exposed to Long-term ASA Treatment. *Life Sci.* **2006,** *79*, 1804–1811.

75. Le, N. A., Walter, M. F. The Role of Hypertriglyceridemia in Atherosclerosis. *Curr. Atheroscler. Rep.* **2007,** *9*(2), 110–115.

76. Le, P. M.; Benhaddou-Andaloussi, A.; Settaf, A.; Cherrah, Y.; Haddad, P. S. The Petroleum Ether Extract of *Nigella sativa* Exerts Lipid Lowering Action in the Rats. *J. Ethnopharmacol.* **2004,** *94*(2–3), 251–259.

77. Lee, A. Y.; Peterson, E. A. Treatment of Cancer-associated Thrombosis. *Blood* **2013,** *122*(14), 2310–2317.

78. Lee, K. W.; Blouin, A.; Lip, G. Y. H. The Role of Omega-3 Fatty Acids in the Secondary Prevention of Cardiovascular Disease. *Q. J. Med.* **2004,** *96*, 465–480.

79. Lee, S. J.; Kavanaugh, A. Autoimmunity, Vasculitis, and Autoantibodies. *J. Allergy Clin. Immunol.* **2006,** *117*(2), 445–450.

80. Li, S.; Yao, X.; Liu, H.; Li, J.; Fan, B. Prediction of T-cell Epitopes Based on Least Squares Support Vector Machines and Amino Acid Properties. *Anal. Chim. Acta* **2007,** *584*, 37–42.

81. Liu, X.; Park, J. H.; Abd El-Aty, A. M.; Assayed, M. E.; Shimoda, M.; Shim, J. H. Isolation of Volatiles from *Nigella sativa* Seeds Using Microwave-assisted Extraction, Effect of Whole Extracts on Canine and Murine CYP1A. *Biomed. Chromatogr.* **2013,** *27*, 938–45.

82. Lutterodt, H.; Luther, M.; Slavin, M.; Yin, J. J.; Parry, J.; Gao, J. M.; Yu. L. Fatty Acid Profile, Thymoquinone Content, Oxidative Stability and Antioxidant Properties of Cold-pressed Black Cumin Seed Oils. *Food Sci. Technol.* **2010,** *43*, 1409–1413.

83. Ma, W.; Lu, Y.; Dai, X.; Liu, R.; Hu, R.; Pan, Y. Determination of Anti-tumor Constitute Mollugin from Traditional Chinese Medicine *Rubia cordifolia*: Comparative Study of Classical and Microwave Extraction Techniques. *Sep. Sci. Technol.* **2009,** *44*, 995–1006.

84. Mabrouk, G. M.; Moselhy, S. S.; Zohny, S. F.; Ali, E. M.; Helal, T. E.; Amin, A. A.; Khalifa, A. A. Inhibition of Methylnitrosourea MNU Induced Oxidative Stress and Carcinogenesis by Orally Administered Bee Honey and Nigella Grains in Sprague Dawely Rats. *J. Exp. Clin. Cancer Res.* **2002,** *21*, 341–346.

85. Machmudah, S.; Murakami, W. K.; Sasaki, M.; Goto, M. Production of Nanofibers by Electrospinning Under Pressurized CO_2. *High Pressure Res.: Int. J.* **2012,** *32*(1), 54–59.

86. Mahgoub, A. A. Thymoquinone Protects Against Experimental Colitis in Rats. *Toxicol. Lett.* **2003,** *143*, 133–143.

87. Majdalawieh, A. F.; Fayyad, M. W. Immunomodulatory and Anti-inflammatory Action of *Nigella sativa* and Thymoquinone: A Comprehensive Review. *Int. Immunopharmacol.* **2015,** *28*, 295–304.

88. Majdalawieh, A., Ro, H. S. Regulation of IkappaBalpha function and NF-kappaB Signaling: AEBP1 is a Novel Proinflammatory Mediator in Macrophages. *Mediators Inflammation* **2010,** *2010*, e823821.

89. Mansi, K. M. S. Effects of Oral Administration of Water Extract of *Nigella sativa* on Serum Concentrations of Insulin and Testosterone in Alloxan-induced Diabetic Rats. *Pak. J. Biol. Sci.* **2005,** *8*, 1152–1156.

90. Mansi, K. M. S. Effects of Oral Administration of Water Extract of *Nigella sativa* on the Hypothalamus Pituitary Adrenal Axis in Experimental Diabetes. *Int. J. Pharm.* **2006,** *2*, 104–109.

91. Mariod, A. A.; Ibrahim, R. M.; Ismail, M.; Ismail, N. Antioxidant Activity and Phenolic Content of Phenolic Rich Fractions Obtained from Black Cumin (*Nigella sativa*) Seed Cake. *Food Chem.* **2009,** *116*, 306–312.

92. Martin-Moreno, J. M.; Soerjomataram, I.; Magnusson, G. Cancer Causes and Prevention: A Condensed Appraisal in Europe in 2008. *Eur. J. Cancer* **2008,** *44*(10), 1390–403.

93. Martos, G.; López-Fandiño, R.; Molina, E. In vitro Digestions and IgE Binding of Proteins from White and Whole Hen's Egg. *Food Allergy Anaphylaxis Meet.* **2011,** *1*, 08.

94. Matsuura, E.; Hughes, G. R.; Khamashta, M. A. Oxidation of LDL and its Clinical Implication. *Autoimmun. Rev.* **2008**, *7*, 558–566.

95. Matthäus, B. Antioxidant Activity of Extracts Obtained from Residues of Different Oilseeds. *J. Agric. Food Chem.* **2002**, *50*, 3444–3452.

96. MeÂzes, M.; Virag, G. Y.; Barta, M.; Abouzeid, A. D. Effect of Lipid Peroxide Oading on Lipid Peroxidation and on the Glutathione and Cytochrome Systems in Rabbits. *Acta Vet. Hung.* **1996**, *44*, 443–450.

97. Mehta, B. K.; Mehta, P.; Gupta, M. A New Naturally Acetylated Triterpene Saponin from *Nigella sativa. Carbohydr. Res.* **2009**, *344*, 149–151.

98. Meral, I.; Kanter, M. Effects of *Nigella sativa* L. and *Urtica dioica* L. on Selected Mineral Status and Hematological Values in CCl_4-Treated Rats. *Biol. Trace Elem. Res.* **2003**, *96*, 263–270.

99. Meral, I.; Yener, Z.; Kahraman, T.; Mert, N. Effect of *Nigella sativa* on Glucose Concentration, Lipid Peroxidation, Antioxidant Defence System and Liver Damage in Experimentally Induced Diabetic Rabbits. *J. Vet. Med.* **2001**, *48*, 593–599.

100. Morikawa, T.; Xu, F.; Ninomiya, K.; Matsuda, H.; Yoshikawa, M. Nigellamines A3, A4, A5 and C, New Dolabellane-type Diterpene Alkaloids, with Lipid Metabolism-promoting Activities from the Egyptian Medicinal Food Black Cumin. *Chem. Pharm. Bull.* **2004**, *52*, 494–497.

101. Mozzafari, F. S.; Ghorbanli, M.; Babai, A.; Farzami Scpehr, M. The Effect of Water Stress on the Seed Oil of *Nigella sativa* L. *J. Essent. Oil Res.* **2000**, *12*, 36–38.

102. Mutabagani, A.; El-Mahdy, S. A. M. A Study of the Anti-inflammatory Activity of *Nigella sativa* L. and Thymoquinone in Rats. *Saudi Pharmacol. J.* **1997**, *5*, 110–113.

103. Nagi, M. N.; Mansour, M. A. Protective effect of Thymoquinone Against Doxoru-bicin-Induced Cardiotoxicity in Rats: A Possible Mechanism of Protection. *Pharmacol. Res.* **2000**, *41*, 283–289.

104. Nehar, S.; Rani, P. HPTLC Studies on Ethanolic Extract of *Nigella sativa* Linn. Seeds and its Phytochemical Standardization. *Ecoscan* **2011**, *1*, 105–108.

105. Nickavar, B.; Mojab, F.; Javidnia, K.; Amoli, M. A. R. Chemical Composition of the Fixed and Volatile Oils of *Nigella sativa* L. *Iran. Z. Naturforsch.* **2003**, *58*, 629–631.

106. Nies, L. K.; Cymbala, A. A.; Kasten, S. L.; Lamprecht, D. G.; Olson, K. L. Complementary and Alternative Therapies for the Management of Dyslipidemia. *Ann. Pharmacother.* **2006**, *40*, 1984–1992.

107. Nikakhlagh, S.; Rahim, F.; Hossein, F.; Aryani, N.; Syahpoush, A.; Brougerdnya, M.; Saki, N. Herbal Treatment of Allergic Rhinitis: The Use of *Nigella sativa. Am. J. Otolaryngol.* **2011**, *32*(5), 402–407.

108. Noguchi, H. Stem Cells for the Treatment of Diabetes. *Endocr. J.* **2007**, *54*(1), 7–16.

109. Nordlie, R. C.; Foster, J. D. Regulation of Glucose Production by the Liver. *Ann. Rev. Nutr.* **1999**, *19*, 379–406.

110. Norwood, A. A.; Tan, M.; May, M.; Tucci, M.; Benghuzzi, H. Comparison of Potential Chemotherapeutic Agents, 5-fluoruracil, Green Tea and Thymoquinone on Colon Cancer Cells. *Biomed. Sci. Instrum.* **2006**, *42*, 350–356.

111. Ohishi, W.; Fujiwara, S.; Cologne, J. B.; Suzuki, G.; Akahoshi, M.; Nishi, N.; Takahashi, I.; Chayama, K. Risk Factors for Hepatocellular Carcinoma in a Japanese Population: A Nested Case-Control Study. *Cancer Epidemiol., Biomarkers Prev.* **2008**, *17*(4), 846–854.

112. Oktay, M.; Gülçin, I.; Küfrevio˘glu, O. I. Determination of in vitro Antioxidant Activity of Fennel (*Foeniculum vulgare*) Seed Extracts. *LWT—Food Sci. Technol.* **2003,** *36*(2), 263–271.

113. Ozkal, S. G.; Salgin, U.; Yener, M. E. Supercritical Carbon Dioxide Extraction of Hazelnut Oil. *J. Food Eng.* **2005,** *69*, 217–223.

114. Parhizkar, S.; Latiff, L. A.; Rahman, S. A.; Dollah, M. A. Metabolic Impact of *Nigella sativa* Extracts on Experimental Menopause Induced Rats. *J. Appl. Pharm. Sci.* **2011,** *1*(9), 38–42.

115. Parhizkar, S.; Latiff, L. A.; Rahman, S. A.; Dollah, M. A. Preventive Effect of *Nigella sativa* on Metabolic Syndrome in Menopause Induced Rats. *J. Med. Plants Res.* **2011,** *5*(8), 478–1484.

116. Perveen, R.; Suleria, H. A. R.; Anjum, F. M.; Butt, M. S.; Pasha, I.; Ahmad, S. Tomato (*Solanum lycopersicum*) Carotenoids and Lycopenes Chemistry; Metabolism, Absorption, Nutrition, and Allied Health Claims—A Comprehensive Review. *Crit. Rev. Food Sci. Nutr.* **2015,** *55*(7), 919–929.

117. Prior, R. L.; Cao, G. Antioxidant Phytochemicals in Fruits and Vegetables. Diet and Health Implications. *Hortic. Sci.* **2000,** *35*, 588–592.

118. Ramadan, M. F. Nutritional Value, Functional Properties and Nutraceuticals Applications of Black Cumin (*Nigella sativa* L.): An Overview. *Int. J. Food Sci. Technol.* **2007,** *42*, 1208–1218.

119. Ramadan, M. F.; Mörsel, J. T. Neutral Lipid Classes of Black Cumin (*Nigella sativa* L.) Seed Oils. *Eur. Food Res. Technol.* **2002,** *214*, 202–206.

120. Ramadan, M. F.; Mörsel, J. T. Oxidative stability of black cumin (*Nigella sativa* L.), coriander (Coriandrum sativum L.) and niger (Guizotia abyssinica Cass.) crude seed oils upon stripping. *European Journal of Lipid Science and Technology*, 2004, *106*, 35–43.

121. Rana, J. S.; Li, T. Y.; Manson, J. E.; Hu, F. B. Adiposity Compared with Physical Inactivity and Risk of Type 2 Diabetes in Women. *Diabetes Care* **2007,** *30*(1), 53–58.

122. Raskin, I.; Ribnicky, D. M.; Komamytsky, S.; Ilic, N.; Poulev, A.; Borisjuk, N.; Brinker, A.; Moreno, D. A.; Ripoll, C.; Yakoby, N.; O'Neal, J. M.; Cornwell, T.; Pastor, I.; Fridlender, B. Plants and Human Health in the Twenty-First Century. *Trends Biotechnol.* **2002,** *20*(12), 522–531.

123. Rastogi, L.; Feroz, S.; Pandey, B. N.; Jagtap, A.; Mishra, K. P. Protection Against Radiationinduced Oxidative Damage by an Ethanolic Extract of *Nigella sativa* L. *Int. J. Rad. Biol.* **2010,** *86*(9), 719–731.

124. Rasul Suleria, H. A.; Sadiq Butt, M.; Muhammad Anjum, F.; Saeed, F.; Batool, R.; Nisar Ahmad, A. Aqueous Garlic Extract and its Phytochemical Profile; Special Reference to Antioxidant Status. *Int. J. Food Sci. Nutr.* **2012,** *63*(4), 431–439.

125. Rates, S. M. K. Plants as Source of Drugs. *Toxicon* **2001,** *39*, 603–613.

126. Retelny, V. S.; Neuendorf, A.; Roth, J. L. Nutrition Protocols for the Prevention of Cardiovascular Disease. *Nutr. Clin. Pract.* **2008,** *23*, 468–476.

127. Rooney, S.; Ryan, M. F. Effects of Alpha-Hederin and Thymoquinone, Constituents of *Nigella sativa*, on Human Cancer Cell Lines. *Anticancer Res.* **2005,** *25*, 2199–204.

128. Routray, W.; Orsat, V. Microwave-Assisted Extraction of Flavonoids: A Review. *Food Bioprocess Technol.* **2012,** *5*, 409–24.

129. Salea, R.; Widjojokusumo, E.; Hartanti, A. W.; Veriansyah, B.; Tjandrawinata, R. R. Supercritical Fluid Carbon Dioxide Extraction of *Nigella sativa* (Black Cumin) Seeds Using Taguchi Method and Full Factorial Design. *Biochem. Compd.* **2013,** 1–7. DOI: 10.7243/2052–9341–1–1.

130. Salem, M. L. Immunomodulatory and Therapeutic Properties of the *Nigella sativa* L. Seed. *Int. J. Immunopharmacol.* **2005,** *5,* 1749–1770.

131. Salem, M. L.; Hossain, M. S. Protective Effect of Black Seed Oil from *Nigella sativa* Against Murine Cytomegalovirus Infection. *Int. J. Immunopharmacol.* **2000,** *22,* 729–740.

132. Salim, E. I.; Fukushima, S. Chemopreventive Potential of Volatile Oil from Black Cumin (*Nigella sativa* L.) Seeds Against Rat Colon Carcinogenesis. *Nutr. Cancer* **2003,** *45*(2), 195–202.

133. Seifried, H. E.; Anderson, D. E.; Fisher, E. I.; Milner, J. A. A Review of the Interaction Among Dietary Antioxidants and Reactive Oxygen Species. *J. Nutr. Biochem.* **2007,** *18,* 567–579.

134. Sen, C. K.; Khanna, S.; Roy, S. Tocotrienols: Vitamin E Beyond Tocopherols. *Life Sci.* **2006,** *78,* 2088–2098.

135. Shahidi, F.; Alasalvar, C.; Liyana-Pathirana, C. M. Antioxidant Phytochemicals in Hazelnut Kernel (*Corylus avellana* L.) and Hazelnut Byproducts. *J. Agric. Food Chem.* **2007,** *55,* 1212–1220.

136. Singh, B. B.; Vinjamury, S. P.; Der-Martirosian, C.; Kubik, E.; Mishra, L. C.; Shepard, N.; Singh, V. J.; Meier, M.; Madhu, S. G. Ayurvedic and Collateral Herbal Treatments for Hyperlipidemia: A Systematic Review of Randomized Controlled Trials and Quasi-Experimental Designs. *Altern. Ther. Health Med.* **2007,** *13,* 22–28.

137. Singh, N.; Verma, M.; Mehta, D.; Mehta, B. K. Two New Lipid Constituents of *Nigella sativa* (Seeds). *Indian J. Chem.* **2005,** *44,* 742–744.

138. Sohail, M.; Rakha, A.; Butt, M. S.; Iqbal, M. J.; Rashid, S. Rice Bran Nutraceutics: A Comprehensive Review. *Crit. Rev. Food Sci. Nutr.* **2017,** *57*(17), 3771–3780.

139. Solati, Z.; Baharin, B. S.; Bagheri, H. Supercritical Carbon Dioxide (SC–CO_2) Extraction of *Nigella sativa* L. Oil Using Full Factorial Design. *Ind. Crops Prod.* **2012,** *36*(1), 519–523.

140. Stepnowski, P.; Synak, E.; Szafranek, B.; Kaczyński, Z. *Separation Techniques* (*Techniki separacyjne*). *Wydawnictwo Uniwersytetu Gdańskiego w Gdańsku,* 2010; pp 69–95.

141. Steppan, C. M.; Bailey, S. T.; Bhat, S.; Brown, E. J.; Banerjee, R. R.; Wright, C. M.; Patel, H. R.; Ahima, R. S.; Lazar, M. A. The Hormone Resistin Links Obesity to Diabetes. *Nature* **2001,** *409*(6818), 307–312.

142. Suleria, H. A. R.; Khalid, N.; Sultan, S.; Raza, A.; Muhammad, A.; Abbas, M. Functional and Nutraceutical Bread Prepared by Using Aqueous Garlic Extract. *Internet J. Food Saf.* **2015,** *17,* 10–20.

143. Suleria, H. A. R.; Addepalli, R.; Masci, P.; Gobe, G.; Osborne, S. A. In vitro Anti-inflammatory Activities of Blacklip Abalone (*Haliotis rubra*) in RAW 264.7 Macrophages. *Food Agric. Immunol.* **2017,** 1–14.

144. Suleria, H. A. R.; Butt, M. S.; Anjum, F. M.; Saeed, F.; Khalid, N. Onion: Nature Protection Against Physiological Threats. *Crit. Rev. Food Sci. Nutr.* **2015**, *55*(1), 50–66.

145. Suleria, H. A. R.; Butt, M. S.; Khalid, N.; Sultan, S.; Raza, A.; Aleem, M.; Abbas, M. Garlic (*Allium sativum*): Diet Based Therapy of 21st Century—A Review. *Asian Pacific J. Trop. Dis.* **2015**, *5*(4), 271–278.

146. Suleria, H.; Butt, M.; Anjum, F.; Ashraf, M.; Qayyum, M.; Khalid, N.; Younis, M. Aqueous Garlic Extract Attenuates Hypercholesterolemic and Hyperglycemic Perspectives; Rabbit Experimental Modeling. *J. Med. Plants Res.* **2013**, *7*(23), 1709–1717.

147. Sultan, M. T.; Butt, M. S.; Ahmad, R. S.; Batool, R.; Naz, A.; Rasul Suleria, H. A. Supplementation of Powdered Black Cumin (*Nigella sativa*) Seeds Reduces the Risk of Hypercholesterolemia. *Funct. Foods Health Dis.* **2011**, *12*, 516–524.

148. Sultan, M. T.; Butt, M. S.; Karim, R.; Ahmad, A. N.; Suleria, H. A. R.; Saddique, M. S. Toxicological and Safety Evaluation of *Nigella sativa* Lipid and Volatile Fractions in Streptozotocin Induced Diabetes Mellitus. *Asian Pacific J. Trop. Dis.* **2014**, *4*, S693–S697.

149. Swamy, S. M.; Huat, B. T. Intracellular Glutathione Depletion and Reactive Oxygen Species Generation are Important in Alpha-Hederin-Induced Apoptosis of P388 Cells. *Mol. Cell. Biochem.* **2003**, *245*(1–2), 127–39.

150. Taka, E.; Mazzio, E. A.; Goodman, C. B.; Redmon, N.; Flores-Rozas, H.; Reams, R.; Darling-Reed, S.; Soliman, K. F. A. Anti-inflammatory Effects of Thymoquinone in Activated BV-2 Microglial Cells. *J. Neuroimmunol.* **2015**, *286*, 5–12.

151. Tapsell, L. C.; Hemphill, I.; Cobiac, L.; Patch, C. S.; Sullivan, D. R.; Fenech, M.; Roodenrys, S.; Keogh, J. B.; Clifton, P. M.; Williams, P. G.; Fazio, V. A.; Inge, K. E. Health Benefits of Herbs and Spices: The Past, the Present, the Future. *Med. J. Austr.* **2006**, *185*(4), 4–24.

152. Tubesha, Z.; Iqbal, S.; Ismail, M. Effects of Hydrolytic Conditions on Recovery of Antioxidants from Methanolic Extracts of *Nigella sativa* Seeds. *J. Med. Plants Res.* **2011**, *5*(14), 3152–3158.

153. Tuck, K. L.; Hayball, P. J. Major Phenolic Compounds in Olive Oil: Metabolism and Health Effects. *J. Nutr. Biochem.* **2002**, *13*, 636–644.

154. Türkdoğan, M. K.; Ozbek, H.; Yener, Z.; Tuncer, I.; Uygan, I.; Ceylan, E. The Role of Antioxidant Vitamins C and E, Selenium and *Nigella sativa* in the Prevention of Liver Fibrosis and Cirrhosis in Rabbits: New Hopes. *DTW-Dtsch Tierarztl Wochenschr* **2001**, *108*, 71–73.

155. Ugochukwu, N. H.; Figgers, C. L. Dietary Caloric Restriction Improves the Redox Status at the Onset of Diabetes in Hepatocytes of Streptozotocin-Induced Diabetic Rats. *Chemico-Biol. Interact.* **2007**, *165*, 45–53.

156. Vats, V.; Yadav, S. P.; Grover, J. K. Effect of *T. foenumgraecum* on Glycogen Content of Tissues and the Key Enzymes of Carbohydrate Metabolism. *J. Ethnopharmacol.* **2003**, *85*, 237–242.

157. Vina, J.; Borras, C.; Gomez-Cabrera, M. C.; Orr, W. C. Part of the Series: From Dietary Antioxidants to Regulators in Cellular Signalling and Gene Expression. Role of Reactive Oxygen Species and (Phyto) Oestrogens in the Modulation of Adaptive Response to Stress. *Free Rad. Res.* **2006**, *40*, 111–119.

158. Vinson, J. A.; Teufel, K.; Wu, N. Red Wine, Dealcoholized Red Wine, and Especially Grape Juice, Inhibit Atherosclerosis in a Hamster Model. *Atherosclerosis* **2001,** *156,* 67–72.

159. Wahle, K. W.; Heys, S. D. Atherosclerosis: Cell Biology and Lipoproteins. *Curr. Opin. Lipidol.* **2008,** *19,* 435–437.

160. Wajs, A.; Bonikowski, R.; Kalemba, D. Composition of Essential Oil from Seeds of *Nigella sativa* L. Cultivated in Poland. *Flavor Fragrance J.* **2008,** *23,* 126–132.

161. Wang, L.; Weller, C. L. Recent Advances in Extraction of Nutraceuticals from Plants. *Trends Food Sci. Technol.* **2006,** *17,* 300–312.

162. Wang, Y.; You, J.; Yu, Y.; Qu, C.; Zhang, H.; Ding, L.; Zhang, H.; Li, X. Analysis of Ginsenosides in Panax ginseng in High Pressure Microwave-Assisted Extraction. *Food Chem.* **2008,** *110,* 161–167.

163. Wild, S.; Roglic, G.; Green, A.; Sicree, R.; King, H. Global Prevalence of Diabetes: Estimates for the Year 2000 and Projections for 2030. *Diabetes Care* **2004,** *27*(5), 1047–1053.

164. Wolf, S. P. Diabetes Mellitus and Free Radicals. *Br. Med. Bull.* **1993,** *49,* 643–649.

165. Wong, S. P.; Leong, L. P.; William-Koh, J. H. Antioxidant Activities of Aqueous Extracts of Selected Plants. *Food Chem.* **2006,** *99,* 775–783.

166. Wong, S.; Baltuch, G. H.; Jaggi, J. L.; Danish, S. F. Functional Localization and Visualization of the Subthalamic Nucleus from Microelectrode Recordings Acquired during DBS Surgery with Unsupervised Machine Learning. *J. Neural Eng.* **2009,** *6,* 2.

167. Yang, C. S.; Landau, J. M.; Huang, M. T.; Newmark, H. L. Inhibition of Carcinogenesis by Dietary Polyphenolic Compounds. *Ann. Rev. Nutr.* **2001,** *21,* 381–406.

168. Yousefi, M.; Barikbin, B.; Kamalinejad, M.; Abolhasani, E.; Ebadi, A.; Younespour, S.; Manouchehrian, M.; Hejazi, S. Comparison of Therapeutic Effect of Topical Nigella with Betamethasone and Eucerin in Hand Eczema. *J. Eur. Acad. Dermatol. Venereol.* **2013,** *27*(12), 1498–1504.

169. Zakir, S.; Allen, J. C.; Sarwar, M.; Nisa, M. U.; Chaudhry, S. A.; Arshad, U.; Javaid, A.; Slam-ud-Din, I. Sweet Potato Glycemic Index in Relation to Serum Glucose Level in Human Participants. *Int. J. Agric. Biol.* **2008,** *10,* 311–315.

170. Zaoui, A.; Cherrah, Y.; Alaoui, K.; Mahassine, N.; Amarouch, H.; Hassar, M. Effects of *Nigella sativa* Fixed Oil on Blood Homeostasis in Rat. *J. Ethnopharmacol.* **2002,** *79,* 23–26.

171. Zaoui, A.; Cherrah, Y.; Lacaille-Dubois, M. A.; Settaf, A.; Amarouch, H.; Hassar, M. Diuretic and Hypotensive Effects of *Nigella sativa* in the Spontaneously Hypertensive Rat. *Therapie* **2000,** *55,* 379–382.

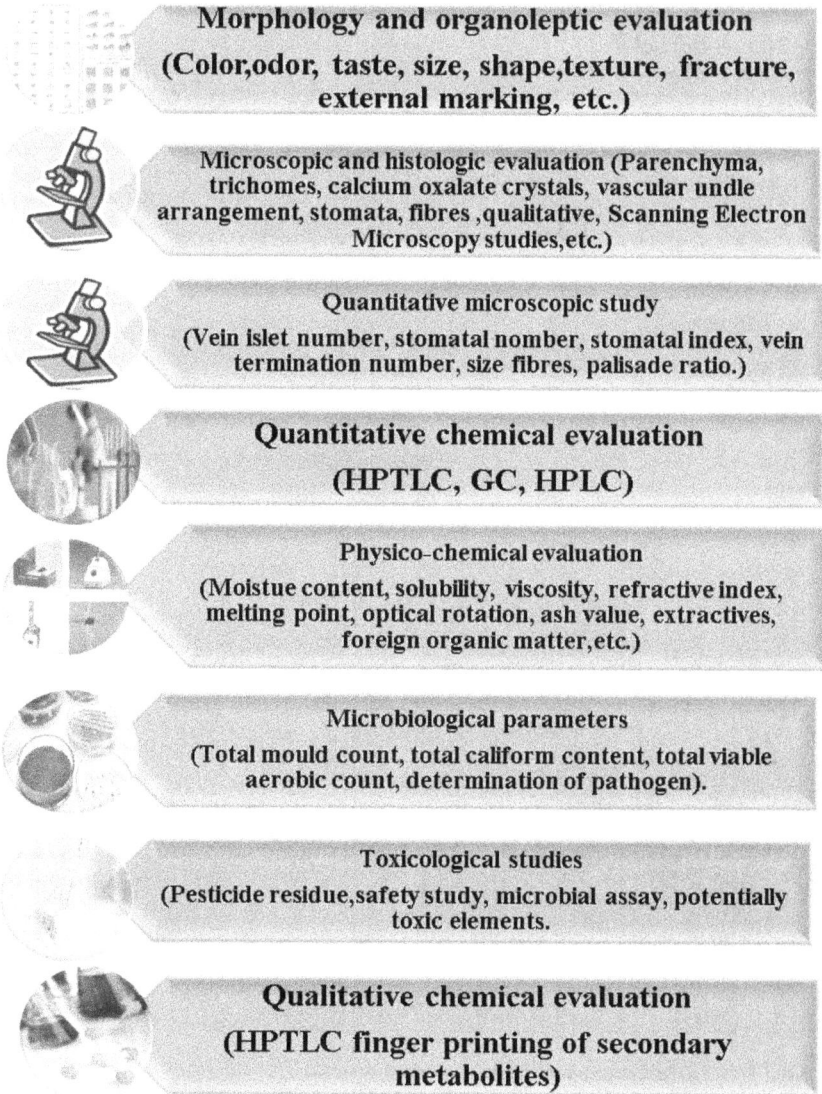

Morphology and organoleptic evaluation
(Color,odor, taste, size, shape,texture, fracture, external marking, etc.)

Microscopic and histologic evaluation (Parenchyma, trichomes, calcium oxalate crystals, vascular undle arrangement, stomata, fibres ,qualitative, Scanning Electron Microscopy studies,etc.)

Quantitative microscopic study
(Vein islet number, stomatal nomber, stomatal index, vein termination number, size fibres, palisade ratio.)

Quantitative chemical evaluation
(HPTLC, GC, HPLC)

Physico-chemical evaluation
(Moistue content, solubility, viscosity, refractive index, melting point, optical rotation, ash value, extractives, foreign organic matter,etc.)

Microbiological parameters
(Total mould count, total caliform content, total viable aerobic count, determination of pathogen).

Toxicological studies
(Pesticide residue,safety study, microbial assay, potentially toxic elements.

Qualitative chemical evaluation
(HPTLC finger printing of secondary metabolites)

FIGURE 1.2   Parameters for standardization and quality evaluation of herbal drugs.

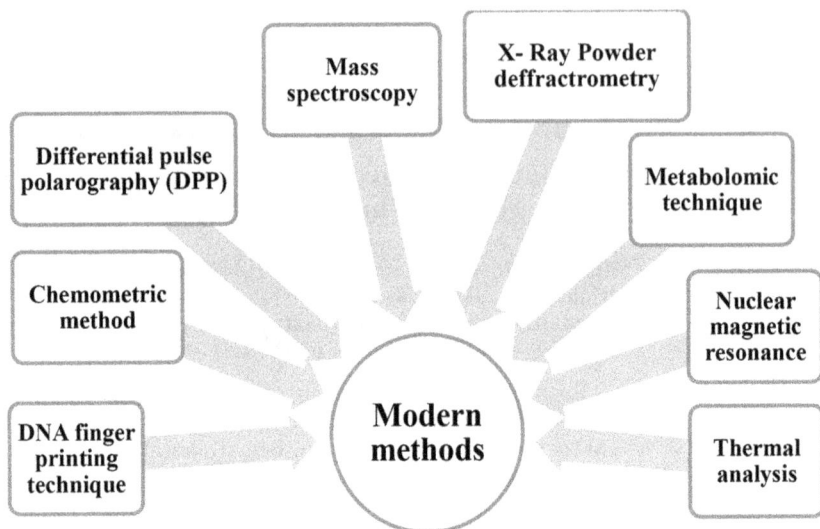

FIGURE 1.4 Modern methods for standardization of herbal drugs.

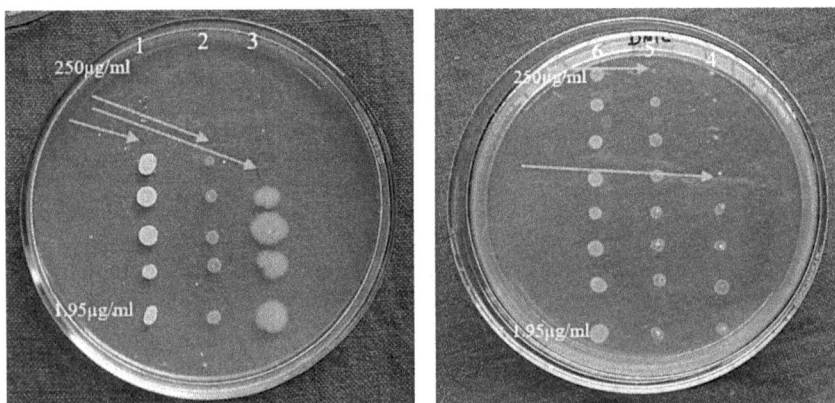

FIGURE 2.3 Minimum inhibitory concentration (MIC) of flindersine against tested bacteria: 1-*Staphylococcus aureus* (62.5 µg/mL); 2-*Staphylococcus epidermidis* (62.5 µg/mL); 3-*Bacillus subtilis* (31.2 µg/mL); 4-*Enterococcus faecalis* (31.2 µg/mL); 5-*Pseudomonas aeruginosa* (250 µg/mL); and 6-*Escherichia coli* (>250 µg/mL).

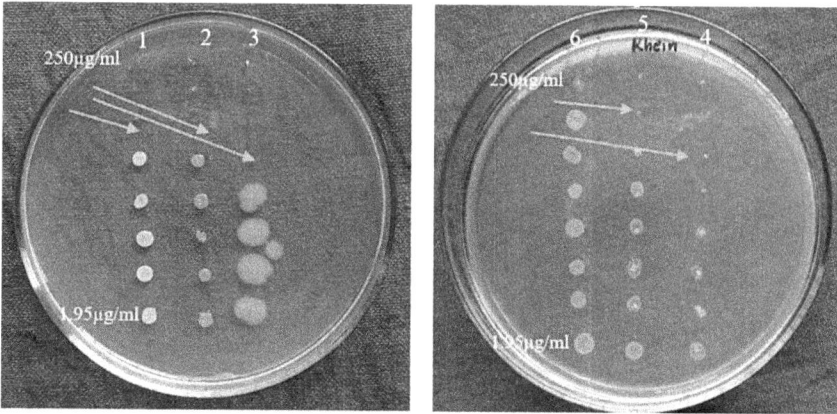

FIGURE 2.4 MIC of rhein against tested bacteria: 1-*Staphylococcus aureus* (62.5 µg/mL); 2-*Staphylococcus epidermidis* (62.5 µg/mL); 3-*Bacillus subtilis* (31.2 µg/mL); 4-*Enterococcus faecalis* (62.5 µg/mL); 5-*Pseudomonas aeruginosa* (125 µg/mL); 6-*Escherichia coli* (>250 µg/mL).

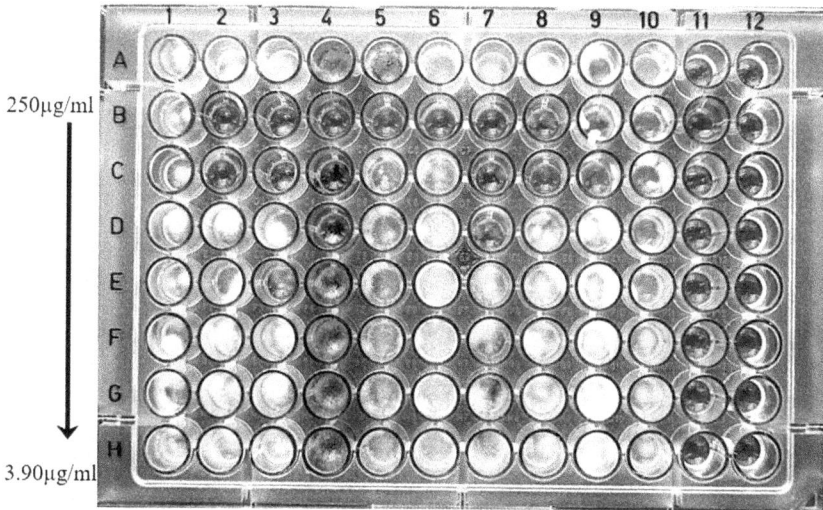

FIGURE 2.5 Antifungal activity of Flindersine: B1 to H1-*Trichophyton mentagrophytes* (>250 µg/mL); B2 to H2-*Epidermophyton flocossum* (125 µg/mL); B3 to H3-*Trichophyton simii* 125 µg/mL); B4 to H4-*Curvularia lunata* (250 µg/mL); B5 to H5-*Aspergillus niger* (>250 µg/mL); B6 to H6-*Botrytis cinerea* (>250 µg/mL); B7 to H7-*Trichophyton rubrum* (296) (62.5 µg/mL); B8 to H8-*Magnoporthe grisea* (125 µg/mL); B9 to H9-*T. rubrum* (57) (>250 µg/mL); B10 toH10-*Scopulariopsis* sp (>250 µg/mL); A1-A10: Control (culture and broth only); A11to H11 and A12 to H12—Blank.

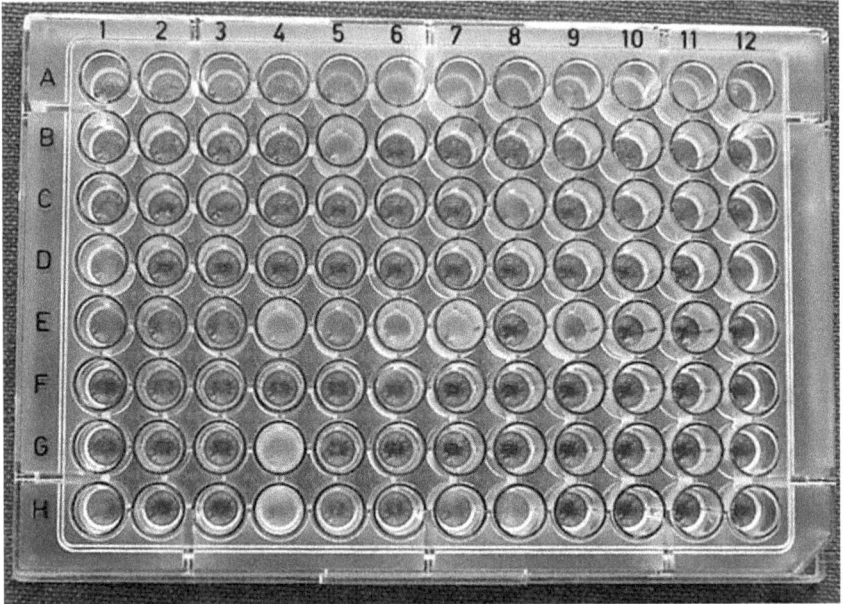

FIGURE 2.6 MIC of rhein as determined using the micro-dilution method against bacteria.

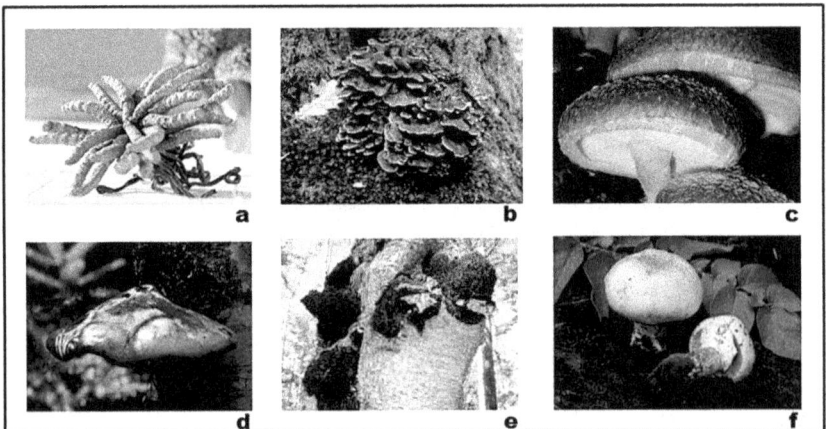

FIGURE 7.1 Some of the common medicinal mushrooms of traditional Chinese medicine (TCM) such as *Ophiocordyceps* (previously *Cordyceps*) *sinensis* (a); *Grifola frondosa* (b); *Lentinula edodes* (c); *Piptoporus betulinus* (d); *Inonotus obliquus* (e); and *Agaricus subrufescens* (f).

FIGURE 7.2 Various paintings of *Ganoderma lucidum* in Chinese literature depicting its wide therapeutic use in traditional Chinese medicine (TCM).

FIGURE 7.3 Morphological forms of *G. lucidum*: young sporocarp (a); antler-shaped fruiting body (b); matured fruiting body (c); and fruiting body in sawdust culture (d).

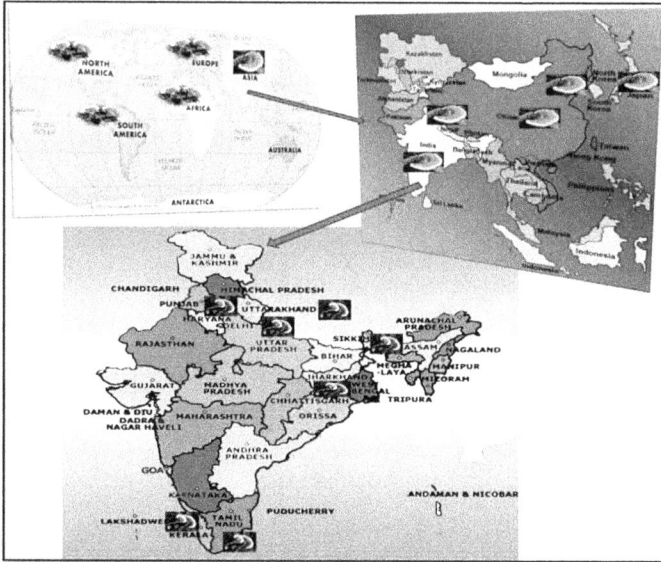

FIGURE 7.4   Geographical distribution of *G. lucidum* throughout the world.

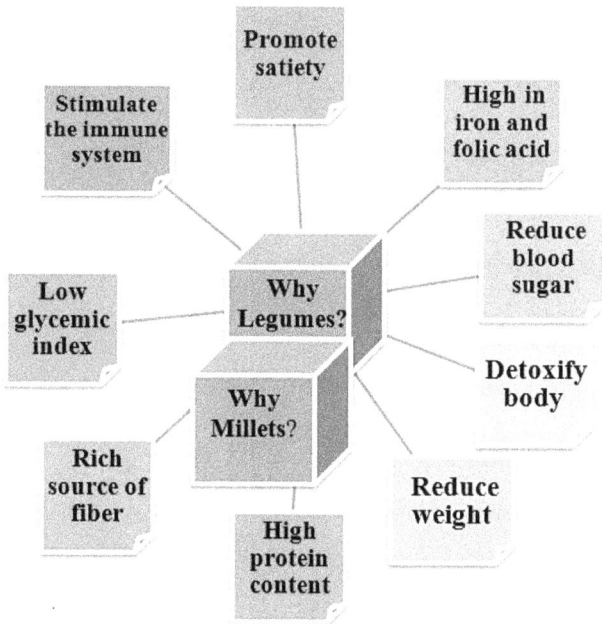

FIGURE 8.1   Health-benefiting properties of legumes and millets.

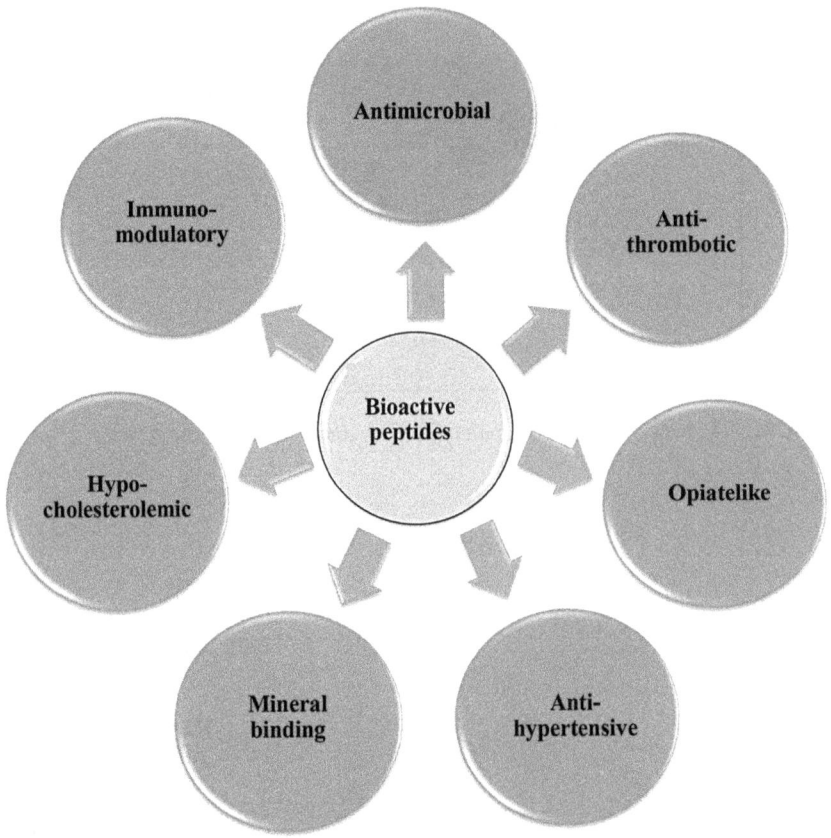

FIGURE 8.3   Biofunctional activity of bioactive peptides.

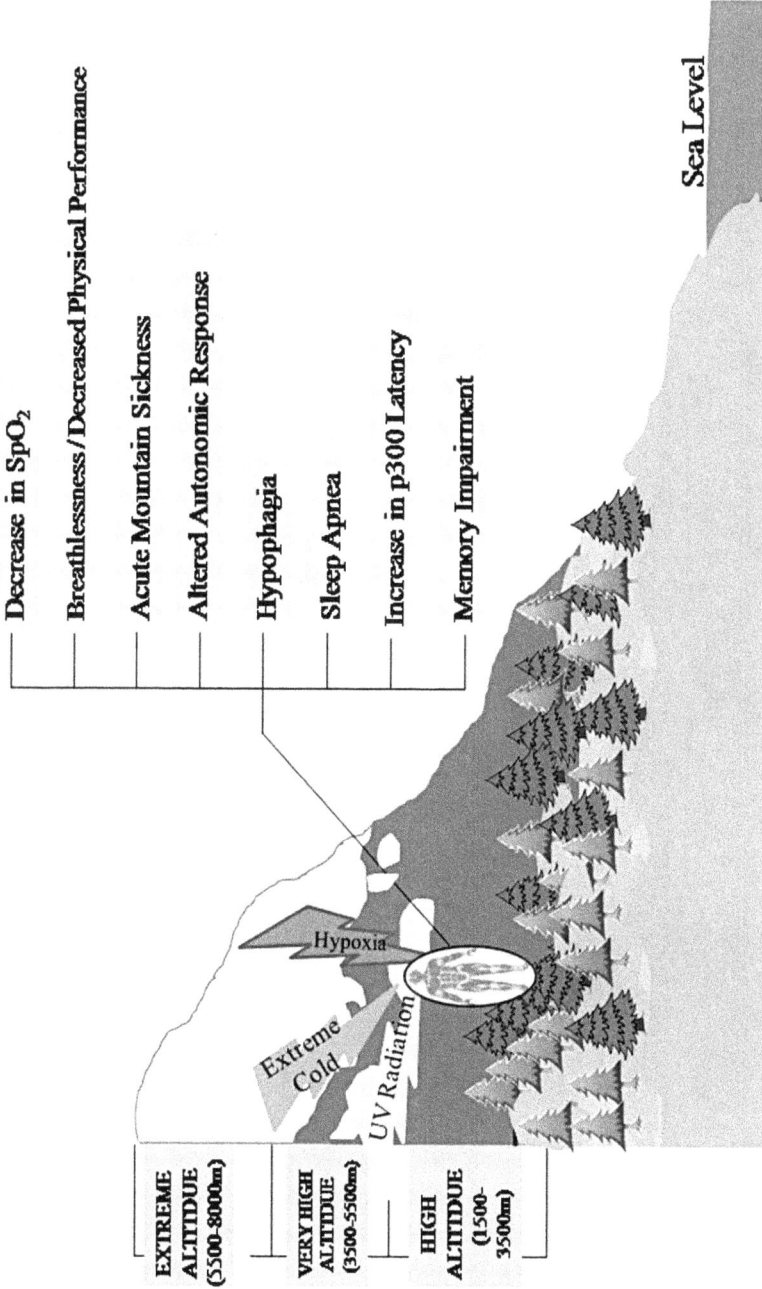

FIGURE 9.1 Adverse effects of high altitude on human health.

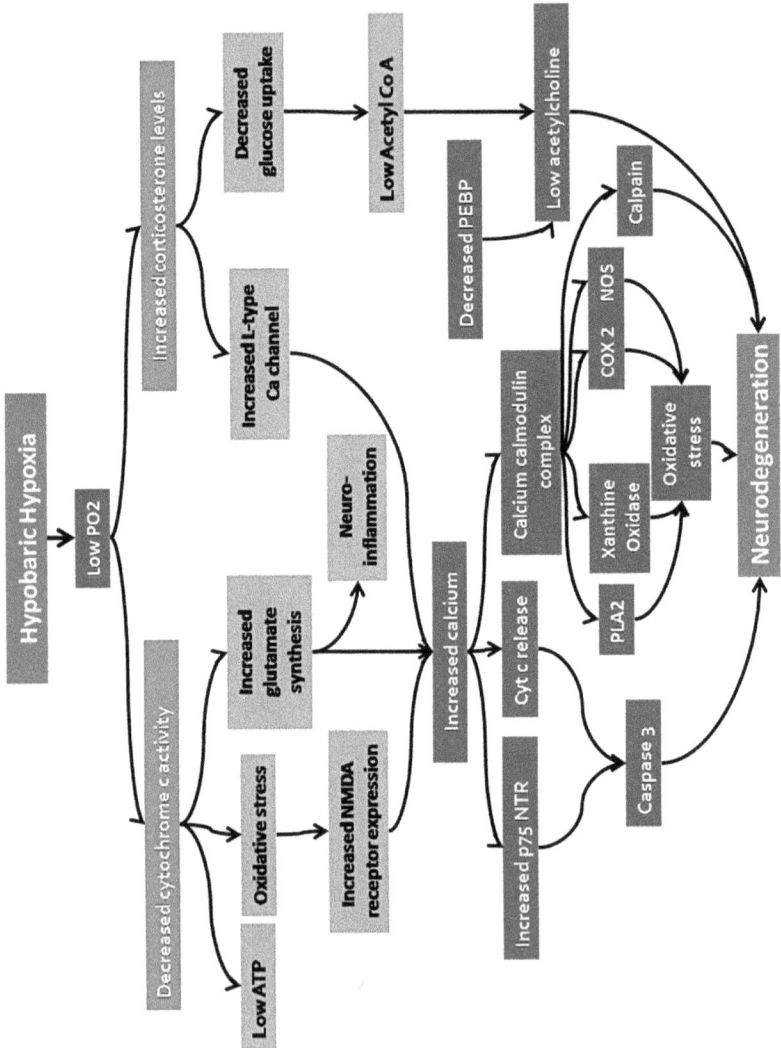

FIGURE 9.2 Molecular mechanisms leading to neurodegeneration in hypobaric hypoxia.

 Ginger (*Zingiber officinale*)	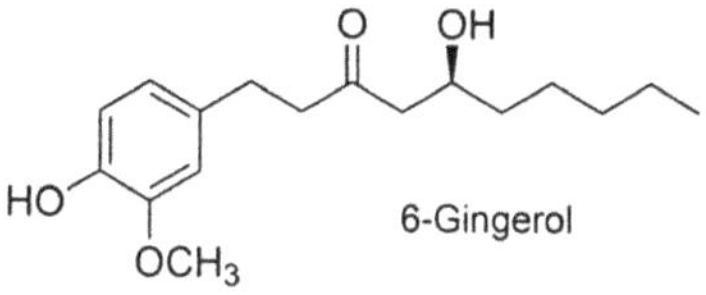 6- Gingerol
 Garlic (*Allium sativum L.*),	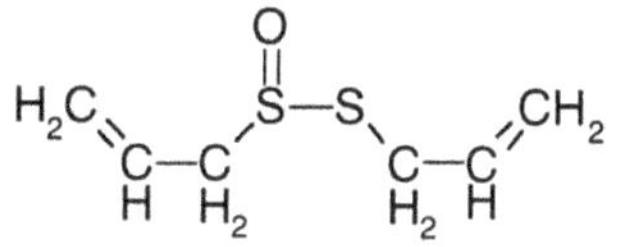 *Allicin*
 Caraway *(Carum carvi)*	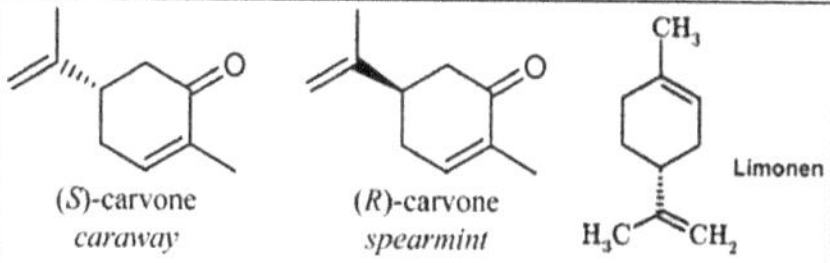 Carvone: (S) and (R); Limonen (right)
 Cumin	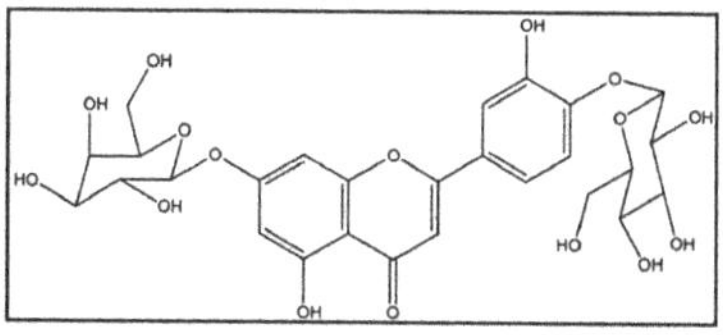 Luteolin: 7-O-beta-D- galactouronide-4'-O-beta-D-glucopyranoside ($C_{21}H_{20}O_{11}$)

FIGURE 10.1   Different types of bioenhancers.

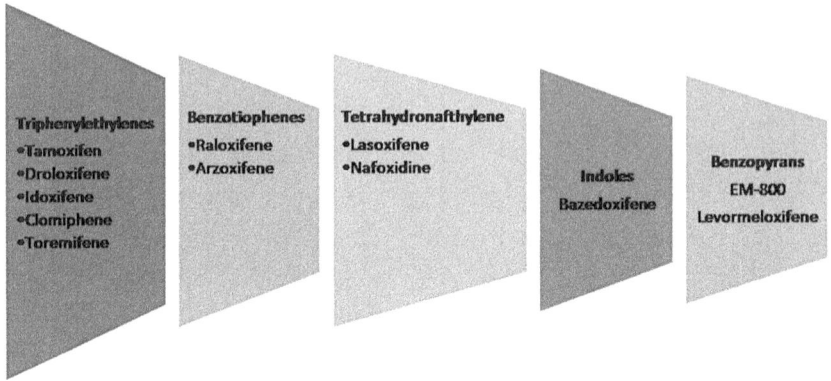

FIGURE 11.1   Chemical classification of selective estrogen receptor modulators (SERMs).

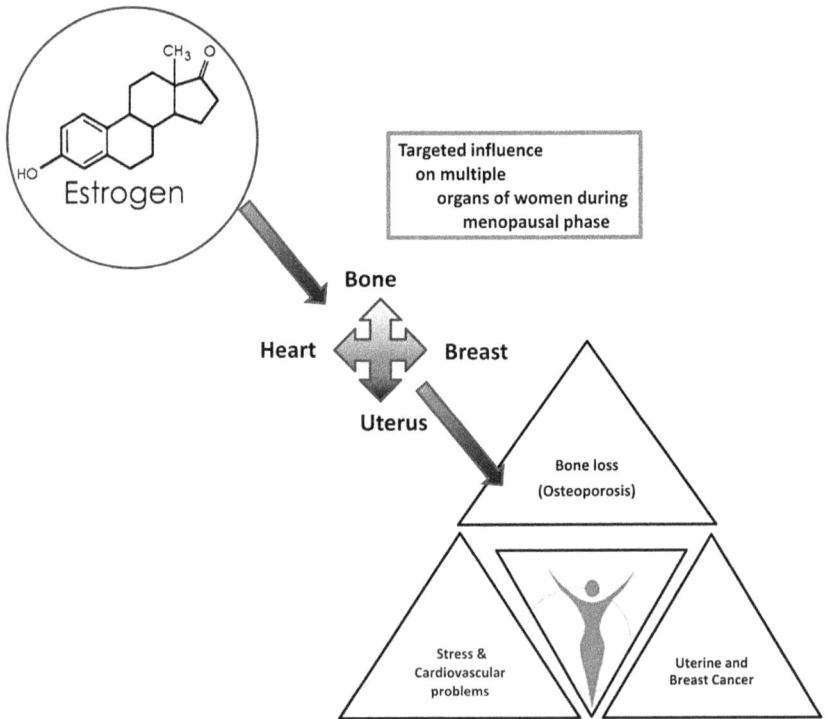

FIGURE 11.2   Discrepancy of estrogen causing menaces among women in early stages of menopause.

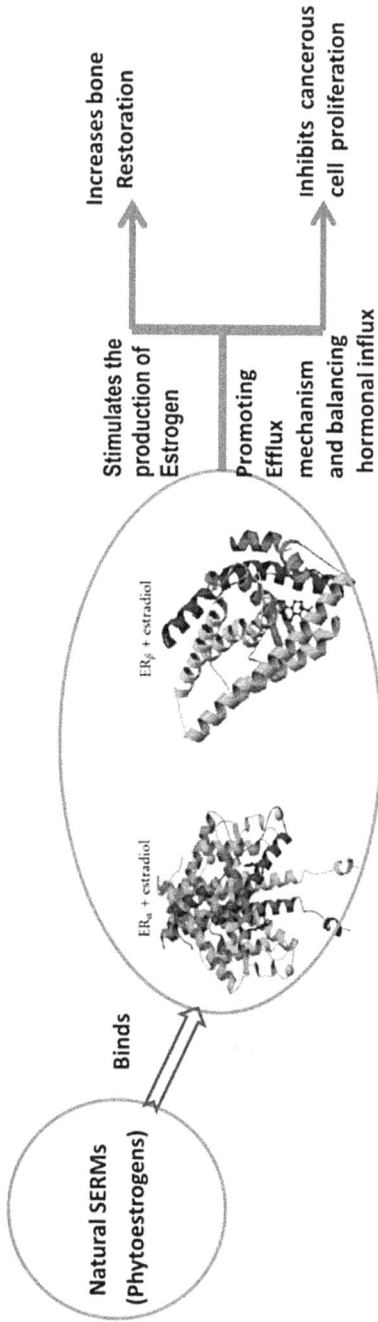

FIGURE 11.3 An outline representing the mechanism of phytoestrogens as natural SERMs that compete with estrogen receptors.

FIGURE 11.6 Flax plant: flower and seed.

FIGURE 11.7   Veldt grape plant.

FIGURE 11.8   Broccoli.

PART III
Therapeutic Attributes of Mushrooms, Cereal Grains, and Legumes

CHAPTER 7

THERAPEUTIC MEDICINAL MUSHROOM (*GANODERMA LUCIDUM*): A REVIEW OF BIOACTIVE COMPOUNDS AND THEIR APPLICATIONS

A. BHARDWAJ and K. MISRA

CONTENTS

ABSTRACT

In view of continuing the legacy of traditional medicine in human health, this chapter presents a review on the various therapeutic uses of the medicinal mushroom, *Ganoderma lucidum* along with the major bioactive constituents that contribute towards its medicinal effects. *G. lucidum* possesses a wide array of pharmaceutical properties such as: anticarcinogenic, antimicrobial (antibacterial and antifungal), antioxidant, and antitumorigenic properties, which could be attributed to the presence of fatty acids, nucleotides, peptides, polysaccharides, proteins, steroids, sterols, terpenoids, and so forth.

7.1 INTRODUCTION

Herbs from the earliest times have provided a plethora of remedies as prophylaxis for the treatment of various diseases. They constitute an important element of herbal medicine.[139] Herbal medicines often employ plants or herbs-derived active ingredients, which are used as a whole or in part to produce pharmaceutical or nutraceutical products. In spite of the enormous advances in modern medicine, they continue to contribute significantly to the health care industry. They have played a pivotal role in disease prevention and treatment not only in developing countries like India (Ayurveda) and China (Traditional Chinese Medicine (TCM)) that have a long traditional use of herbal medicines, but also these are becoming popular in Western nations like Europe (Germany, Italy, and France) and United States.[13]

In recent times, herbal medicines as a part of TCM are being pursued meticulously by the researchers in academia and industry and manufactured globally. About 80% of TCM derived from plants or fungi (mushrooms) are used in accordance with methodical and self-sufficient theories practiced to ameliorate several human health conditions. TCM include thousands of years old accumulated human trials with well-documented herbal formulations that are archived in Chinese medical literature from the earliest known '*Shen Nong Ben Tsao Jing*' to the more contemporary Chinese "*Materia Medica*." In these texts, the use of medicinal mushrooms in traditional therapies for treating various diseases is most noteworthy.[76]

Medicinal mushrooms are macroscopic fungi, mostly Basidiomycetes and Ascomycetes, with therapeutic properties that are used in the form of extracts or powder for prevention, alleviation, or treatment of diseases and/ or for their nutritional values.[82] Some of these include *Agaricus subrufescens* (Almond mushroom), *Ganoderma lucidum* (Lingzhi), *Grifola frondosa* (Hen of the woods/Maitake), *Inonotus obliquus* (Chaga), *Lentinula edodes* (Shiitake), *Ophiocordyceps* (previously Cordyceps) species and *Piptoporus betulinus* (Birch polypore/bracket), among numerous others (Fig. 7.1).

FIGURE 7.1 (See color insert.) Some of the common medicinal mushrooms of traditional Chinese medicine (TCM) such as *Ophiocordyceps* (previously *Cordyceps*) *sinensis* (a); *Grifola frondosa* (b); *Lentinula edodes* (c); *Piptoporus betulinus* (d); *Inonotus obliquus* (e); and *Agaricus subrufescens* (f).

There are various commercial preparations that are being marketed as tablets, elixirs, or capsules extensively in Oriental countries and increasingly in the USA and Europe as nutraceuticals from these medicinal mushrooms.[9,140,167] However, reading thoroughly the TCM history, *G. lucidum* is probably the most widely consumed and highly revered medicinal mushroom with deep-rooted medicinal significance.[9,103,159,168] Currently, several formulations made from the morphological forms like mycelium, fruiting bodies and spores of *G. lucidum* are available commercially as dietary supplements and functional foods due to its numerous pharmaceutical and nutraceutical properties.[9,159,165] Approximately, 5000 t of GL are produced a worldwide annually worth 2.5 billion US$.[9,77,200]

In view of continuing the legacy of traditional medicine in human health, the current chapter presents the review on the various therapeutic uses of the medicinal mushroom, *G. lucidum* along with the major bioactive constituents that contribute towards its medicinal effects.

7.2 HISTORY AND TRADITIONAL USAGE

Held high in esteem, *G. lucidum* has been regarded as a renowned mushroom that provides a historical record of more than two millennia.[9,15,97,166] It is usually called as "*Lingzhi*" in China, "*Youngzhi*" in Korea, "*Reishi*" in Japan, and simply "*Ganoderma*" in USA.[26] However, according to some researchers, the species *G. lucidum* has been used with reference to Chinese Lingzhi, globally.[9,49,66] In earlier times, this rare and thus highly prized fungus was reserved only for the royal class and was referred as the "Elixir of Life".[66,85]*G. lucidum* has also been commonly referred to as the "mushroom of immortality," "ten-thousand-year mushroom," "mushroom of spiritual potency," and "the spirit plant".[9,156]

This fungus has been included as a medicinal mushroom in the ancient Chinese text of "*Shen Nong Ben Cao Jing*" also known as "*Classics of the Materia Medica.*" This Chinese text is solely devoted to the description of medicinal herbs and their benefits. Besides, it has also been cited in the *Supplement to Classic of Materia Medica* (502–536 AD) and the *Ben Cao Gang Mu* (the first pharmacopoeia in China, 1590 AD) among the superior tonics (*shang pin*).[9,159] Several researchers also advocate that this mushroom was only recently included in the *Pharmacopoeia of the People's Republic of China* (2000), though it is a widely used by Asian physicians and herbalists. According to these texts, it acts to replenish *qi*, ease the mind, and relieves cough and asthma, and is recommended for dizziness, insomnia, palpitation, and shortness of breath.[9,66,85,159]Figure 7.2 illustrates a wide therapeutic use of *G. lucidum* in TCM as a depicted from various paintings of Chinese literature.

7.3 ORIGIN, DISTRIBUTION, AND CLASSIFICATION

G. lucidum (P. Curtis, 1881) was first recognized by William Curtis as the only type species under the genus *Ganoderma* (established by Karsten in 1881) based on the material collected from hazel (*Corylus avellana*)

FIGURE 7.2 (See color insert.) Various paintings of *Ganoderma lucidum* in Chinese literature depicting its wide therapeutic use in traditional Chinese medicine (TCM).

in Peckham Common, London, UK.[45,49,112] Formerly, it was known as *Polyporus lucidus* (Curtis) Fr. 1821 with the basionym, *Boletus lucidus* Curtis 1781.[45,112] Presently, it is included in the family *Ganoderma taceae* and the order *Polyporales*.[45,97]

Ganoderma species as a suggested by several researchers are cosmopolitan lignin-cellulosic wood-decaying fungi that occur on conifers and hardwoods.[51,112] The various morphological forms of *G.lucidum* are shown in Figure 7.3.

FIGURE 7.3 (See color insert.) Morphological forms of *G. lucidum*: young sporocarp (a); antler-shaped fruiting body (b); matured fruiting body (c); and fruiting body in sawdust culture (d).

The species *G. lucidum* grows as parasite or saprotroph in tropical, subtropical, and temperate geographical regions including North and South America, Europe, Africa, and Asia.[124] In Asia, it is found in China, Korea, India, Nepal, and Japan, where it grows primarily on plum trees. It is also found on stumps, generally, near the soil surface and occasionally on soils arising from buried roots.[120] In India, it is found in both Southern (e.g., Kerala and Tamil Nadu) and Northern (e.g., Gharwal Himalaya, Uttarakhand, and Assam) regions.[27] The geographical distribution of *G. lucidum* throughout the world has been given in Figure 7.4.

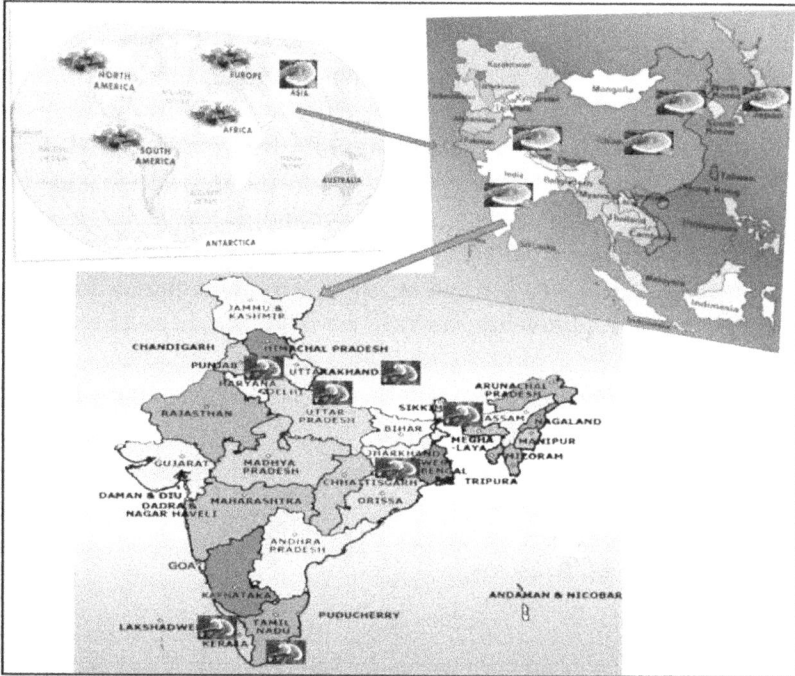

FIGURE 7.4   (See color insert.) Geographical distribution of *G. lucidum* throughout the world.

Ganoderma is typified by the species of *G. lucidum* (Curtis) and is referred as "*G. lucidum* complex".[112] However, different authors have considered various criterions in its identification and nomenclature.[45,194] Besides, out-crossing over generations and different geographical origins have led to great genetic heterogeneity among the *Ganoderma* species and consequences are varied morphological traits even within the same species with an uncertain nomenclature.[45] During the taxonomic studies, some authors have considered mainly host–specificity, geographical distribution, and morphology of basidiomes, while others only the chlamydospore characters;[29,45,148] whereas few others considered the chemotaxonomic traits[9,149] or described a phylogeny with low resolution to certain clades.[95,179,194] Recently, Richter et al. (2015) suggested that a more stable taxonomy for the genus *Ganoderma* could be established by integrating morphological, chemotaxonomic, and molecular approaches.[112] Therefore, throughout the literature, numerous morphological characteristics and taxonomic studies are available to mostly differentiate the European *G. lucidum* and the *Lingzhi* fungus.[16,49,161,163,180,194] However, it is still unclear which name has priority and should be assigned to the *Lingzhi* type specimen.[49]

In order to assist with the clarification of the species, Hawksworth had proposed that a new name be assigned to the European "*G. lucidum*" and that the well-known Chinese *Lingzhi* should retain the name *G. lucidum*; while Cao et al.[16] proposed that the original Eurasian *G. lucidum* maintain its name whereas *Lingzhi*, predominantly from East Asia, be renamed *Ganoderma lingzhi* and referred to as *Lingzhi*. The latter appears to have been partially accepted.[9,32,149,179] Thus, we may conclude that *Ganoderma* species show vast species diversity being cosmopolitan in distribution.

7.4 MAJOR BIOACTIVE COMPOUNDS IN *GANODERMA LUCIDUM*

In today's research, isolation, characterization, and bio-efficacy studies of bioactive principles have attained significant importance.[25,133] The literature provides numerous evidence on *G. lucidum* supporting its richness in various bioactive principles including amino acids, nucleotides and nucleosides, organic germanium, peptidoglycans, phenols, polysaccharides, steroids, triterpenoids, vitamins, and so forth, each possessing remarkable medicinal effects.[9,26,64,159] However, among these, the polysaccharides and the triterpenoids are considered as the major physiologically active constituents.[12,159,195] Some of the major bioactive principles (with examples) in *G. lucidum* are listed in Table 7.1.

TABLE 7.1   The Major Bioactive Principle Compounds Found in *Ganoderma lucidum*.

Chemical constituents	Examples	Reference
Nucleobases, nucleosides, and nucleotides	NUCLEOBASES: Adenine, cytosine, 2'-deoxyuridine, 2'-deoxyadenosine, guanine, hypoxanthine, thymine, uracil, xanthine, etc. NUCLEOSIDES: Adenosine, cytidine, guanosine, inosine, thymidine, uridine, etc. Nucleotides: 5'-adenosine monophosphate, 5'-cytosine monophosphate, 5'-guanosine monophosphate, 5'-inosine monophosphate; 5'-uridine monophosphate; 5'-xanthosine monophosphate, etc.	[22, 53, 67, 72, 91, 97, 105, 120, 133, 166, 170]
Polyphenolic compounds	Catechin, chlorogenic acid, cinnamic acid, coumarin, ellagic acid, ferulic acid, gallic acid, genistein, kaempferol, morin, myricetin, p-coumaric acids, p-hydroxybenzoic, quercetin, rutin, syringic acid, tannic acid, ursolic acid, vanillin, etc.	[25, 30, 33, 47, 63, 72, 84, 111, 118, 119, 125, 181]
Poly-saccharides	β-1–3-D glucans; β-1–3-D glucans; proteoglycan (GLPG, GLIS, PGY, GLPP), etc.	[12, 62, 120, 134, 154, 158, 166]
Proteins, peptides, and amino acids	Ling Zhi-8; Ganodermin; GLP; antihypertensive related proteins (cystathionine α- synthase-like protein, DEAD/DEAH box helicase-like protein, paxillin-like protein and α/β hydrolase-like protein); LZP-1, LZP-2, LZP-3; Se-GL, Lectins (GLL-F, GLL-M); peptidoglycans (GLIS); peptidoglycans, etc.	[12, 64, 66, 91, 97, 102, 103, 133, 154, 159]

TABLE 7.1 (Continued)

Chemical constituents	Examples	Reference
	ENZYMES:	
	α-galactocidase; Ribonucleases; Laccases; Proteinase A inhibhitor; Lignolytic enzymes, etc.	
	Amino acids:	
	L-alanine, glycine, L- isoleucine, L-leucine, L-methionine, L-phenylalanine, L-proline, L-serine, L-threonine, L-tryptophan, L-tyrosine, valine, and so forth.	
Triterpenes	Ganoderic acids (GA-A, AM1, B, C1, C2, D, DM, F, G, H, K, Me, MK, S, T, TR, Y); GAs (A, B, D); Ganoderols (A, B); Ganoderiol F; Ganoderal A; Ganodermanontriol; Me Ganoderates (D, G); Lucideric acids, and so forth.	[23, 58, 64, 66, 133, 159, 174, 195]
Vitaminsandminerals	Vitamin C, vitamin E, β-carotene, calcium, copper, germanium, iron, magnesium, manganese, phosphorous, potassium, selenium, silica, sodium, sulfur, zinc, and so forth.	[53, 120, 133, 159]
Volatile organic compounds(VOCs)	Choline; betaine; tetracosanoic acid; stearic acid; palmitic acid; ergosta-7, 22- dien-3-ol, nonadecanoic acid; behenic acid; tetracosane; hentriacontane; ergosterol; β-sitosterol; pyrophosphatidic acid; sterols [oleic acid, cyclooctasulfur, ergosterol peroxide (5,8-epidioxy-ergosta-6,22E-dien-3-ol]; cerebrosides [(4E',8E)-N-D-2'-hydroxystearoyl-1-O- D-glucopyranosyl-9-methyl-4–8-sphingadienine, (4E,8E)- N-D-2'-hydroxypamitoyl-1-O–D-glucopyranosyl-9- methyl-4–8-sphingadienine]; trans-anethol, R-()-linalool, S-(+)-carvone, α-bisabolol; 1-octen-3-ol, ethanol, hexanal, 1-hexanol; sesquirosefuran; 3-octanol; 3-octanone; aldehydesphenols; 3-Methyl; 2-Butanamine, 2-propanamine; 3-Methyl butanal, and so forth.	[23, 35, 90, 96, 103, 120, 133, 144, 151, 199]

7.4.1 NUCLEOBASES, NUCLEOSIDES, AND NUCLEOTIDES

The nucleosides are naturally occurring compounds, which are involved in various biological processes consisting of a sugar and a heterocyclic nucleobase (purine or pyrimidine). The nucleotides occur as monomeric building blocks of polymeric nucleic acids (DNA and RNA) and contain an additional phospho-ester group in comparison to nucleosides.[116] The nucleosides and nucleotides play a pivotal role in the maintenance of functions of the bone marrow hematopoietic cells, intestinal mucosa, and the brain, which have a limited *de novo* synthesis of purine and pyrimidine bases.[183]

Several studies support the presence of these essential bioactive principles in *G. lucidum*. For example, Kasahara et al.[65] performed bioguided fractionation to isolate adenosine from *G. lucidum* by monitoring analgesic activity. The isolated adenosine reduced spontaneous motor activity, elevated pain threshold, prolonged death time induced by caffeine, and relaxed skeletal muscle in mice. In another study, adenosine and its derivative 5'-deoxy-5'-methylsulfonyl adenosine from *G. lucidum* were also found to inhibit platelet aggregation.[120,133,166]

Huang et al. (1999) designed a study to determine total nucleoside (s) in sporophores of *G. lucidum* produced in different Asian localities by UV-spectrophotometry. They found that the contents of total nucleosides in sporophore were found to be higher in *G. lucidum* of Taiwan, Jiangsu, and Guangdong origin than from China and Vietnam. Thin layer chromatography (TLC) identification of adenine, adenosine, uracil, and uridine in sporophore of *G. lucidum* was also performed.[53] In another study, a high-performance liquid chromatography-diode array detector-mass spectrometry (HPLC–DAD-MS) analytical method was employed for detection of nucleosides and nucleobases in *G. lucidum* fruiting bodies by Gao et al.[35] The method qualitatively identified six nucleosides namely adenosine, cytidine, guanosine, inosine, thymidine, uridine, and five nucleobases namely adenine, guanine, hypoxanthine, thymine, and uracil.[97,133]

The nucleobases cytosine, uracil, and xanthine have also been detected in *G. lucidum* of Korean and Chinese origin using NMR-based metabolomics by Wen et al.[171] Chen et al. (2012) developed an effective zwitterionic hydrophilic interaction chromatographic (ZIC-HILIC) method for the simultaneous detection and quantification of 16 nucleosides and nucleobases in *Ganoderma* of different species and origins including *G. lucidum*.[22] Peng et al. (2014) performed HPLC-DAD method for the simultaneous determination of 13 nucleosides and nucleobases including 2'-deoxyuridine, 2'-deoxyadenosine adenine, adenosine, cytidine, cytosine, guanosine, hypoxanthine, inosine, thymidine, thymine, uracil, and uridine in *Ganoderma* and related species.[105] Besides, studies have shown that the Indian variety of *G. lucidum* also possesses substantial amounts of nucleosides as evident from the study by Kirar et al.[72] and Khan et al.[67]

Kirar et al.[72] employed high performance thin layer chromatography (HPTLC) for the determination of adenine, adenosine, and uracil content in the aqueous and alcoholic extracts of *G. lucidum* of Indian Himalayas.[72] Khan et al.[67] developed an HPLC method for routine quality control quantification of nucleosides adenine and adenosine in the herbal extracts. They quantified adenine (0.16%) and adenosine (0.14%) in *G. lucidum* fruiting body extract of Indian origin.[67]

7.4.2 POLYPHENOLIC COMPOUNDS

Polyphenols, mainly the flavonoids and phenolics are the most important and widely occurring groups of secondary metabolites.[60,115,164] These have been proven to exhibit a wide range of pharmacological activities such as antiallergic, anticancer, anti-inflammatory, antimicrobial and so forth.[60,68,115] Most of these biological activities might be attributable to the antioxidative activity of polyphenols,[60] which they possess due to the presence of the conjugated ring structures and the hydroxyl groups.[68] The polyphenolic compounds have the potential to act as antioxidants by scavenging or stabilizing free radicals involved in the oxidative process through hydrogenation or complexing with oxidizing species.[68]

Phenolic compounds and flavonoids from mushrooms exhibit potential antioxidant activity.[33,63] Several studies have been conducted to evaluate the antioxidant potential of *G. lucidum* in terms of total phenolic and flavonoids content either in whole or crude extracts[72] or extracts from fruiting body,[25] mycelium[119], and spores.[48] However, fewer studies have been carried out to detect and quantify individual flavonoids as well as phenolic compounds in *G. lucidum*. Some of these have been discussed in this section.

Liu et al.[84] were identified and isolated Kaempferol and genistein from the ethanolic extract of *G. lucidum* using column chromatography and HPLC. Heleno et al.[47] carried a study to compare the antioxidant activity of phenolic and polysaccharidic extracts from fruiting body, spores, and mycelium, obtained in three different culture media, of *G. lucidum* from

Northeast Portugal using HPLC coupled to photodiode array detection. The phenolic extract of *G. lucidum* fruiting body was the most potent and revealed the highest content in total phenolics and phenolic acids such as p-hydroxybenzoic, p-coumaric acids, and cinnamic acid among all.[47] Heleno et al. (2013) also compared the antimicrobial (antibacterial and antifungal) and demalinizing activities of acetylated glucuronide derivatives of p-hydroxybenzoic acid and cinnamic acids (two compounds identified in *G. lucidum*)[48] to the parent acids and also a methanolic extract of *G. lucidum*. The synthesized acetylated glucuronide derivatives of p-hydroxybenzoic and cinnamic acids revealed even higher antimicrobial activity than the standards. However, the methanolic extract of *G. lucidum* was the only one with demelanizing activity against *A. niger*.[47] Sheikh et al. (2014) reported phenolic compounds (chlorogenic acid, rutin, vanillin, and cinnamic and gallic acids) during HPLC profiling of ethanolic extract from *G. lucidum*.[129]

Recently, Reis et al. (2015) reported induction of autophagy in human gastric adenocarcinoma cells by a methanolic extract of *G. lucidum* that was found to contains substantial amounts of phenolic compounds such as p-hydroxybenzoic acid (123 ± 9 µg/g extract), p-coumaric acid (80 ± 6 µg/g extract), and cinnamic acid (59 ± 6 µg/g extract). Autophagy is a "self-digestive" process whereby, a cell is able to survive under nutrient (s) deprivation conditions. However, recent studies suggest that it may also be regarded as a cell death mechanism since the cells exposed to prolonged stress and continuous autophagy eventually "die off." Although the methanolic extract induced autophagy, yet it was observed that the phenolics were not responsible for such an observation.[111] The presence of phenolic compounds in *G. lucidum* namely protocatechuic acid, p-hydroxybenzoic acid, catechin, chlorogenic acid, vanillic acid, syringic acid, p-coumaric acid, rutin, and t-cinnamic acid was also confirmed by during the HPLC analysis of three mushrooms including *G. lucidum*. The study showed that *G. lucidum* had highest phenolic content, as well as individual phenolic compounds and also exhibited the highest antioxidant activity.[181] Saltarelli et al. (2015) performed a comparative analysis between *G. lucidum* extracts (Gl-4 and Gl-5) and *G. resinaceum* for their antioxidant and antiproliferative activities on U937 cells. The study demonstrated highest total phenol content in the Gl-4 isolate of *G. lucidum* and had highest antioxidant activity as measured by DPPH (2, 2-diphenyl-1-picrylhydrazyl) free radical scavenging activity and Fe^{+2} chelating activity. The major flavonoids detected included morin, myricetin, quercetin, and rutin.[118]

Furthermore, a recent publication reported the isolation of a novel tannic acid from the fruiting body aqueous extract of *G. lucidum*. It was extracted by TLC and confirmed by HPLC. The extracted tannic acid was found to possess antioxidant activity in vitro along with antihypergly-cemic, hypolipedemic, and renal protective effects against streptozotocin (STZ)-induced diabetic nephropathy in rats.[30]

7.4.3 POLYSACCHARIDES

Polysaccharides from *G. lucidum* represent structurally diverse biological macromolecules with wide-ranging physicochemical properties.[12,33,103,134,159] Approximately, more than 200 types of polysaccharides have been isolated from the fruiting body, spores, and mycelia, or separated from the broth of a submerged liquid culture of *G. lucidum*.[33,64,103,120,133,154,159,166] Polysac-charides are long chain sugar molecules joined together by glycosidic bonds.[64] The major bioactive *G. lucidum* polysaccharides (GLPS) include (1.3), (1.6)-α/β-glucans, glycoproteins, and water soluble hetero-polysac-charides with arabinose, fucose, galactose, glucose, mannose, and xylose combined in different proportions and types of glycosidic linkages, as well as peptide bonds.[33,64,103,154,166] However, amongst these polysaccharides, β-1–3-D and β-1–6-D glucans are the most active biologically[154] and have been characterized with novel antitumor activity against oncogenesis and tumor metastasis.[184]

The β-D-glucans consist of a linear backbone of β -(1,3)- linked D-glucopyranosyl groups with varying degrees of branching from the C6 position. In addition to water-soluble β -D glucans, these also exist with hetero-polysaccharide chains of β -D-glucans–protein complexes, galac-tose, mannose, uronic acid, and xylose that are present at 10–50% in dry *G. lucidum*.[12,120,133,159,166] The biological activity of these glucans is determined by their water solubility, molecular size and form.[133,159] The effectiveness of β-glucans is also influenced by the number of lateral branches in the main chain, the length of the lateral chain and the ratio of the number of bonds (1.4) to (1.6) and (1.3).[133,159] For example, their 1,3-linked back-bone, relatively small side chains and an organized helical structure are beneficial for the immunostimulation.[12,64,159]

Besides, various bioactive peptidoglycans have also been isolated such as *G. lucidum* proteoglycan (GLPG; with antiviral activity), *G. lucidum* immunomodulating substance (GLIS), PGY (a water soluble glycopeptide), *G. lucidum* polysaccharide peptide (GLPP), and a fucose-containing glycoprotein fraction (F3).[120,159,166] They are usually produced by fungal mycelia cultured in fermenters and can differ in their sugar and peptide compositions and molecular weight (e.g., ganoderns A, B, and C).[159] GLPS are reported to exhibit a broad range of bioactivities, including anti-inflammatory, hypoglycemic, anti-ulcer, antitumorigenic, and immunomodulatory effects.[12,134,159]

GLPS are believed to exert antitumor activities through enhancement of host-mediated immunity indirectly, via the stimulation of natural killer cells, T cells, B cells, and macrophage-dependent immune system responses rather than a direct cytotoxic effect.[64,103,135,154] *Ganoderma* β-D-glucans have been reported to have higher antitumor activity and are better absorbed orally than the commercially available synthetic β-D-glucans.[184] Numerous refined polysaccharide preparations extracted from *G. lucidum* are now marketed as an over-the-counter treatment for chronic diseases, including cancer and liver diseases.[159] Therefore, the selection of an extraction method is very crucial to meet the market demands as it is hypothesized that the polysaccharides extracted from different parts of *G. lucidum* induce different immune responses with varying immune potencies.[64]

The extraction procedures are applied according to the characteristics (e.g., molecular weight, solubility) of the target polysaccharide (s) and depend especially on the cell wall structure. Most polysaccharides are extractable with hot water, or acidic, saline and dilute alkali solutions, or with dimethyl sulfoxide.[33,159,184] After initial extraction by water, the compounds can be further extracted by ethanol and sodium hydroxide; the purified molecules are less soluble in water and become more soluble in alkali.[184] The hot water extraction yields water-soluble polysaccharides; on the other hand, extraction with alkaline solution yields water-insoluble ones.[33] Other techniques such as microwave, ultrasonic, ultrasonic/microwave, and enzymatic treatments are also used, which could promote the breakage of the cell wall and increase the yield of the extracted polysaccharides.[33]

7.4.4 PROTEINS, PEPTIDES, AND AMINO ACIDS

At present, many bioactive proteins with profound therapeutic effects have been isolated from *G. lucidum*[103,120,133,159,166] and characterized by chromatographic or electrophoretic techniques.[54,102,103]

The protein, Ling Zhi-8 (LZ-8) was the first one to be isolated by Kino et al. (1989) from the mycelia of *G. lucidum*.[54,120,133,154,166] This 12 kDa polypeptide consisted of 110 amino-acid residues with an acetylated amine ending.[133,154] As shown by sequencing studies, it was found to be similar to the variable region of the immunoglobulin heavy chain in its sequence.[120,166] Major biological properties of LZ-8 included immunomodulatory, mitogenic[12,133,154,159] and immunosuppressive activities.[12,159]

Owing to the immunomodulatory effect of LZ-8, its candidature as a novel anticancer drug has been evaluated by several researchers. And it has been found that LZ-8 enhance the efficacy of DNA vaccine by activating dendritic cells and promoting innate and adaptive immune responses through toll-like receptor 4 (TLR-4) in vivo and therefore, could be used as an immunoadjuvant to a DNA vaccine model for cancer therapy.[81,102]

Ganodermin, 15-kDa is another protein isolated from *G. lucidum* fruiting bodies.[12,97,103,159] The isolation procedure utilized chromatography on DEAE-cellulose, Affi-gel blue gel, CM-Sepharose, and Superdex G-75.[12,102] During the study, ganodermin expressed an antifungal activity by inhibiting the mycelia growth of *Botrytis cinerea, Fusarium oxysporum,* and *Physalospora piricola*.[12,103,133] However, it was devoid of hemagglutinating, deoxyribonuclease, ribonuclease, and protease inhibition activities.[102,103] Recently, Ansor et al. (2013) following purification by HPLC and identification by SDS-PAGE and MALDI TOF/TOF MS discovered four antihypertensive -related proteins from *G. lucidum* mycelia. These included cystathionine α- synthase-like protein, DEAD/DEAH box helicase-like protein, paxillin-like protein, and α/β hydrolase-like protein. These proteins are suggested in lowering blood pressure level through the inhibition of angiotensin converting enzyme (blood pressure regulatory enzyme) activity.[3] Three more bioactive proteins (LZP-1, LZP-2, and LZP-3) were found from the fruiting body and spores of *G. lucidum* with mitogenic activity by Ye et al. (2002) using a gel filtration chromatography for fractionation and SDS-PAGE.[54,103]

Besides these, polysaccharide-protein complexes with reported bioactivities have also been isolated from GLPP, a *G. lucidum* polysaccharide-protein complex with an average molecular weight of 5.13×10^5 has been reported with the protective neutralizing effects on the mice peritoneal macrophages injured by reactive oxygen species (ROS) both in vitro and in vivo.[102,154] It consisted of 16 different kinds of amino acids and the polysaccharides; fructose, galactose, glucose, rhamnose, and xylose both linked together by β-glycosidic linkages.[102,154]

Later in 2004, Sun et al. isolated a low-molecular-weight peptide (GLPS) from water extraction of the fungus[102,154,159] and investigated its antioxidant activity.[102] They concluded that GLPS is the major antioxidant component besides the polysaccharides, polysaccharide-peptide complex, and phenolic components of *G. lucidum*.[102,154] It has shown to play an important role in the inhibition of lipid peroxidation in vivo, due to its antioxidant, metal chelating, and free radical scavenging activities.[158,163] In an investigation by Cao and Lin (2006), GLPS demonstrated antitumor and potential antiangiogenesis effects of *G. lucidum*.[102,103,154]

Proteins as enzymes have also been isolated and characterized by various types of column chromatography as well as electrophoretic techniques.[54,102,103] Sripuan et al. (2003) isolated α- galactosidase from the fruiting body of wild *G. lucidum*. This enzyme hydrolyzed p-nitrophenyl-α-D-galactopyranoside as well as melibiose, raffinose and stachyose[102,154] by using DEAE-Sephadex column gel filtration and Con A-Sepharose affinity chromatography. In another study, Wang and Ng (2006) described the isolation of a ligninolytic enzyme from *G. lucidum* fresh fruiting bodies with potent inhibitory effect against HIV-1 reverse transcriptase.[102,161] In addition, there are previous reports on the isolation and characterization of laccase isoenzymes through anion exchange chromatography, preparative gel electrophoresis, and electroelution.[97,102,154] The electrophoretic patterns of the laccase isoenzymes have been applied in the identification of various *Ganoderma* species.[54]

In 2004, a ribonuclease with a molecular mass of 42 kDa and with an N-terminal sequence different from other mushroom ribonucleases was also isolated from the fresh fruiting bodies of *G. lucidum* by Wang et al.[12] In another study, a purified 38 kDa enzyme, proteinase A inhibitor, it was found to be more specific against yeast proteinase A than other proteinases and had remarkable heat stability.[12]

An interesting fact has been highlighted that *G. lucidum* is able to biotransform inorganic selenium into organic selenium, which is stored in a water-soluble protein, Se-GL. This glycoprotein in its native state was identified as a monomer of 36,600 Da and had remarkable free radical scavenging activity in vitro.[102,154]

In addition, a variety of amino acids have also been found in *G. lucidum*.[97,103] Interestingly, spores of *G. lucidum* have been reported to contain higher amounts of amino acids compared to the fruiting bodies.[54] Various techniques for isolation and characterization of amino acids in *G. lucidum* such as Cu-Sephadex G-25 column chromatography, capillary electrophoresis, and amino acid analyzer have been employed.[54] For example, Mau et al. (2001) detected 15 free amino acids including L-isoleucine, L-leucine, L-methionine, L-phenylalanine, L-tyrosine, and L-tryptophan using HPLC after derivatization.[91] Whereas, Kim et al. (2009) analyzed amino acids including L-alanine, glycine, valine, L-proline, L-serine, and L-threonine and so forth. from the fruiting bodies of *G. lucidum*.[70]

The fungus *G. lucidum* has also been reported to contain a proteoglycan with a carbohydrate: protein ratio of 10.4:1 as isolated from its cultivated mycelia by Liu and coworkers.[84] This proteoglycan demonstrated antiviral activities against herpes simplex virus 1 and 2. According to this research group, the proteoglycan inhibited viral replication by interfering with the early events of viral adsorption and entry into the target cells.[12,97,154] Another proteoglycan, GLIS was isolated from *G. lucidum* fruiting body with a carbohydrate:protein ratio of 11.5:1. It stimulated mouse spleen lymphocyte proliferation and activation wherein, most of the activated lymphocytes were B cells. It also caused a slight increase in interleukin-2 (IL-2) and enhanced the expression of protein kinase C-α, and protein kinase C-γ in B cells.[102] Besides, several glycans such as ganoderans A, B, and C have been isolated from the fruiting body of *G. lucidum*. These peptidoglycans have been found to have hypoglycemic activity.[102,133,154]

Lectins were also isolated from the fruiting body (GLL-F) and mycelium (GLL-M) of *G. lucidum* using affinity chromatography on the BSM-Toy pearl. Sanodiya et al.[120] purified and characterized a novel 114 kDa hexameric lectin (glycoprotein) from the fruiting body of the fungus. It consisted of 9.3% neutral sugar and demonstrated hemagglutinating activity on pronase-treated human erythrocytes.[97,102,159]

7.4.5 TRITERPENES

Triterpenes are considered as one of the major group of compounds that contribute to the various bioactivities of *G. lucidum*.[9,64,66,159] These constitute a subclass of naturally occurring terpenes and usually have complex lanostane-based chemical structures that are highly oxidized.[9,64,159] The triterpenes of *G. lucidum* include ganoderic acids (GAs) (A, AM1, B, C1, C2, D, DM, F, G, H, K, Me, Mk, S, T, TR, Y), ganoderenic acids (A, B, D), ganoderols (A, B), ganoderiol F, ganodermanontriol, ganoderal A, Me ganoderates (D, G), and lucideric acids.[23,58,133,159,195] Among these, GAs are the most abundantly occurring triterpenes.[9] These bitter tasting triterpenes consists of four cyclic and two linear isoprenes.[64] GA-A and GA-B were the first triterpenes isolated from *G. lucidum* by Kubota et al. in 1982.[159] Since then, the occurrence of more than 200 triterpenes has been reported from the spores, fruiting body, and mycelia of *G. lucidum*.[9,64,66,159,162,174,183] Studies have shown that the spores have considerably higher GA contents than other parts.[103,166] Also, it has been found that with geographic distribution, the triterpene composition of *G. lucidum* fruiting body varies considerably.[166]

Although, appreciable variation in triterpene contents exist among the different parts and growing stages of the fungus still they are considered as marker compounds of *G. lucidum*.[159] Accordingly, several researchers have used the triterpene profiles during the nomenclature and taxonomic studies of *G. lucidum*. The triterpene content has also been used as a measure of the quality of different *Ganoderma* species.[19,159] *G. lucidum* triterpenes (GLT) have numerous health benefits such as anti-inflammatory, antitumorigenic, anti-HIV, antihypertensive, antiangiogenic, immunomodulatory, antiandrogenic, antioxidant, and antimicrobial activities.[64,66,159,174]

7.4.6　VOLATILE ORGANIC COMPOUNDS (VOCS)

Enrichment of fungi with numerous volatile organic compounds (VOCs) imparts them several bioactivities such as antioxidant, anti-inflammatory, and antimicrobial.[90,96] Several electrophoretic and chromatographic techniques have been employed to characterize the VOCs in *G. lucidum*. Common chromatographic techniques such as gas chromatography (GC) and TLC were used to detect fatty acids in *G. lucidum*.[54] Later in 2014, Stojković et al. used the technique of GC with flame ionization detection (GC-FID)/capillary column to determine fatty acids in *G. lucidum* from different origins.[144] Ziegenbein et al. (2006) detected trans-anethol, R-()-linalool, S-(+)-carvone, and α-bisabolol as the major volatiles from the hydrodistillates and solvent extracts of *G. lucidum* fruiting body.[199]

Chen et al. (2010) using headspace solid-phase microextraction combined with gas chromatography-mass spectrometry (HS-SPME-GC-MS) enabled detection of fifty-eight volatile compounds in *G. lucidum* mycelium. The main volatile flavor compounds included 1-octen-3-ol, ethanol, hexanal, 1-hexanol, sesquirosefuran, 3-octanol, and 3-octanone.[23] Similar chromatographic technology (HS-SPME-GC-MS) was used to detect volatile aroma compounds in *G. lucidum* from Turkey. They detected acids, alcohols, aldehydes, phenols, L-Alanine, D-Alanine, 3-Methyl, 2-Butanamine, 2-Propanamine, and identified 1-Octen-3-ol and 3-Methyl butanal as the major aroma compounds.[151] C-19 fatty acids were also detected in the ethanolic extract of *G. lucidum* spores by Gao and coworkers.[35] During their research, 2-naphthyl esters of nonadecanoic and cis-9-nonadecenoic acids isolated by multiple column chromatography and preparative HPLC and characterized by ^{1}H- and ^{13}C-NMR and MS spectral data from the *G. lucidum* spores were identified as the bioactive constituents responsible for the antitumor activity.[35]

Ganoderma lucidum also contains sterols, oleic acid, cyclooctasulfur, ergosterol peroxide (5,8-epidioxy-ergosta-6,22E-dien-3-ol), and the cerebrosides (4E′,8E)-N-D-2′-hydroxystearoyl-1-O-D-glucopyranosyl-9-methyl-4–8-sphingadienine and (4E,8E)-N-D-2′-hydroxypamitoyl-1-O-D-glucopyranosyl-9-methyl-4–8-sphingadienine.[103,120,133] The compounds oleic acid and cyclooctasulfur isolated from cultured broth of *G. lucidum* were found to inhibit histamine release, which is an essential activity for the treatment of inflammation, allergies, and anaphylactic shock. The spores themselves contain the alkaloids such as choline and betaine, tetracosanoic acid, stearic acid, palmitic acid, ergosta-7, 22- dien-3-ol, nonadecanoic acid, behenic acid, tetracosane, hentriacontane, ergosterol and β-sitosterol, and the lipid pyrophosphatidic acid.[120]

7.4.7 OTHER CONSTITUENTS

There are several researches that support the presence of various vitamins and minerals in *G. lucidum* besides polysaccharides, proteins and peptides, triterpenes, and VOCs. Vitamins such as vitamin C and E and β-carotene have been detected in the spores of *G. lucidum* using RP-HPLC.[54] Among the minerals, selenium and germanium detected in fruiting bodies of *G. lucidum* are particularly worth mentioning. These have been known to enhance the anticancer effect.[133,159] Germanium although not an essential element yet has been credited with immunopotentiating, antitumor, anti-oxidant, and antimutagenic activities, at low doses.[159] As reported by Du et al. (2008), *G. lucidum* can also biotransform 20–30% of inorganic sele-nium present in the growth substrate into selenium-containing proteins.[159] Other elements detected in log-cultivated *G. lucidum* fruiting bodies included phosphorous, silica, sulfur, potassium, calcium, magnesium iron, sodium, zinc, copper, and manganese in traces. Heavy metals such as lead, cadmium, and mercury were also detected in traces.[120,159]

7.5 THERAPEUTIC EFFECTS

The presence of numerous bioactive principles bestows *G. lucidum* with various therapeutic effects. Some of the prime bioactivities of *G. lucidum* are summarized in Table 7.2.

TABLE 7.2 Various Therapeutic Effects of *G. lucidum*.

Therapeutic effect (S)	Principle bioactive compounds	References
Antiangiogenic effects	Polysaccharides, polysaccharide-peptides, triterpenoids, ethanolic extract, and so forth.	14, 71, 99, 170, 175, 185
Anticancer effects	Polysaccharides, triterpenoids, and so forth.	11, 52, 57, 59, 64, 77, 83, 111, 125, 126, 152, 161, 182
Anti-inflammatory effects	Triterpenes, polysaccharides, and so forth.	24, 28, 51, 62, 136
Antimetastatic effects	Triterpenes, polysaccharides, and so forth.	1, 24, 20, 58, 64, 86, 172, 173
Antimicrobial effects (Antibacterial, antifungal, and antiviral)	Triterpenes, polysaccharides, proteins, proteolytic enzymes, lectins, essential oil, and so forth.	2, 5, 10, 17, 31, 47, 50, 61, 98, 106, 120, 127, 132, 133, 157
Antioxidative effects	Triterpenes, polysaccharides, and so forth.	64, 75, 138
Anti-stress, antihypoxic, and adaptogenic effects	Polysaccharides, methanol extract, and so forth.	88, 104, 110, 131, 193
Antitumor effects	Polysaccharides, polysaccharide-proteins, proteoglycans, triterpenoids, phenolic fractions, organic germanium, aqueous extract, methanol extract, and so forth.	14, 36, 46, 48, 78, 79, 80, 120, 145, 155, 163, 169, 175, 178, 182, 187, 198
Immunomodulatory effects	Polysaccharides, proteins, triterpenoids, methanol extract, and so forth.	4, 9, 34, 69, 89, 120, 133, 141, 147, 192
Neuroprotective effects	Triterpenoids, polysaccharides, and so forth.	9, 73, 123, 188, 191

7.5.1 MAJOR BIOACTIVITIES

The literature on several in vitro and in vivo studies on *G. lucidum* describes its various bioactivities and affirms its therapeutic benefits. Some of the major bioactivities of *G. lucidum* have been briefly described in this section.

7.5.1.1 ANTITUMOR ACTIVITY

The antitumor activity of *G. lucidum* is mainly attributable to polysaccharides followed by the triterpenoids. It has been suggested that the antitumor potential of polysaccharides seems to be highly correlated with their chemical composition and configuration, as well as their physical properties.[100] As reported in literature, the structural features such as $(1\rightarrow3)$-β-linkages in the main chain of the glucan and additional $(1\rightarrow6)$-β-branch points;[93,165] presence of glucose and mannose subunits and triple helical confirmation of $(1\rightarrow3)$-β-glucans;[187] lower level of branching and greater water solubility of β-glucans[34] are considered important factors for their immunostimulating activity besides higher molecular weight.[94,165] According to Lowe et al. (2001), the antitumor potential of *Ganoderma* polysaccharides is usually related to their immunomodulatory activity, since polysaccharides have a large molecular weight; therefore, these compounds cannot penetrate cells, but they bind to immune cell receptors. It has been confirmed that there are fungal pattern-recognition molecules for the innate immune system. However, the mechanism by which the innate immune system recognizes and responds to fungal cell wall carbohydrates is a very complex and multifactorial process.[87]

Ganopoly is a polysaccharide aqueous extract of *G. lucidum* well-known for its antitumor activity. It has demonstrated antitumor activity against sarcoma-180 in mice and indicated activation of murine macrophages, resulting in release of cytokines such as interferon-γ (IFN-γ) and tumor necrosis factor-α (TNF-α), nitric oxide along with interleukins (IL-1β and IL-6) and other mediators, and activated nuclear factor kappa B (NF-κB). Furthermore, the polysaccharide had caused significant cytotoxicity in the human tumor cell lines such as human caucasian cervical epidermoid carcinoma (CaSki), human cervical cancer (SiHa), human hepatoma (Hep 3B), human hepatocellular liver carcinoma (HepG2), human colon carcinoma (HCT 116), and human colon adenocarcinoma grade II (HT29) cells in vitro, with marked apoptotic effects observed in CaSki, HepG2, and HCT 116 cells.[37] Other studies showed that Ganopoly could enhance the immune response in patients with advanced-stage cancer, which could be an approach for overcoming immunosuppressive effects of chemotherapy/radiotherapy.[36]

Tsai et al. (2012) established the relationship between the structural features of *G*. GLPS and their immunomodulating mechanisms. Two different bioactive components (GLPS-SF1 and GLPS-SF2) were isolated from GLPS, which exhibited distinct bioactivities. The GLPS-SF1 consisted of heterogeneous glycopeptides. The sugar part was composed of glucose and mannose in a molar ratio of 4:1, with main sugar 1,6-glucan linkage. The GLPS-SF1 showed diverse immune stimulating properties in human monocytes, NK cells, and T lymphocytes as well as induced cytokine expression via TLR4-dependent signaling in mouse macrophages. The GLPS-SF2 were acidic oligosaccharides with the $[-\alpha\text{-}1,4\text{-Glc}-(\beta\text{-}1,4\text{-GlcA})_3-]$ n structure. The GLPS-SF2 were shown to have immunomodulating activities specific to NK cells and T lymphocytes by inducing the IL-2 cytokine expression and cell proliferation.[155]

Liang et al. (2014) demonstrated that GLPS reduced cell viability on human colorectal (HCT)-116 cells in a time- and dose-dependent manner, which in turn induced cell apoptosis. The observed apoptosis was characterized by morphological changes, DNA fragmentation, mitochondrial membrane potential decrease, S-phase population increase, and caspase-3 and caspase-9 activation. Furthermore, inhibition of c-Jun N-terminal kinase (JNK) by SP600125 (an anthrapyrazolone inhibitor of JNK) led to a dramatic decrease of the GLPS-induced apoptosis. Besides, GLPS up-regulated the expression of Bax/Bcl-2, caspase-3, and poly (ADP-ribose) polymerase (PARP). These results demonstrated that apoptosis stimulated by GLPS in HCT-116 cells was associated with activation of mitochondrial and mitogen-activated protein kinase (MAPK) pathways. These data provide evidence for the possible mechanism of GLPS-mediated apoptosis.[80]

In another study, Liang et al. (2013) also investigated the cytotoxic and apoptotic effect of GLPS on HCT-116 human colon cancer cells and the molecular mechanisms involved. Treatment of HCT-116 cells with various concentrations of GLPS (0.625–5 mg/mL) resulted in a significant decrease in cell viability. This study showed that the antitumor activity of GLPS was related to cell migration inhibition, cell morphology changes, intracellular Ca^{2+} elevation, and lactate dehydrogenase (LDH) release. Also, increase in the levels of caspase-8 activity was involved in GLPS-induced apoptosis. Western blotting indicated that *Fas* and caspase-3 protein expressions were upregulated after exposure to GLPS. This investigation demonstrated anticancer activities of GLPS against HCT-116 human colon cancer cell line by triggering intracellular calcium release and the death receptor pathway.[79]

In a most recent study, the antitumor activity both in vitro and in vivo of a polysaccharide obtained from *G. lucidum* was investigated on HL-60 acute myeloid leukemia cells. The prime focus was on the effect of the polysaccharide on MAPK pathways. It was found that GLPS blocked the extracellular signal-regulated kinase/MAPK signaling pathway, simultaneously activated p38 and JNK MAPK pathways, and therefore, regulated their downstream genes and proteins, including p53, c-myc, c-fos, c-jun, Bcl-2, Bax, cleaved caspase-3, and cyclin D1. As a result, cell cycle arrest and apoptosis of HL-60 cells were induced. Therefore, GLPS exerted antitumor activity via MAPK pathways in HL-60 acute leukemia cells.[178]

Polysaccharide-protein or—peptide complexes have also been described as having antitumor properties. GLPP, potentially inhibited human lung carcinoma cell line (PG), proliferation in vitro, and reduced the xenograft (of the PG cell line) in an albino laboratory-bred strain of the house mouse (BALB/c) nude mice in vivo. This compound proved to have an antiangiogenic activity which can be the basis for its antitumor effects.[14]

A well-known proteoglycan from *G. lucidum*, GLIS exhibited an effective antitumor effect by increasing both humoral and cellular immune activities.[186] The *G. lucidum* polysaccharide-peptide conjugate also demonstrated antitumor potential in different studies. For example, it significantly inhibited tumor growth in a murine sarcoma 180 model, and inhibited proliferation of human umbilical vein endothelial cells (HUVECs) by inducing cell apoptosis and decreasing the expression of secreted vascular endothelium growth factor (VEGF) in human lung cancer cells.[14,78]

Besides, GLPS, triterpenoids especially GAs also possess potential antitumor activity. GAs such as GA-U, GA-V, GA-W, GA-X, and GA-Y were found to exhibit cytotoxicity against hepatoma cell in vitro more than 20 years ago.[153,175] Recently, the cytotoxic and antiproliferative effects of many GAs against tumor cells have been demonstrated in numerous investigations.[175] GA-D was found to inhibit the proliferation of HeLa human cervical carcinoma, induce cell cycle arrest at G2/M phase, and trigger cell apoptosis.[182] GA-F, GA-K, GA-B, and GA-AM1 also exhibited cytotoxicity on HeLa human cervical carcinoma cells with half maximal inhibitory concentration (IC50) values of 15–20 µM.[183] GA-F inhibited primary solid tumor growth in the spleen and liver metastasis and secondary metastatic tumor growth in the liver in intrasplenic Lewis lung carcinoma (LLC) in implanted mice.[71] GA-E showed significant cytotoxic activity against Hep G2, G15, and P-388 tumor cells.[175]

Tang et al. (2006) reported that GA-T had higher cytotoxicity to 95-D cell line than to normal cell lines, and it induced apoptosis of metastatic lung tumor cells through an intrinsic pathway related to mitochondrial dysfunction and p53 expression, indicating that it could be a potentially useful chemotherapeutic agent.[150] GA-Me, a lanostane triterpenoid purified from *G. lucidum* mycelia, effectively inhibited both tumor growth and lung metastasis of LLC in C57BL/6 mice by increasing expressions of IL-2 and IFN-γ and upregulating expression of NF-κB, which might be involved in the production of IL-2,[160] and tumor invasion through downregulating matrix metalloproteinase 2/9 (MMP2/9) gene expression.[20] It was found that GA-Me depressed the same viability of tumor cells at much lower concentrations than against normal cells.[20] Thus, it might be an effective potential therapeutic drug for prevention and treatment of tumors.

Inhibition of DNA polymerase and post-translational modification of oncoproteins has also been reported to contribute to the antitumor activity of *G. lucidum*. The organic germanium in *G. lucidum* may also contribute to its antitumor activity as reported by Chiu et al. 2000.[120]

7.5.1.2 ANTICANCER ACTIVITY

Throughout the literature, triterpenoids and the polysaccharides from *G. lucidum* have been considered as the major anticancer bioactive principles. In accordance with the various in vitro and in vivo anticancer studies conducted on *G. lucidum*, there are several contributing mechanisms. In case of triterpenoids, these include antiangiogenic, anti-inflammatory, antimetastatic, antioxidative, cell-cycle arrest, and direct cytotoxic effects, besides immunomodulation.[11,64] On the other hand, there are three mechanisms (antiangiogenic activity, antioxidative activity, and immunomodulation activity) by which GLPS exert their anticancer effects.[64] The anticancer activities of *G. lucidum* are presented in Table 7.2 and the contributing bioactivities (mechanisms) are detailed in this section.

7.5.1.2.1 Antiangiogenic activity

Angiogenesis besides being a normal physiological process involving new blood vessels formation from the existing ones is also an essential step in tumor transition. Thus, as suggested by Sanodiya et al. (2009) and Xu et al. (2011), antiangiogenic therapy might be an important component in cancer therapy.[120,176] The GLPP was isolated from the fruiting body of *G. lucidum*. It directly inhibited HUVECs proliferation in vitro, but had no direct effect on PG cell (a human lung carcinoma cell line) proliferation. However, the serum collected from tumor-bearing mice after GLPP treatment showed inhibition of PG cell proliferation. Chick chorioallantoic membrane (CAM) assay revealed that GLPP and GLPP treated serum showed antiangiogenic effect. Furthermore, GLPP noticeably reduced the xenograft (human lung carcinoma cell PG) in BALB/c nude mice in vivo, suggesting that GLPP has antiangiogenic effects.[14,171,176,185] The same group of investigators also found antiproliferative effect of GLPP on HUVECs due to its proapoptotic action, mediated by the reduction of Bcl-2 expression; increase of Bax expression and downregulation of VEGF secretion.[14,176,185]

Antiangiogenic effects of *G. lucidum* have also been demonstrated by an ethanolic extract of *G. lucidum* fruiting body extract in a dose-dependent manner and supported by the inhibition of inducible NO production.[142,185] Stanley et al. (2005) demonstrated that *G. lucidum* inhibited the early event of angiogenesis in prostate cancer-3 (PC-3) cells. This action was mediated through the modulation of MAPK and Akt signaling pathways that resulted in downregulation of VEGF and TGF-β1 from PC-3 cells.[143,185]

Triterpenoid, another active ingredient of *G. lucidum*, has also been found to possess antitumor and antiangiogenic activities. Angiogenesis induced by Matrigel was inhibited by triterpenoids fraction of *G. lucidum* and a reduction of primary splenic tumor size as well as suppression of secondary metastasis in intrasplenic implant mice with LLC cells was seen.[71,176,185] GA-F was further identified as the active component of the triterpenoid fraction, as it was responsible for the antimetastasis activity via inhibition of tumor-induced angiogenesis.[71,171] Recently, Nguyen et al. (2015) found the higher inhibitory effect of ganoderic acid F on the formation of capillary-like structures of HUVECs among other triterpenoids isolated from Vietnamese *G. lucidum*.[99]

In addition, culture soybean extracts with *G. lucidum* mycelia produced a cultivated product called genistein combined polysaccharide, which inhibited angiogenesis in the CAM assay as well as in the mouse dorsal air-sac model following implantation of colon-26 tumor cells.[92,171,176] Some of the studies that support antiangiogenic activity of *G. lucidum* are mentioned in Table 7.2.

7.5.1.2.2 *Antimetastatic activity*

Cancer metastasis is a complex process that involves the spread of malignant cells from a primary tumor to distant sites. It is common in the late stages of cancer and is the major cause of death.[24,39] Studies have shown that several key proteins involved in cancer metastasis are regulated by GLT.[64]

Jiang et al. (2008) found that GA-A and GA-H suppressed invasive behavior of breast cancer cells by inhibiting activating protein-1 (AP-1) and NF-κB signaling and suppressing secretion of urokinase plasminogen activator (uPA).[58] A recent study demonstrated inhibition of migration and adhesion of highly metastatic breast cancer cells by a ganoderiol A-enriched *G. lucidum* extract via suppression of the focal adhesion kinase (FAK)-SRC-paxillin cascade pathway.[24]

Liu et al. (2008) isolated GA, GA-DM from the ethanol extracts and showed that it blocked osteoclastogenesis, which has a high incidence of bone metastasis.[84] GA-Me effectively inhibited the invasive behavior of highly metastatic lung and breast cancer cells via inhibition of MMP2/9 gene expression, respectively[20] that degrade the extracellular matrix and can promote cancer metastasis.[64] GA-T also inhibited colon cancer cell migration, invasion, and adhesion in vitro through p53 dependent inhibition of MMP expression.[20]

In addition, Ganodermanontriol also exerted its effect on anti-invasiveness of breast cancer cells by inhibiting secretion of uPA and expression of its receptor (uPAR).[59] Furthermore, lucidenic acids -A, -B, -C, and -N were found to possess potential anti-invasive activity in hepatoma cells by suppressing phorbol-12-myristate-13-acetate (PMA)-induced MMP-9 activity.[172] Among them, lucidenic acid B (LAB) might inhibit cell invasion by suppressing the phosphorylation of extracellular signal-regulated kinase (ERK1/2) and by reducing activation protein (AP)-1 and NF-κB DNA-binding activities.[173]

Oral administration of standardized *G. lucidum* extract (GLE), containing 6% chemically characterized triterpenes and 13.5% polysaccharides, suppressed breast-to-lung cancer metastases in vivo through the downregulation of genes responsible for cell invasiveness.[1,86] These studies which highlight the antimetastatic activities of *G. lucidum* have been mentioned in Table 7.2.

7.5.1.2.3   Antioxidative activity

Oxidative stress is known to be a major contributor to increased cancer risk. Free radicals and ROS generated as by-products of metabolic processes and exposure to some exogenous chemicals can damage proteins and DNA within cells leading to oxidative stress. Oxidative stress can be countered by antioxidative enzymes and repair mechanisms. However, excess oxidative stress can overwhelm the innate protective systems leading to a variety of physiological disorders including cancer.[64]

Studies suggest that the cancer-inducing oxidative damage might be prevented or limited by antioxidants.[146] A group of researchers showed that triterpenes extracted from *G. lucidum* exhibit antioxidative activity in vitro by directly scavenging free radicals, increasing the activities of antioxidant enzymes in blood and tissues and reducing radiation-induced oxidative DNA damage in mice splenocytes.[137,138] Nevertheless, polysaccharides from *G. lucidum* have also been reported with antioxidative activity. Kao et al. (2013) demonstrated that dietetic treatment using a *Ganoderma* mycelium-derived polysaccharide extract could suppress aberrant crypt foci formation in rat colon by reducing ROS induced oxidative damage.[64] In addition, inhibition of iron-induced lipid peroxidation in rat brain homogenates; inactivation of hydroxyl radicals and superoxide anions and reduction of UV-induced DNA strand breakage in differentiated HL-60 (human promyelocytic leukaemia) cells by an amino-polysaccharide *G. lucidum* fraction (G009) has also been reported.[75]

7.5.1.2.4   Cytotoxic and cytostatic activities

The normal cellular division is a self-regulated process during cell cycle with restricted controls. However, when these controls are bypassed, deregulation of the cell cycle triggers aberrant cell reproduction, leading to cancer formation.[24] Cycle-phase-specific anticancer drugs have either cytotoxic or cytostatic activities. Cytotoxic drugs result in programmed cell death (apoptosis and autophagy) whereas, cytostatic drugs cause cell cycle arrest to halt the rapid proliferation of cancer cells at a particular phase.[158]

Yue et al. (2008) indicated that the GLT treatment regulated expression of 14 proteins in human cervical carcinoma cells, which play important roles in cell proliferation, cell cycle, oxidative stress, and apoptosis.[182] In addition, Thyagarajan et al. (2010) demonstrated that GLT extract suppressed proliferation of human colon cancer cells as well as inhibited tumor growth in a xenograft model, which was associated with the cell cycle arrest at the G0/G1 phase and induction of the programmed cell death Type II–autophagy.[152]

Radwan et al. (2015) recently demonstrated that GA-A induced apoptosis in lymphoma cells through caspase-3 and -9, and enhanced HLA class II-mediated antigen presentation and CD4[+] T-cell recognition.[107] Shang et al. in 2009 isolated a novel polysaccharide from Se-enriched *G. lucidum* (SeGLP-2B-1) that suppressed proliferation of different cancer cell lines in vitro. Induction of mitochondria-mediated apoptosis by this polysaccharide was also demonstrated.[125,126]

7.5.1.2.5   *Immunomodulatory activity*

Ganoderma lucidum contains several substances with potent immuno-modulatory action. These comprise of polysaccharides (β-D-glucans), proteins (Ling Zhi-8), and triterpenoids.[120,133] GLPS have been demonstrated to modulate and improve immune functions in both human studies and animal models.[9,38,160] Literature also reports that the administration of *G. lucidum* extracts resulted in increased secretion of cytokines from immune cells, which lead to increased cellular activity and survival of immune cells related to innate (macrophages) and adaptive immunity (lymphocytes).[4,9,89] In addition, it has been demonstrated that the polysaccharides from this fungus enhanced the expression of major histocompatibility complex (MHC) in a melanoma cell line, which improved antigen presentation and thus promoted viral and cancer immunity.[9,147] Other major studies related to immunomodulatory activity are mentioned in Table 7.2.

7.5.1.2.6 Anti-inflammatory activity

Inflammation has been colligated as a causative agent in nearly 20% of cancers.[24,64] Chronic over-expression of inflammatory cytokines such as VEGF, IL-6, and TNF-α could promote carcinogenesis.[64] Anti-inflammatory activity in human colon carcinoma cells was demonstrated by a triterpene-enriched ethanol extract from *G. lucidum* on exposure to pro-inflammatory stimuli.[51] Whereas, another ethanol extract from the mycelium of *G. lucidum* showed anti-inflammatory activity against carrageenan-induced (acute) and formalin-induced (chronic) inflammation in mouse and phorbol ester–induced mouse skin inflammation.[74] GLT extract showed anti-inflammatory effects in macrophages, which were mediated by the inhibition of NF-κB and AP-1 signaling pathways.[28] In addition, colitis-associated carcinogenesis in mice was also prevented by GLTextract.[136]

Besides, GLPS have also shown significant anti-inflammatory activity in acute and chronic inflammation in in vivo models.[62] Liu et al. (2008) evaluated an extract of *G. lucidum* polysaccharides (EORP) and concluded that it attenuated lipopolysaccharide (LPS)-induced expression of adhesion molecule and monocyte adherence by ERK phosphorylation suppression and NF-κB activation in vitro and in vivo.[84]

7.5.1.3   NEUROPROTECTIVE ACTIVITY

The neuroprotective effect of *G. lucidum* has been supported by the work carried out by Zhang et al. (2011) and Zhao et al. (2012), wherein, a mixture of triterpenoid compounds in *G. lucidum*, including methyl GA-A, methyl GA-B, GA-S1, and GA-TQ promoted neuronal survival and reduced fatigue.[188,191] In addition, this fungus has also been examined for the treatment of neurological diseases. Researchers demonstrated that long-term consumption of *G. lucidum* could diminish the progression of Alzheimer's disease,[73,196] maybe due to the promotion of neuritogenesis and reduction of senescence of the neurons.[123]

In addition to triterpenes, the polysaccharides found in *G. lucidum* are also believed to confer a neurological benefit. Although little scientific evidence exists and further work is required to support this claim, yet *G. lucidum* has been used for its analgesic and muscle relaxing properties since ancient times.[188] On the other hand, it was demonstrated that the aqueous extract (containing polysaccharides) of *G. lucidum* exhibited an antidepressant-like effect and reduced anxiety-type behavior in rats, a finding which is yet to be replicated in studies involving human subjects.[9]

7.5.1.4 ANTIMICROBIAL ACTIVITY

Ganoderma species are known to produce various compounds such as polysaccharides, triterpenoids, lectins, proteins, proteolytic enzymes, and essential oils with potential antimicrobial activity.[132] The various antimicrobial activities of *G. lucidum* including antibacterial, antifungal, and antiviral are discussed in this section.

7.5.1.4.1 Antibacterial activity

GLPS have been found effective against plant pathogens (*Erwinia carotovora*, *Penicillium digitatum*, and *Botrytis cinerea*) and harmful food spoiling microorganisms (*Bacillus cereus*, *Bacillus subtilis*, *Escherichia coli*, *Aspergillus niger*, and *Rhizopus nigricans*).[5] Extract from *G. lucidum* has also demonstrated bacteriostatic activity against *Helicobacter pylori*, a causative agent for the formation of gastric ulcers and gastric cancer.[133] Quereshi et al. (2010) demonstrated the antimicrobial activity of various solvent extracts of *G. lucidum* against the bacteria such as *E. coli*, *S. aureus*, *K. pneumoniae*, *B. subtilis*, *S. typhi*, and *P. aeruginosa*.[106] Heleno et al. (2013) showed antibacterial and antifungal properties of *G. lucidum* extract. The extract showed a higher activity against *S. aureus* and *B. cereus* than the antibiotics ampicillin and streptomycin.[47] In a recent study, Vazirian et al. (2014) found that addition of purified triterpenoids and steroids from the crude fractions (hexane and chloroform) of *G. lucidum* enhanced their antimicrobial activity against the yeast, *C albicans*; gram-positive (*S. aureus*, *B. subtilis*) and gram-negative (*P. aeruginosa*, *E. coli*) bacteria.[157]

7.5.1.4.2 Antiviral activity

GA isolated from fruiting bodies of *G. lucidum* exhibited antiviral activity for example, against HIV and Epstein-Barr virus.[31] Water soluble substances (GLhw and GLlw) and methanol soluble substances (GLMe-1–8) isolated from fruiting bodies inhibited replication of influenza-A virus.[133] Polysaccharides demonstrated a direct action towards hepatitis B virus (HBV) by inhibiting DNA polymerase. Ganodermadiol exhibited activity against herpes simplex virus type 1.[10] In another study, *G. lucidum* water extract inhibited proliferation of virus del papiloma humano (VPH) transformed cells.[50]

7.5.1.4.3 Antifungal activity

There are several documented scientific studies regarding the antibacterial and antiviral activity of *G. lucidum*. However, there are only a few investigations on its antifungal activity. Wang et al. (2005) isolated an antifungal protein, Ganodermin from *G. lucidum* mycelium that inhibited mycelial growth of *Botrytis cinerea, Fusarium oxysporum*, and *Physalospora piricola*.[162] Heleno et al. (2013) showed the antifungal properties of *G. lucidum* extract against *Trichoderma viride* which exceeded that of standard antibiotics.[48,133] *T. viride* causes green mold disease of mushrooms during their cultivation and is also known for producing allergic symptoms, especially in the immune-compromised individuals.

Besides, studies exploring the effect of *G. lucidum* have also been carried out against the fungal pathogen, *C. albicans*. *Candida species* are the most common opportunistic pathogenic fungi that constitute the normal microbiota of an individual's mucosal oral cavity, gastrointestinal tract, and vagina.[2,121] As reported in the literature, these species are the major causative agents of nosocomial infections. These ubiquitous fungi are known to cause superficial to invasive infections; mainly in immune-compromised patients.[40]

There are some studies on the activity of *G. lucidum* against the planktonic forms of *Candida* although fewer in number. For example, Nayak et al. (2010) examined varying concentrations of toothpaste containing *G. lucidum* in vitro for its antifungal properties against *C. albicans*. The toothpaste exhibited antifungal properties against the tested organism and the MIC value of *C. albicans* was found to be less than 2.0 mg/ml.[98] A similar observation was made by Adwan et al. (2012) wherein, the anti-*Candida* activity of *G. lucidum* formulated toothpaste was evaluated against 45 oral and nonoral *C. albicans* isolates along with other eight toothpaste samples.

The study supported the use of herbal-based toothpaste against the dental ailments caused by *C. albicans*.[2]

In another study, nine *Ganoderma species* including *G. lucidum*, *G. chalceum*, and *G. stipitatum* of Indian origin (Maharashtra) were examined for their antimicrobial activity and potential against *C. albicans*. Three different terpenoids (sesquiterpene, diterpene, and triterpene) extracts and seven standard antibiotics were considered. The sesquiterpenoid extracts were found highly effective against both gram-positive, negative bacteria, and *C. albicans*, as compared to the standard antibiotics. *C. albicans* showed resistance against diterpene and triterpene extracts. Therefore, the study clearly indicated that *Ganoderma species* especially, *G. lucidum*, *G. chalceum*, and *G. stipitatum* have potential antimicrobial properties.[130] Moreover, the antimicrobial effects of ethanolic and methanolic extracts of *G. lucidum* have also been tested against *C. glabrata* and *C. albicans*. The highest inhibitory activity was determined against *C. glabrata* than *C. albicans*.[17]

7.5.1.5 ANTISTRESS, ANTIHYPOXIC, AND ADAPTOGENIC EFFECTS

Stress is a state of threatened homeostasis that produces a different physiological as well as pathological changes depending on severity, type, and duration of stress.[56] Hypoxia, whereas, is defined as an oxygen deprivation at the tissue level. As reported in the literature, exposure to extreme physiological stress induces various deleterious effects at cellular level. Recent findings suggest that there is an increased cellular oxidative stress with a consequent imbalance between excess ROS generated and antioxidant defense mechanism due to enhanced metabolic rate. These harmful ROS cause oxidative damage to various biomolecules and often leads to mitochondrial and muscular structure damage. Thereby, compromising cell integrity and reducing capacity to maintain cellular energy levels. These results in muscular atrophy and muscle fatigue leading to stress-induced decreased performance of an individual.[42]

In a recent study, the effect of GLPS on exhaustive exercise-induced oxidative stress in skeletal muscle tissues of mice was designed. After 28 days of GLPS administration, the mice performed an exhaustive swimming exercise. The superoxide dismutase (SOD), glutathione peroxidase

(GPX), catalase (CAT) activities, and malondialdehyde (MDA) levels in the skeletal muscle were also determined. The results showed that GLPS could increase antioxidant enzymes activities and decrease the MDA levels in the skeletal muscle of mice and thus, supported protective effects of GLPS supplementation against exhaustive exercise-induced oxidative stress.[193]

In addition to these, the antihypoxic effect of *G. lucidum* has also been studied by several research groups. For example, the effects of polysaccharide extract isolated from *GlPS* on rat cortical neuronal cultures exposed to hypoxia/reoxygenation (H/R) were studied in vitro. *GlPS* increased neuron viability following H/R and also significantly reduced MDA content and ROS production besides, increasing manganese SOD (Mn-SOD) activity. Moreover, NF-κB translocation induced by H/R was blocked. These findings suggested that *GlPS* might be useful in treating H/R-induced oxidative stress and Mn-SOD play a critical role in the neuroprotective effect of *GlPS* against H/R injury.[189] A similar observation was made with *Ganoderma* total sterol (GS) and its main components (GS (1)). GS (1) demonstrated the more potent protective effect on neurons compared with GS at the same doses (0.01, 0.1, and 1 μg/ml).[190] Later on, Zhou et al. [197] employed two models namely, middle cerebral artery occlusion (MCAO) in Sprague–Dawley (SD) rats and oxygen and glucose deprivation (OGD) in primary cultured rat cortical neurons to mimic ischemia-reperfusion (I/R) damage, in vivo and in vitro, respectively and neurological functional deficits were assessed at 24 h after I/R. The results showed that oral administration of water-soluble GLPS (100, 200, and 400 mg/kg) significantly reduced cerebral infarct area, attenuated neurological functional deficits, and reduced neuronal apoptosis in ischemic cortex. In OGD model, GLSP (0.1, 1, and 10 μg/ml) effectively reduced neuronal cell death and relieved cell injury. Moreover, GLPS decreased the percentage of apoptotic neurons, relieved neuronal morphological damage, suppressed over-expression of active caspase-3, caspase-8 and caspase-9 and Bax, and inhibited the reduction of Bcl-2 expression.[197]

It has been found that supplementation with adaptogenic preparations increase the physical and mental performance of an individual during the exposure to stressful environments[42] and also provide non-specific resistance to the body.[117,128] Various single and polyherbal preparations have been found useful to increase stress tolerance during the stressful situations.[41,43] These include those derived from *Aloe vera*,[54] *Cordyceps sinensis*,[101] *Ginkgo biloba*,[6,44] *Hippophae rhamnoides*,[117] *Rhodiola crenulata*,[6,44] *Rhodiola imbricata*,[41] *Valeriana wallichii*,[128] and so forth.

The ancient Chinese literature also supports the adaptogenic potential of *G. lucidum*, but, scientific researches considering this effect is limited. Some of these are discussed in this section. Pawar et al.[104] evaluated the adaptogenic effect of methanolic extract of *G. lucidum* fruiting bodies against swimming endurance followed by post swimming antifatigue and motor coordination and hypoxic stress tolerance test in mice. Oral administration (100, 300, and 500 mg/kg/day) of test extract showed dose dependent significant enhancement in the abovementioned parameters. The 100 mg/kg dose produced a considerable increase in hypoxia tolerance time. Whereas, concomitant treatment with the extract at doses 300 and 500 mg/kg showed significant increase.[104]

In another recent study, Luo et al. (2014) evaluated antihypoxia effects of Gl-PS in mice under the anoxic condition of ordinary pressure. The antifatigue effects were determined by exhaustive swimming times of mice, and biochemical parameters related to fatigue. The results demonstrated that GlPS prolonged survival and exhaustive swimming times, decreased blood lactic acid (BLA) and blood urea nitrogen (BUN) contents, and increased the liver and muscle glycogen contents of mice. Thus, GlPS had antihypoxia and antifatigue effects.[88]

Besides these, recently Rossi et al. (2014) have also showed that after a brief three-month supplementation of the athletes with *G. lucidum* and *Ophiocordyceps sinensis (combination) capsules*, the testosterone/cortisol ratio changed in a statistically significant manner, thereby, protecting the athletes from nonfunctional overreaching (NFO) and the overtraining syndrome (OTS). Besides, an increased scavenger capacity of free radicals in the athletes' serum after the race was observed thus, protecting the athletes from oxidative stress.[113] Therefore, as evident from the literature, *G. lucidum* exhibit substantial antistress, antioxidative, and adaptogenic properties.

7.6 CASE STUDIES

7.6.1 ANTICANDIDA BIOFILM ACTIVITY OF G. LUCIDUM

There are various human fungal pathogens that cause life threatening diseases, *Candida* species being one of them. As mentioned earlier (Section 5.1.4), it causes superficial to invasive infections primarily, in immunecompromised patients. There are various species of *Candida* but about 90% of *Candida* infections are caused by five major species such as *C. albicans, C. glabrata, C. tropicalis, C. parapsilosis,* and *C. krusei.*[40] The pathogenicity of *Candida* species is associated with their ability to form biofilms onto non-biotic surfaces of medical devices like catheters, dentures, and other surgical instruments in addition to the host tissues (mucosal). Biofilms are clusters of yeast cells embedded in an extracellular polymeric matrix that contribute primarily towards propagation of *Candida* pathogenicity[109,122] and impart them resistance to antifungal agents and thus, support their persistence or survival.[122]

Most of the available antifungal agents (amphotericin B, nystatin, fluconazole, and so forth.) are either ineffective or required in very high concentrations to inhibit *Candida* biofilm growth. The high doses of these drugs are often associated with marked side effects due to the toxicity. Moreover, biofilms may act as reservoirs and release the infective cells into the body. Even if the free cells are removed by antiobiotic treatment, the reservoirs remain unaffected and are able to cause reinfection.[110] This has opened avenues for the application of herbal products in the treatment of *Candidal* infections. There are several publications that support the use of natural products to evaluate interference *in C. albicans* biofilms.[121] However, its potential anti-biofilm activity against *Candida species* has remained untapped.

Considering these facts, in the laboratory at Defense Institute of Physiology (Delhi, India) authors evaluated various extracts (aqueous, methanolic, and ethyl acetate) of *G. lucidum* mycelium and the fruiting body against two *Candida species*; *C. albicans* and *C. glabrata* biofilms using XTT [2,3-Bis-(2-methoxy-4-nitro-5-sulfophenyl)-2H-tetrazolium-5-carboxanilide] reduction assay. The extracts reduced the adhesion, biofilm formation as well as the mature biofilm of the two test pathogens. The aqueous extract of *G. lucidum* mycelium was most potent among all the extracts.[8]

7.6.2 ADAPTOGENIC ACTIVITY OF G. LUCIDUM

Currently, in the laboratory at Defense Institute of Physiology (Delhi, India), authors evaluated the adaptogenic potential of the aqueous mycelium extract of *G. lucidum*[7] using cold, hypoxia, and restraint (C-H-R) animal model. The C-H-R model is the only passive multiple stress (cold, hypoxia, and restraint) animal model among other stress models.[108] In this model, the experimental animals in restraint state are exposed to cold (5°C) and hypoxia (428 mmHg) equivalent to an altitude of 4572 m.[128] This model evaluates endurance promotion of an adaptogen by thermo-regulation under stressful conditions (cold, hypoxia, and immobilization) via rectal temperature (T_{rec}), which is an indirect indicator of physical and mental performance capacity, the metabolic status of the animal and hypo-thalamic and higher brain function.[108] Therefore, C-H-R model is an ideal stress model to study the adaptogenic potential of any herbal preparation against the multifactorial pathophysiology that is associated with high altitude.

Authors selected three doses (50, 100, and 150 mg/kg body wt.) of the extract. After the C-H-R exposure, the rats were taken out when they attained T_{rec} of 23°C. The time is taken (min) to reach termination point that is, T_{rec} 23°C during the exposure was used as a measure of endurance. The dose 100 mg/kg b. wt. provided maximum resistance (93.55%) to cold-hypoxia-restraint (C-H-R) induced hypothermia by delaying fall in $T_{rec} = 23°C$ and was found comparable with other reported potent adap-togenic agents. And therefore, it was regarded as the dose with highest adaptogenic potential. However, 50 mg/kg body weight dose showed significant results as observed from hematological and biochemical [CAT, LDH, lipid peroxidation (MDA), reduced glutathione (GSH) SOD] param-eters in comparison to 100 and 150 mg/kg b. wt. doses. Overall, authors concluded that *G. lucidum* has protective adaptogenic efficacy against C-H-R exposure. Therefore, it was concluded that the aqueous extract of *G. lucidum* could be used as a protective herbal remedy against high alti-tude induced pathologies.

7.7 SUMMARY

G. lucidum is one of the most widely consumed and highly revered medicinal mushrooms with deep-rooted medicinal significance.It has been reported to possess a wide array of pharmaceutical properties such as anticarcinogenic, antimicrobial (antibacterial and antifungal), antioxi-dant, antitumorigenic properties, and so forth, which could be attributed to the presence of numerous bioactive principles it harbors, such as fatty acids, nucleotides, peptides, polysaccharides, proteins, steroids, sterols, terpenoids, and so forth. At present, *G. lucidum* has led to an establishment of 2.5 billion US$ industry, considering its commercial preparations. However, there are still a lot of scopes to endorse its potential use in food and pharmaceutical industry as dietary supplements and therapeutics and this necessitates further research.

In view of continuing the legacy of traditional medicine in human health, this chapter presents a review on the various therapeutic uses of the medicinal mushroom, *G. lucidum* along with the major bioactive constituents that contribute towards its medicinal effects.

7.8 ACKNOWLEDGMENT

The authors are thankful to the Director DIPAS (DRDO) for the constant support and encouragement.

KEYWORDS

- adaptogen
- antiangiogenic
- antibacterial
- antihypoxia
- anti-inflammatory
- antimetastatic
- antioxidative
- *Candida species*
- *Candida albicans*
- *Candida glabrata*
- chromatography
- cold-hypoxia-restraint
- cytotoxic
- flavonoids
- ganoderic acid
- *Ganoderma lucidum*
- hypoxia
- immunomodulatory
- *Ling zhi*
- medicinal mushrooms
- nucleobases
- nucleosides
- phenolics
- polysaccharides
- *Reishi*
- selenium
- **Traditional Chinese medicine**
- triterpenoids
- volatile organic compounds

REFERENCES

1. Adamec, J.; Jannasch, A.; Dudhgaonkar, S.; Jedinak, A.; Sedlak, M.; Sliva, D. Development of a New Method for Improved Identification and Relative Quantification of Unknown Metabolites in Complex Samples: Determination of a Triterpenoid Metabolic Fingerprint for the in Situ Characterization of *Ganoderma* Bioactive Compounds. *J. Sep. Sci.* **2009,** *32*(23–24), 4052–4058.

2. Adwan, G.; Salameh, Y.; Adwan, K.; Barakat, A. Assessment of Antifungal Activity of Herbal and Conventional Toothpastes Against Clinical Isolates of *Candida albicans. Asian Pac. J. Trop. Biomed.* **2012,** *2*(5), 375–379.

3. Ansor, N. M.; Abdullah, N.; Aminudin, N. Anti-Angiotensin Converting Enzyme (ACE) Proteins from Mycelia of *Ganoderma lucidum* (Curtis) P. Karst. *BMC Complementary Altern. Med.* **2013,** *13*(1), 1.

4. Bach, J. P.; Deuster, O.; Balzer-Geldsetzer, M.; Meyer, B.; Dodel, R.; Bacher, M. The Role of Macrophage Inhibitory Factor in Tumorigenesis and Central Nervous System Tumors. *Cancer* **2009,** *115*(10), 2031–2040.

5. Bai, D.; Chang, N. T.; Li, D. H.; Liu, J. X.; You, X. Y. Antiblastic Activity of *Ganoderma lucidum*Polysaccharides. *Acta. Agric. Bor. Sin.* **2008,** *S1*.

6. Basnyat, B.; Murdoch, D. R. High-Altitude Illness. *Lancet* **2003,** *361*(9373), 1967–1974.

7. Bhardwaj, A.; Sharma, P.; Mishra, J.; Rakhee., Suryakumar, G.; Misra, K. Abstracts of Papers, 4th International Conference and Exhibition on Pharmacognosy, Phytochemistry and Natural Products, Sao Paulo, Brazil, August 29–31, 2016; *Nat. Prod. Chem. Res.* **2016,** *4*, 5(Suppl); DOI: 10.4172/2329–6836.C1.011.

8. Bhardwaj, A.; Gupta, P.; Kumar, N.; Mishra, J.; Kumar, A.; Misra, K. *Int. J. Med. Mushrooms* **2017,** *in press*.

9. Bishop, K. S.; Kao, C. H.; Xu, Y.; Glucina, M. P.; Paterson, R. R. M.; Ferguson, L. R. From 2000 Years of *Ganoderma lucidum* to Recent Developments in Nutraceuticals. *Phytochemistry* **2015,** *114*, 56–65.

10. Bisko, N. A.; Mitropolskaya, N. Y. Some Biologically Active Substances from Medicinal Mushroom *Ganoderma lucidum* (W. Curt.: Fr.) P. Karst. (Aphyllophoromycetideae). *Int. J. Med. Mushrooms* **2003,** *5*(3).

11. Boh, B. *Ganoderma lucidum*: A Potential for Biotechnological Production of Anti-Cancer and Immunomodulatory Drugs. *Recent Pat. Anti-Cancer Drug Discovery* **2013,** *8*(3), 255–287.

12. Boh, B.; Berovic, M.; Zhang, J.; Zhi-Bin, L. *Ganoderma lucidum* and its Pharmaceutically Active Compounds. *Biotechnol. Ann. Rev.* **2007,** *13*, 265–301.

13. Calixto, J. B., Efficacy, Safety, Quality Control, Marketing and Regulatory Guidelines for Herbal Medicines (Phytotherapeutic Agents). *Braz. J. Med. Biol. Res.* **2000,** *33*(2), 179–189.

14. Cao, Q. Z.; Lin, Z. B. Antitumor and Anti-Angiogenic Activity of *Ganoderma lucidum* Polysaccharides Peptide. *Acta Pharmacol. Sin.* **2004,** *25*, 833–838.

15. Cao, Q. Z.; Lin, Z. B. *Ganoderma lucidum* Polysaccharides Peptide Inhibits the Growth of Vascular Endothelial Cell and The Induction of Vegf in Human Lung Cancer Cell. *Life Sci.* **2006,** *78*(13), 1457–1463.

16. Cao, Y.; Wu, S. H.; Dai, Y. C. Species Clarification of the Prize Medicinal *Gano-derma* Mushroom "Lingzhi". *Fungal Diversity* **2012,** *56*(1), 49–62.

17. Celik, G. Y.; Onbasli, D.; Altinsoy, B.; Alli, H. In Vitro Antimicrobial and Antioxidant Properties of *Ganoderma lucidum* Extracts Grown in Turkey. *Eur. J. Med. Plants* **2014,** *4*(6), 709.

18. CFR - Ferreira, I.; A Vaz, J.; Vasconcelos, M. H.; Martins, A. Compounds from Wild Mushrooms with Antitumor Potential. *Anti-Cancer Agents in Med. Chem. (Formerly Current Medicinal Chemistry-Anti-Cancer Agents)* **2010,** *10*(5), 424–436.

19. Chen, D. H.; Shiou, W. Y.; Wang, K. C.; Huang, S. Y.; Shie, Y. T.; Tsai, C. M.; Shie, J. F.; Chen, K. D., Chemotaxonomy of Triterpenoid Pattern of HPLC of *Ganoderma lucidum* and *Ganoderma tsugae*. *J. Chin.Chem. Soc.* **1999,** *46*(1), 47–51.

20. Chen, N. H.; Liu, J. W.; Zhong, J. J. Ganoderic Acid Me Inhibits Tumor Invasion Through Down-Regulating Matrix Metalloproteinases 2/9 Gene Expression. *J. Pharmacol.Sci.* **2008,** *108*(2), 212–216.

21. Chen, N. H.; Liu, J. W.; Zhong, J. J. Ganoderic Acid T Inhibits Tumor Invasion in Vitro and in Vivo Through Inhibition of Mmp Expression. *Pharmacol. Rep.* **2010,** *62*(1), 150–163.

22. Chen, Y.; Bicker, W.; Wu, J.; Xie, M.; Lindner, W. Simultaneous Determination of 16 Nucleosides and Nucleobases by Hydrophilic Interaction Chromatography and its Application to the Quality Evaluation of *Ganoderma*. *J. Agric. Food. Chem.* **2012,** *60*(17), 4243–4252.

23. Chen, Z. J.; Yang, Z. D.; Gu, Z. X. Determination of Volatile Flavor Compounds in *Ganoderma lucidum* by HSSPME-GC-MS.*Food Res. Develop.* **2010,** *2*, 43.

24. Cheng, S.; Sliva, D. *Ganoderma lucidum* for Cancer Treatment: We are Close but Still not There. *Int. Cancer Ther.* **2015,** *14*(3), 249–257.

25. Ćilerdžić, J.; Vukojević, J.; Stajić, M.; Stanojković, T.; Glamočlija, J. Biological Activity of *Ganoderma lucidum*Basidiocarps Cultivated on Alternative and Commercial Substrate. *J. Ethnopharmacol.* **2014,** *155*(1), 312–319.

26. Deepalakshmi, K.; Mirunalini, S. Therapeutic Properties and Current Medical Usage of Medicinal Mushroom: *Ganoderma lucidum*. *Int. J. Pharm. SciRes.* **2011,** *2*(8), 1922.

27. Devi, R. Diversity of Wild Edible Mushrooms in Indian Subcontinent and Its Neighboring Countries. *Recent Adv. Biol.Med.* **2015,** *1*, 69–75.

28. Dudhgaonkar, S.; Thyagarajan, A.; Sliva, D. Suppression of the Inflammatory Response by Triterpenes Isolated from the Mushroom *Ganoderma lucidum*. *Int. Immunopharmacol.* **2009,** *9*(11), 1272–1280.

29. Ekandjo, L. K.; Chimwamurombe, P. M. Traditional Medicinal Uses and Natural Hosts of the *Genus Ganoderma* in North-Eastern Parts of Namibia. *J. Pure Appl. Microbiol.* **2012,** *6*(3), 1139–1146.

30. Elhussainy, E.; Elzawawy, N.; Shorbagy, S. Novel Tannic Acid from *Ganoderma lucidum* Fruiting Bodies Extract Ameliorates Early Diabetic Nephropathy in Strepto-zotocin Induced Diabetic Rats. *Int. J. Pharm.Sci. Res.* **2016,** *7*(1), 62.

31. Eo, S. K.; Kim, Y. S.; Lee, C. K.; Han, S. S. Possible Mode of Antiviral Activity of Acidic Protein Bound Polysaccharide Isolated from *Ganoderma lucidum*on Herpes Simplex Viruses. *J. Ethnopharmacol.* **2000,** *72*(3), 475–481.

32. Fatmawati, S.; Kondo, R.; Shimizu, K. Structure—Activity Relationships of Lanostane-Type Triterpenoids from *Ganoderma lingzhi*as α-Glucosidase Inhibitors. *Bioorg. Med. Chem. Lett.* **2013,** *23*(21), 5900–5903.

33. Ferreira, I. C.; Barros, L.; Abreu, R. Antioxidants in Wild Mushrooms. *Curr. Med. Chem.* **2009,** *16*(12), 1543–1560.

34. Ferreira, I. C.; Heleno, S. A.; Reis, F. S.; Stojkovic, D.; Queiroz, M. J. R.; Vasconcelos, M. H.; Sokovic, M.Chemical Features of *Ganoderma* Polysaccharides with Antioxidant, Antitumor and Antimicrobial Activities. *Phytochemistry* **2015,** *114*, 38–55.

35. Gao, P.; Hirano, T.; Chen, Z.; Yasuhara, T.; Nakata, Y.; Sugimoto, A. Isolation and Identification of C-19 Fatty Acids with Anti-Tumor Activity from the Spores of *Ganoderma lucidum*(Reishi Mushroom). *Fitoterapia* **2012,** *83*(3), 490–499.

36. Gao, Y.; Gao, H.; Chan, E.; Tang, W.; Xu, A.; Yang, H.; Huang, M.; Lan, J.; Li, X.; Duan, W. Antitumor Activity and Underlying Mechanisms of Ganopoly, the Refined Polysaccharides Extracted from *Ganoderma lucidum*, in Mice. *Immunol. Invest.* **2005,** *34*(2), 171–198.

37. Gao, Y.; Tang, W.; Dai, X.; Gao, H.; Chen, G.; Ye, J.; Chan, E.; Koh, H. L.; Li, X.; Zhou, S. Effects of Water-Soluble *Ganoderma lucidum* Polysaccharides on the Immune Functions of Patients with Advanced Lung Cancer. *J.Med. Food* **2005,** *8*(2), 159–168.

38. Gao, Y.; Zhou, S.; Jiang, W.; Huang, M.; Dai, X. Effects of Ganopoly® (A *Ganoderma lucidum*Polysaccharide Extract) on the Immune Functions in Advanced-Stage Cancer Patients. *Immunol. Invest.* **2003,** *32*(3), 201–215.

39. Geiger, T. R.; Peeper, D. S. Metastasis Mechanisms. *Biochimica et Biophysica Acta (BBA)-Rev. Cancer* **2009,** *1796*(2), 293–308.

40. Guinea, J.Global Trends in the Distribution of Candida Species Causing Candidemia. *Clin. Microbiol. Infect.* **2014,** *20*(s6), 5–10.

41. Gupta, V.; Lahiri, S.; Sultana, S.; Kumar, R. Mechanism of Action of *Rhodiola imbricata Edgew* During Exposure to Cold, Hypoxia and Restraint (C–H–R) Stress Induced Hypothermia and Post Stress Recovery in Rats. *Food Chem. Toxicol.* **2009,** *47*(6), 1239–1245.

42. Gupta, V.; Lahiri, S.; Sultana, S.; Tulsawani, R.; Kumar, R. Anti-Oxidative Effect of *Rhodiola imbricata*Root Extract in Rats During Cold, Hypoxia and Restraint (C–H–R) Exposure and Post-Stress Recovery. *Food Chem. Toxicol.* **2010,** *48*(4), 1019–1025.

43. Gupta, V.; Saggu, S.; Tulsawani, R.; Sawhney, R.; Kumar, R. Dose Dependent Adaptogenic and Safety Evaluation of *Rhodiola imbricata Edgew*, a High Altitude Rhizome. *Food Chem. Toxicol.* **2008,** *46*(5), 1645–1652.

44. Hackett, P. H.; Roach, R. C. High-Altitude Illness. *New Engl. J. Med.* **2001,** *345*(2), 107–114.

45. Hapuarachchi, K.; Wen, T.; Deng, C.; Kang, J.; Hyde, K. Mycosphere Essays-1: Taxonomic Confusion in the *Ganoderma lucidum*Species Complex. *Mycosphere* **2015,** *6*(5), 542–559.

46. Harhaji-Trajković, L. M.; Mijatović, S. A.; Maksimović-Ivanić, D. D.; Stojanović, I. D.; Momčilović, M. B.; Tufegdžić, S. J.; Maksimović, V. M.; Marjanovi, Ž. S.;

Stošić-Grujičić, S. D. Anticancer Properties of *Ganoderma lucidum*Methanol Extracts in Vitro and in Vivo. *Nutr. Cancer* **2009,** *61*(5), 696–707.

47. Heleno, S. A.; Ferreira, I. C.; Esteves, A. P.; Ćirić, A.; Glamočlija, J.; Martins, A.; Soković, M.; Queiroz, M. J. R. Antimicrobial and Demelanizing Activity of *Ganoderma lucidum* Extract, P-Hydroxybenzoic and Cinnamic Acids and Their Synthetic Acetylated Glucuronide Methyl Esters. *Food Chem. Toxicol.* **2013,** *58*, 95–100.

48. Heleno, S. A; Barros, L.; Martins, A.; Queiroz, M. J. R; Santos-Buelga, C.; Ferreira, I. C. Fruiting Body, Spores and in Vitro Produced Mycelium of *Ganoderma lucidum*from Northeast Portugal: A Comparative Study of the Antioxidant Potential of Phenolic and Polysaccharidic Extracts. *Food Res. Int.* **2013,** *46*(1), 135–140.

49. Hennicke, F.; Cheikh-Ali, Z.; Liebisch, T.; Maciá-Vicente, J. G.; Bode, H. B.; Piepenbring, M. Distinguishing Commercially Grown *Ganoderma lucidum*from Ganoderma Lingzhi from Europe and East Asia on the Basis of Morphology, Molecular Phylogeny, and Triterpenic Acid Profiles. *Phytochemistry* **2016,** *127*, 29–37.

50. Hernandez-Marquez, E.; Lagunas-Martinez, A.; Bermudez-Morales, V. H.; Burgete-Garcfa, A. I.; Leon-Rivera, I.; Montiel-Arcos, E.; Garcia-Villa, E.; Gariglio, P.; Ondarza-Vidaurreta, R. N. Inhibitory Activity of Lingzhi or Reishi Medicinal Mushroom, *Ganoderma lucidum*(Higher Basidiomycetes) on Transformed Cells by Human Papillomavirus (HPV). *Int. J.Med. Mushrooms* **2014,** *16*(2), 179–187.

51. Hong, S. G.; Jung, H. S. Phylogenetic Analysis of *Ganoderma* Based on Nearly Complete Mitochondrial Small-Subunit Ribosomal DNA Sequences. *Mycologia* **2004,** *96*(4), 742–755.

52. Huang, C. Y.; Chen, J. Y. F.; Wu, J. E.; Pu, Y. S.; Liu, G. Y.; Pan, M. H.; Huang, Y. T.; Huang, A. M.; Hwang, C. C.; Chung, S. J. Ling-Zhi Polysaccharides Potentiate Cytotoxic Effects of Anticancer Drugs Against Drug-Resistant Urothelial Carcinoma Cells. *J. Agric. Food. Chem.,* **2010,** *58*(15), 8798–8805.

53. Huang, H.; Jiang, Y., The Quantitative Determination and TLC Identification of Total Nucleosides in Sporophore of *Ganoderma lucidum*(Leyss. ex Fr.) Karst. *J. Plant Resour. Environ.* **1999,** *9*(3), 61–62.

54. Huie, C. W.; Di, X. Chromatographic and Electrophoretic Methods for Lingzhi Pharmacologically Active Components. *J. Chromatogr. B* **2004,** *812*(1), 241–257.

55. Imray, C.; Grieve, A.; Dhillon, S. Cold Damage to the Extremities: Frostbite and Non-Freezing Cold Injuries. *Postgrad. Med. J.* **2009,** *85*(1007), 481–488.

56. Jaggi, A. S.; Bhatia, N.; Kumar, N.; Singh, N.; Anand, P.; Dhawan, R. Review on Animal Models for Screening Potential Anti-Stress Agents. *Neurol. Sci.* **2011,** *32*(6), 993–1005.

57. Jedinak, A.; Thyagarajan-Sahu, A.; Jiang, J.; Sliva, D. Ganodermanontriol, a Lanostanoid Triterpene from *Ganoderma lucidum*, Suppresses Growth of Colon Cancer Cells Through Ss-Catenin Signaling. *Int. J. Oncol.* **2011,** *38*(3), 761–767.

58. Jiang, J.; Grieb, B.; Thyagarajan, A.; Sliva, D. Ganoderic Acids Suppress Growth and Invasive Behavior of Breast Cancer Cells by Modulating Ap-1 and Nf-Kappa B Signaling. *Int. J. Mol. Med.* **2008,** *21*(5), 577.

59. Jiang, J.; Jedinak, A.; Sliva, D. Ganodermanontriol (GDNT) Exerts its Effect on Growth and Invasiveness of Breast Cancer Cells Through the Down-Regulation of CDC20 and uPA. *Biochem. Biophys. Res. Commun.* **2011,** *415*(2), 325–329.

60. Jing, L.; Ma, H.; Fan, P.; Gao, R.; Jia, Z. Antioxidant Potential, Total Phenolic and Total Flavonoid Contents of *Rhododendron* Anthopogonoides and its Protective Effect on Hypoxia-Induced Injury in PC12 Cells. *BMC Complementary Altern.Med.* **2015,** *15*(1), 287.

61. Jonathan, S.; Awotona, F. Studies on Antimicrobial Potentials of Three *Ganoderma* Species. *Afr. J. Biomed. Res.* **2010,** *13*(2), 131–139.

62. Joseph, S.; Sabulal, B.; George, V.; Antony, K.; Janardhanan, K. Antitumor and Anti-Inflammatory Activities of Polysaccharides Isolated from *Ganoderma lucidum*. *Acta Pharma.* **2011,** *61*(3), 335–342.

63. Kalogeropoulos, N.; Yanni, A. E.; Koutrotsios, G.; Aloupi, M. Bioactive Microcon-stituents and Antioxidant Properties of Wild Edible Mushrooms from the Island of Lesvos, Greece. *Food Chem. Toxicol.* **2013,** *55*, 378–385.

64. Kao, C.; Jesuthasan, A. C.; Bishop, K. S.; Glucina, M. P.; Ferguson, L. R. Anti-Cancer Activities of *Ganoderma lucidum*: Active Ingredients and Pathways. *Funct. Foods Health Dis.* **2013,** *3*(2), 48–65.

65. Kasahara, Y.; Hikino, H. Central Actions of Adenosine, a Nucleotide of *Ganoderma lucidum*. *Phytother. Res.* **1987,** *1*(4), 173–176.

66. Keypour, S.; Rafati, H.; Riahi, H.; Mirzajani, F.; Moradali, M. F. Qualitative Analysis of Ganoderic Acids in *Ganoderma lucidum*from Iran and China by RP-HPLC and Electrospray Ionisation-Mass Spectrometry (ESI-MS). *Food Chem.* **2010,** *119*(4), 1704–1708.

67. Khan, M. S.; Parveen, R.; Mishra, K.; Tulsawani, R.; Ahmad, S. Determination of Nucleosides in Cordyceps Sinensis and *Ganoderma lucidum*by High Performance Liquid Chromatography Method. *J.Pharm.Bioallied Sci.* **2015,** *7*(4), 264.

68. Khan, R. A.; Khan, M. R.; Sahreen, S.; Ahmed, M. Evaluation of Phenolic Contents and Antioxidant Activity of Various Solvent Extracts of Sonchus Asper (L.) Hill. *Chem. Central J.* **2012,** *6*(1), 1.

69. Kim, H. W.; Shim, M. J.; Choi, E. C.; Kim, B. K. Inhibition of Cytopathic Effect of Human Immunodeficiency Virus-1 by Water-Soluble Extract of *Ganoderma lucidum*. *Arch. Pharmacal Res.* **1997,** *20*(5), 425–431.

70. Kim, M. Y.; Lee, S. J.; Ahn, J. K.; Kim, E. H.; Kim, M. J.; Kim, S. L.; Moon, H. I.; Ro, H. M.; Kang, E. Y.; Seo, S. H. Comparison of Free Amino Acid, Carbohydrates Concentrations in Korean Edible and Medicinal Mushrooms. *Food Chem.* **2009,** *113*(2), 386–393.

71. Kimura, Y.; Taniguchi, M.; Baba, K. Antitumor and Antimetastatic Effects on Liver of Triterpenoid Fractions of *Ganoderma lucidum*: Mechanism of Action and Isolation of an Active Substance. *Anticancer Res.* **2001,** *22*(6A), 3309–3318.

72. Kirar, V.; Mehrotra, S.; Negi, P. S.; Nandi, S. P.; Misra, K. HPTLC Fingerprinting, Antioxidant Potential and Antimicrobial Efficacy of Indian Himalayan Lingzhi: *Ganoderma lucidum*. *Int. J. Pharma.Sci. Res.* **2015,** *6*(10), 4259.

73. Lai, C. S. W.; Yu, M. S.; Yuen, W. H.; So, K. F.; Zee, S. Y.; Chang, R. C. C. Antagonizing β-Amyloid Peptide Neurotoxicity of the Anti-Aging Fungus*Ganoderma lucidum*. *Brain Res.* **2008,** *1190*, 215–224.

74. Lakshmi, B.; Ajith, T.; Jose, N.; Janardhanan, K. Antimutagenic Activity of Methanolic Extract of *Ganoderma lucidum*and its Effect on Hepatic Damage Caused by Benzo [a] Pyrene. *J. Ethnopharmacol.* **2006,** *107*(2), 297–303.

75. Lee, J. M.; Kwon, H.; Jeong, H.; Lee, J. W.; Lee, S. Y.; Baek, S. J.; Surh, Y. J. Inhibition of Lipid Peroxidation and Oxidative DNA Damage by *Ganoderma lucidum*. *Phytother. Res.* **2001**, *15*(3), 245–249.

76. Lee, K. H. Research and Future Trends in the Pharmaceutical Development of Medicinal Herbs from Chinese Medicine. *Public Health Nutr.* **2000**, *3*(4a), 515–522.

77. Li, J.; Zhang, J.; Chen, H.; Chen, X.; Lan, J.; Liu, C. Complete Mitochondrial Genome of the Medicinal Mushroom *Ganoderma lucidum*. *PloS One* **2013**, *8*(8), e72038.

78. Li, L.; Lei, L.; Yu, C. Changes of Serum Interferon-Gamma Levels in Mice Bearing S-180 Tumor and the Interventional Effect of Immunomodulators. *Nan fang yi ke da xue xue bao (J.South. Med. Univ.)* **2008**, *28*(1), 65–68.

79. Liang, Z.; Guo, Y. T.; Yi, Y. J.; Wang, R. C.; Hu, Q. L.; Xiong, X. Y. *Ganoderma lucidum* Polysaccharides Target a Fas/Caspase Dependent Pathway to Induce Apoptosis in Human Colon Cancer Cells. *Asian Pac. J. Cancer Prev. (APJCP)* **2013**, *15*(9), 3981–3986.

80. Liang, Z.; Yi, Y.; Guo, Y.; Wang, R.; Hu, Q.; Xiong, X. Chemical Characterization and Antitumor Activities of Polysaccharide Extracted from *Ganoderma lucidum*. *Int. J. Mol. Sci.* **2014**, *15*(5), 9103–9116.

81. Lin, C. C.; Yu, Y. L.; Shih, C. C.; Liu, K. J.; Ou, K. L.; Hong, L. Z.; Chen, J. D.; Chu, C. L. Novel Adjuvant Ling Zhi-8 Enhances the Efficacy of DNA Cancer Vaccine by Activating Dendritic Cells. *Cancer Immunol. Immunother.* **2011**, *60*(7), 1019–1027.

82. Lindequist, U.; Kim, H. W.; Tiralongo, E.; Van Griensven, L. Medicinal Mushrooms. *Evidence-Based Complementary Altern.Med.: eCAM,* **2014**, *2014*.

83. Liu, J.; Kurashiki, K.; Fukuta, A.; Kaneko, S.; Suimi, Y.; Shimizu, K.; Kondo, R. Quantitative Determination of the Representative Triterpenoids in the Extracts of *Ganoderma lucidum*with Different Growth Stages Using High-Performance Liquid Chromatography for Evaluation of Their 5α-Reductase Inhibitory Properties. *Food Chem.* **2012**, *133*(3), 1034–1038.

84. Liu, S. Y.; Wang, Y.; He, R. R.; Qu, G. X.; Qiu, F. Chemical Constituents of Ganoderma Lucidum (Leys. ex Fr.) Karst [J]. *J. Shenyang Pharm. Univ.* **2008**, *3*, 004.

85. Liu, Y.; Liu, Y.; Qiu, F.; Di, X.Sensitive and Selective Liquid Chromatography—Tandem Mass Spectrometry Method for the Determination of Five Ganoderic Acids in *Ganoderma lucidum* and its Related Species. *J. Pharm. Biomed. Anal.* **2011**, *54*(4), 717–721.

86. Loganathan, J.; Jiang, J.; Smith, A.; Jedinak, A.; Thyagarajan-Sahu, A.; Sandusky, G. E.; Nakshatri, H.; Sliva, D.The Mushroom *Ganoderma lucidum* Suppresses Breast-to-Lung Cancer Metastasis Through the Inhibition of Pro-Invasive Genes. *Int. J. Oncol.,* **2014**, *44*(6), 2009–2015.

87. Lowe, E.; Rice, P.; Ha, T.; Li, C.; Kelley, J.; Ensley, H.; Lopez-Perez, J.; Kalbfleisch, J.; Lowman, D.; Margl, P. The (1→ 3)-β-D-Linked Heptasaccharide is the Unit Ligand for Glucan Pattern Recognition Receptors on Human Monocytes. *Microb. Infect.* **2001**, *3*(10), 789–797.

88. Luo, L.; Cai, L. M.; Hu, X. J. Evaluation of the Anti-Hypoxia and Anti-Fatigue Effects of *Ganoderma lucidum* Polysaccharides. *Appl. Mech. Mater. Trans. Tech. Publ. Ltd.* **2014**, *522–524*, 303–306.

89. Mantovani, A.; Sica, A. Macrophages, Innate Immunity and Cancer: Balance, Tolerance, and Diversity. *Curr. Opin. Immunol.* **2010**, *22*(2), 231–237.

90. Marcos-Arias, C.; Eraso, E.; Madariaga, L.; Quindós, G. In Vitro Activities of Natural Products Against Oral Candida Isolates from Denture Wearers. *BMC Complementary Altern. Med.* **2011,** *11*(1), 1.

91. Mau, J. L.; Lin, H. C.; Chen, C. C. Non-Volatile Components of Several Medicinal Mushrooms. *Food Res. Int.* **2001,** *34*(6), 521–526.

92. Miura, T.; Yuan, L.; Sun, B.; Fujii, H.; Yoshida, M.; Wakame, K.; Kosuna, K. I. Isoflavone Aglycon Produced by Culture of Soybean Extracts with Basidiomycetes and its Anti-Angiogenic Activity. *Biosci. Biotechnol.Biochem.* **2002,** *66*(12), 2626–2631.

93. Miyazak, T.; Nishijima, M. Structural Examination of a Water Soluble Antitumor Polysaccharide of *Ganoderma lucidum. Chem. Pharm. Bull.* **1981,** *29*, 3611–3616.

94. Mizuno, T.; Ykohlui, P.; Kinoshita, T.; Zhuang, C.; Ito, H.; Mayuzumi, Y. Antitumor Activity and Chemical Modification of Polysaccharides from *Niohshimeji* Mushroom, *Tricholma giganteum. Biosci. Biotechnol. Biochem.* **1996,** *60*(1), 30–33.

95. Moncalvo, J. M.; Wang, H. F.; Hseu, R. S. Gene Phylogeny of the *Ganoderma lucidum*Complex Based on Ribosomal DNA Sequences. Comparison with Traditional Taxonomic Characters. *Mycol. Res.* **1995,** *99*(12), 1489–1499.

96. Morath, S. U.; Hung, R.; Bennett, J. W. Fungal Volatile Organic Compounds: Review with Emphasis on Their Biotechnological Potential. *Fungal Biol. Rev.* **2012,** *26*(2), 73–83.

97. Nahata, A. *Ganoderma lucidum*: Potent Medicinal Mushroom with Numerous Health Benefits. *Pharm. Anal. Acta* **2013,** *4*(10), e159. DOI:10.4172/2153–2435.1000e159.

98. Nayak, A.; Nayak, R. N.; Bhat, K. Antifungal Activity of a Toothpaste Containing *Ganoderma lucidum*Against *Candida Albicans*—An in Vitro Study. *J. Int. Oral. Health* **2010,** *2*(2), 51–57.

99. Nguyen, V. T.; Tung, N. T.; Cuong, T. D.; Hung, T. M.; Kim, J. A.; Woo, M. H.; Choi, J. S.; Lee, J. H.; Min, B. S. Cytotoxic and Anti-Angiogenic Effects of Lanostane Triterpenoids from *Ganoderma lucidum. Phytochem. Lett.* **2015,** *12*, 69–74.

100. Ooi, V. E. C.; Liu, F. Review of Pharmacological Activities of Mushroom Polysaccharides. *Int. J.Med.Mushrooms* **1999,** *1*(3), 195–206.

101. Pal, M.; Bhardwaj, A.; Manickam, M.; Tulsawani, R.; Srivastava, M.; Sugadev, R.; Misra, K. Protective Efficacy of the Caterpillar Mushroom, Ophiocordyceps Sinensis (Ascomycetes), from India in Neuronal Hippocampal Cells Against Hypoxia. *Int. J. Med. Mushrooms* **2015,** *17*(9), 829–840.

102. Paliya, B. S.; Verma, S.; Chaudhary, H. S. Major Bioactive Metabolites of the Medicinal Mushroom: *Ganoderma lucidum. Int. J. Pharm. Res.* **2014,** *6*(1), 13.

103. Paterson, R. R. M. *Ganoderma*: Therapeutic Fungal Biofactory. *Phytochemistry* **2006,** *67*(18), 1985–2001.

104. Pawar, V. S.; Shivakumar, H. Adaptogenic (Antistress) Activity of Methanolic Extract of *Ganoderma lucidum*Against Physical and Hypoxic Stress in Mice. *Pharmacology-online* **2011,** *2*, 989–995.

105. Peng, J.; Peng, Q.; Lin, L.; Dong, W.; Liu, T.; Xia, X.; Yang, D. Simultaneous Determination of 13 Nucleosides and Nucleobases in *Ganoderma lucidum*and Related Species by HPLC-DAD. *Asian J. Chem.* **2014,** *26*(12), 3477.

106. Quereshi, S.; Pandey, A.; Sandhu, S. Evaluation of Antibacterial Activity of Different *Ganoderma lucidum*Extracts. *Dep. Biol. Sci. R. D. Univ. Jabalpur. Centre Sci. Res.*

Develop. People's Group Bhanpur, Bhopal-462037 (M.P.), India; *People's J. Sci. Res.* **2010**, *3*(1), 9–13.

107. Radwan, F. F.; Hossain, A.; God, J. M.; Leaphart, N.; Elvington, M.; Nagarkatti, M.; Tomlinson, S.; Haque, A. Reduction of Myeloid-Derived Suppressor Cells and Lymphoma Growth by a Natural Triterpenoid. *J. Cell. Biochem.* **2015**, *116*(1), 102–114.

108. Ramachandran, U.; Divekar, H.; Grover, S.; Srivastava, K. New Experimental Model for the Evaluation of Adaptogenic Products. *J. Ethnopharmacol.* **1990**, *29*(3), 275–281.

109. Ramage, G.; Saville, S. P.; Thomas, D. P.; Lopez-Ribot, J. L. Candida Biofilms: An Update. *Eukaryot. Cell.* **2005**, *4*(4), 633–638.

110. Raut, J. S.; Shinde, R. B.; Chauhan, N. M.; Mohan Karuppayil, S. Terpenoids of Plant Origin Inhibit Morphogenesis, Adhesion, and Biofilm Formation by *Candida albicans*. *Biofouling* **2013**, *29*(1), 87–96.

111. Reis, F. S.; Lima, R. T.; Morales, P.; Ferreira, I. C.; Vasconcelos, M. H. Methanolic Extract of *Ganoderma lucidum*Induces Autophagy of Ags Human Gastric Tumor Cells. *Molecules* **2015**, *20*(10), 17872–17882.

112. Richter, C.; Wittstein, K.; Kirk, P. M.; Stadler, M. An Assessment of the Taxonomy and Chemotaxonomy of *Ganoderma*. *Fungal Diversity* **2015**, *71*(1), 1–15.

113. Rossi, P.; Buonocore, D.; Altobelli, E.; Brandalise, F.; Cesaroni, V.; Iozzi, D.; Savino, E.; Marzatico, F. Improving Training Condition Assessment in Endurance Cyclists: Effects of *Ganoderma lucidum and*Ophiocordyceps Sinensis Dietary Supplementation. *Evidence-Based Complementary Altern. Med.* **2014**, *2014*, Article ID 979613.

114. Ruan, W.; Lim, A. H. H.; Huang, L. G.; Popovich, D. G. Extraction Optimization and Isolation of Triterpenoids from *Ganoderma lucidum*and Their Effects on Human Carcinoma Cell Growth. *Nat. Prod. Res.* **2014**, *28*(24), 2264–2272.

115. Saeed, N.; Khan, M. R.; Shabbir, M. Antioxidant Activity, Total Phenolic and Total Flavonoid Contents of Whole Plant Extracts*Torilis leptophylla* L. *BMC Complementary Altern. Med.* **2012**, *12*(1), 1.

116. Saenger, W. Structure and Function of Nucleosides and Nucleotides. *Angew. Chem. Int. Ed. Engl.* **1973**, *12*(8), 591–601.

117. Saggu, S.; Kumar, R. Modulatory Effect of Sea Buckthorn Leaf Extract on Oxidative Stress Parameters in Rats During Exposure to Cold, Hypoxia and Restraint (C-H-R) Stress and Post Stress Recovery. *J. Pharm. Pharmacol.* **2007**, *59*(12), 1739–1745.

118. Saltarelli, R.; Ceccaroli, P.; Buffalini, M.; Vallorani, L.; Casadei, L.; Zambonelli, A.; Iotti, M.; Badalyan, S.; Stocchi, V. Biochemical Characterization and Antioxidant and Antiproliferative Activities of Different *Ganoderma* Collections. *J.Mol. Microbiol. Biotechnol.* **2015**, *25*(1), 16–25.

119. Saltarelli, R.; Ceccaroli, P.; Iotti, M.; Zambonelli, A.; Buffalini, M.; Casadei, L.; Vallorani, L.; Stocchi, V. Biochemical Characterization and Antioxidant Activity of Mycelium of *Ganoderma lucidum*from Central Italy. *Food Chem.* **2009**, *116*(1), 143–151.

120. Sanodiya, B. S.; Thakur, G. S.; Baghel, R. K.; Prasad, G.; Bisen, P. *Ganoderma lucidum*: Potent Pharmacological Macrofungus. *Curr. Pharm. Biotechnol.* **2009**, *10*(8), 717–742.

121. Sardi, J.; Scorzoni, L.; Bernardi, T.; Fusco-Almeida, A.; Giannini, M. M. Candida Species: Current Epidemiology, Pathogenicity, Biofilm Formation, Natural Antifungal Products and New Therapeutic Options. *J. Med. Microbiol.* **2013,** *62*(1), 10–24.

122. Seneviratne, C.; Silva, W.; Jin, L.; Samaranayake, Y.; Samaranayake, L., Architectural Analysis, Viability Assessment and Growth Kinetics of *Candida Albicans* and *Candida Glabrata* Biofilms. *Arch. Oral Biol.* **2009,** *54*(11), 1052–1060.

123. Seow, S. L. S.; Naidu, M.; David, P.; Wong, K. H.; Sabaratnam, V. Potentiation of Neuritogenic Activity of Medicinal Mushrooms in Rat Pheochromocytoma Cells. *BMC Complementary Altern. Med.* **2013,** *13*(1), 157.

124. Shamaki, B.; Sandabe, U. K.; Fanna, I. A.; Adamu, O. O.; Geidam, Y.; Umar, I.; Adamu, M. S. Proximate Composition, Phytochemical and Elemental Analysis of Some Organic Solvent Extract of the Wild Mushroom *Ganoderma lucidum*. *J. Nat. Sci. Res.* **2012,** *2*(4), 24–35.

125. Shang, D.; Li, Y.; Wang, C.; Wang, X.; Yu, Z.; Fu, X., A Novel Polysaccharide from Se-Enriched *Ganoderma lucidum* Induces Apoptosis of Human Breast Cancer Cells. *Oncol. Rep.* **2011,** *25*(1), 267.

126. Shang, D.; Zhang, J.; Wen, L.; Li, Y.; Cui, Q., Preparation, Characterization, and Antiproliferative Activities of the Se-Containing Polysaccharide Seglp-2b-1 from Se-Enriched *Ganoderma lucidum*. *J. Agric. Food Chem.* **2009,** *57*(17), 7737–7742.

127. Sharifi, A.; Naseri, M. H.; Jahedi, S.; Sarkary, B.; Rooz, S. S. K.; Khosravani, S. M.; Kalantar, E. Antimicrobial Potentials of Crude Fractions of Polysaccharides of *Ganoderma spp*. *Afr. J. Microbiol. Res.* **2012,** *6*(39), 6817–6821.

128. Sharma, P.; Kirar, V.; Meena, D. K.; Suryakumar, G.; Misra, K. Adaptogenic Activity of *Valeriana wallichii* Using Cold, Hypoxia and Restraint Multiple Stress Animal Model. *Biomed. Aging Pathol.* **2012,** *2*(4), 198–205.

129. Sheikh, I. A.; Vyas, D.; Ganaie, M. A.; Dehariya, K.; Singh, V. HPLC Determination of Phenolics and Free Radical Scavenging Activity of Ethanolic Extracts of Two Polypore Mushrooms. *Int. J. Pharm. Pharm. Sci.* **2014,** *6*(2), 679–684.

130. Shekhar, R.; Gauri, B.; Jitendra, G.; Sandhya, A.; Hiralal, B. Antimicrobial Activity of Terpenoid Extracts from *Ganoderma* Samples. *Int. J. Pharm. Life Sci. (IJPLS)* **2010,** *1*(4), 234–240.

131. Shi, Y.; Cai, D.; Wang, X.; Liu, X. Immunomodulatory Effect of *Ganoderma lucidum* Polysaccharides (GLP) on Long-Term Heavy-Load Exercising Mice. *Int. J. Vitam. Nutr. Res. (Internationale Zeitschrift für Vitamin- und Ernahrungsforschung. Journal international de vitaminologie et de nutrition)* **2012,** *82*(6), 383–90.

132. Shikongo, L.; Chimwamurombe, P.; Lotfy, H. Antimicrobial Screening of Crude Extracts from the Indigenous *Ganoderma lucidum* Mushrooms in Namibia. *Afr. J. Microbiol. Res.* **2013,** *7*(40), 4812–4816.

133. Siwulski, M.; Sobieralski, K.; Golak-Siwulska, I.; Sokół, S.; Sękara, A. *Ganoderma lucidum* (Curt.: Fr.) Karst.–Health-Promoting Properties. A Review. *Herba Pol.* **2015,** *61*(3), 105–118.

134. Skalicka-Wozniak, K.; Szypowski, J.; Los, R.; Siwulski, M.; Sobieralski, K.; Glowniak, K.; Malm, A. Evaluation of Polysaccharides Content in Fruit Bodies and Their Antimicrobial Activity of Four *Ganoderma lucidum* (W Curt.: Fr.) P. Karst. Strains Cultivated on Different Wood Type Substrates. *Acta Soc. Bot. Pol.* **2012,** *81*(1).

135. Sliva, D. *Ganoderma lucidum* (Reishi) in Cancer Treatment. *Integr. Cancer Ther.* 2003, *2*(4), 358–364.
136. Sliva, D.; Loganathan, J.; Jiang, J.; Jedinak, A.; Lamb, J. G.; Terry, C.; Baldridge, L. A.; Adamec, J.; Sandusky, G. E.; Dudhgaonkar, S. Mushroom *Ganoderma lucidum*Prevents Colitis-Associated Carcinogenesis in Mice. *PloS One* **2012**, *7*(10), e47873.
137. Smina, T.; De, S.; Devasagayam, T.; Adhikari, S.; Janardhanan, K. *Ganoderma lucidum* Total Triterpenes Prevent Radiation-Induced DNA Damage and Apoptosis in Splenic Lymphocytes in Vitro. *Mutat. Res./Genet. Toxicol. Environ. Mutagen.* **2011**, *726*(2), 188–194.
138. Smina, T.; Mathew, J.; Janardhanan, K.; Devasagayam, T. Antioxidant Activity and Toxicity Profile of Total Triterpenes Isolated from *Ganoderma lucidum* (Fr.) P. Karst Occurring in South India. *Environ. Toxicol. Pharmacol.* **2011**, *32*(3), 438–446.
139. Smith, G. Herbs in Medicine. *Practice* **2004**, *154*, 439–441.
140. Smith, J. E.; Rowan, N. J.; Sullivan, R. Medicinal Mushrooms: Rapidly Developing Area of Biotechnology for Cancer Therapy and Other Bioactivities. *Biotechnol. Lett.* **2002**, *24*(22), 1839–1845.
141. Song, B. J.; Zhu, X. J.; Wei, L. N. Effect of *Ganoderma lucidum*Spores on Immune Modulation and Inhibiting Tumor in Mice [J]. *J. Harbin Med. Univ.* **2010**, *5*, 014.
142. Song, Y. S.; Kim, S. H.; Sa, J. H.; Jin, C.; Lim, C. J.; Park, E. H. Anti-Angiogenic and Inhibitory Activity on Inducible Nitric Oxide Production of the Mushroom *Ganoderma lucidum*. *J. Ethnopharmacol.* **2004**, *90*(1), 17–20.
143. Stanley, G.; Harvey, K.; Slivova, V.; Jiang, J.; Sliva, D. *Ganoderma lucidum* Suppresses Angiogenesis Through the Inhibition of Secretion of VEGF and TGF-β1 from Prostate Cancer Cells. *Biochem. Biophys.Res. Commun.* **2005**, *330*(1), 46–52.
144. Stojković, D. S.; Barros, L.; Calhelha, R. C.; Glamočlija, J.; Ćirić, A.; Van Griensven, L. J.; Soković, M.; Ferreira, I. C. Detailed Comparative Study Between Chemical and Bioactive Properties of *Ganoderma lucidum*from Different Origins. *Int. J. Food Sci. Nutr.* **2014**, *65*(1), 42–47.
145. Suarez-Arroyo, I. J.; Rosario-Acevedo, R.; Aguilar-Perez, A.; Clemente, P. L.; Cubano, L. A.; Serrano, J.; Schneider, R. J.; Martínez-Montemayor, M. M. Antitumor Effects of *Ganoderma lucidum* (Reishi) in Inflammatory Breast Cancer inin Vivo and in Vitro Models. *PloS One* **2013**, *8*(2), e57431.
146. Sun, J.; Chu, Y. F.; Wu, X.; Liu, R. H. Antioxidant and Antiproliferative Activities of Common Fruits. *J. Agric. Food. Chem.* **2002**, *50*(25), 7449–7454.
147. Sun, L. X.; Lin, Z. B.; Li, X. J.; Li, M.; Lu, J.; Duan, X. S.; Ge, Z. H.; Song, Y. X.; Xing, E. H.; Li, W. D.Promoting Effects of *Ganoderma lucidum* Polysaccharides on B16F10 Cells to Activate Lymphocytes. *Basic Clin. Pharmacol. Toxicol.* **2011**, *108*(3), 149–154.
148. Sun, S. J.; Gao, W.; Lin, S. Q.; Zhu, J.; Xie, B. G.; Lin, Z. B. Analysis of Genetic Diversity in *Ganoderma* Population with a Novel Molecular Marker SRAP. *Appl. Microbiol. Biotechnol.* **2006**, *72*(3), 537–543.
149. Sun, X.; Wang, H.; Han, X.; Chen, S.; Zhu, S.; Dai, J. Fingerprint Analysis of Poly-saccharides from Different *Ganoderma* by HPLC Combined with Chemometrics Methods. *Carbohydr. Polym.* **2014**, *114*, 432–439.

150. Tang, W.; Liu, J. W.; Zhao, W. M.; Wei, D. Z.; Zhong, J. J. Ganoderic Acid T from *Ganoderma lucidum*Mycelia Induces Mitochondria Mediated Apoptosis in Lung Cancer Cells. *Life Sci.* **2006,** *80*(3), 205–211.

151. Taskin, H.; Kafkas, E.; Çakiroglu, Ö.; Büyükalaca, S. Determination of Volatile Aroma Compounds of *Ganoderma lucidum*by Gas Chromatography Mass Spectrometry (HS-GC/MS). *Afr. J. Tradit., Complementary Altern. Med.* **2013,** *10*(2), 353–355.

152. Thyagarajan, A.; Jedinak, A.; Nguyen, H.; Terry, C.; Baldridge, L. A.; Jiang, J.; Sliva, D., Triterpenes from *Ganoderma lucidum*Induce Autophagy in Colon Cancer Through the Inhibition of p38 Mitogen-Activated Kinase (p38 MAPK). *Nutr. Cancer* **2010,** *62*(5), 630–640.

153. Toth, J.; Bang, L.; Beck, J. P.; Ourisson, G. Cytotoxic Triterpenes from *Ganoderma lucidum* (Polyporaceae): Structures of Ganoderic Acids UZ. *J. Chem.Res. Synop.* **1983,** *12*, 110–115.

154. Trigos, Á.; Suárez Medellín, J. *Metabolitos biológicamente activos del género Ganoderma: tres décadas de investigación mico-química* (Biologically Active Metabolites of the Genus *Ganoderma*: Three Decades of Myco-Chemical Research). *Revista mexicana de micología (Mex. J.Mycol.)* **2011,** *34*, 63–83.

155. Tsai, C. C.; Yang, F. L.; Huang, Z. Y.; Chen, C. S.; Yang, Y. L.; Hua, K. F.; Li, J.; Chen, S. T.; Wu, S. H. Oligosaccharide and Peptidoglycan of *Ganoderma lucidum* Activate the Immune Response in Human Mononuclear Cells. *J.Agric. Food Chem.* **2012,** *60*(11), 2830–2837.

156. Valverde, M. E.; Hernández-Pérez, T.; Paredes-López, O. Edible Mushrooms: Improving Human Health and Promoting Quality Life. *Int. J. Microbiol.* **2015,** *2015*.

157. Vazirian, M.; Faramarzi, M. A.; Ebrahimi, S. E. S.; Esfahani, H. R. M.; Samadi, N.; Hosseini, S. A.; Asghari, A.; Manayi, A.; Mousazadeh, S. A.; Asef, M. R. Antimicrobial Effect of the Lingzhi or Reishi Medicinal Mushroom, *Ganoderma lucidum* (Higher Basidiomycetes) and its Main Compounds. *Int. J. Med.Mushrooms* **2014,** *16*(1).

158. Villasana, M.; Ochoa, G.; Aguilar, S. Modeling and Optimization of Combined Cytostatic and Cytotoxic Cancer Chemotherapy. *Artif. Intell. Med.* **2010,** *50*(3), 163–173.

159. Wachtel-Galor, S.; Yuen, J.; Buswell, J. A.; Benzie, I. F. *Ganoderma lucidum* (Lingzhi or Reishi). Chapter 9. In*Herbal Medicine: Biomolecular and Clinical Aspects.* 2nd ed.;Benzie, I.F.F., Wachtel-Galor, S.,Eds.; CRC Press/Taylor and Francis: Boca Raton (FL), 2011.

160. Wang, G.; Zhao, J.; Liu, J.; Huang, Y.; Zhong, J.J.; Tang, W. Enhancement of IL-2 and IFN-γ Expression and NK Cells Activity Involved in the Anti-Tumor Effect of Ganoderic Acid Me in Vivo. *Int. Immunopharmacol.* **2007,** *7*(6), 864–870.

161. Wang, J.; Zhang, L.; Yu, Y.; Cheung, P. C. Enhancement of Antitumor Activities in Sulfated and Carboxymethylated Polysaccharides of *Ganoderma lucidum*. *J. Agric. Food. Chem.* **2009,** *57*(22), 10565–10572.

162. Wang, M.; Lamers, R. J. A.; Korthout, H. A.; van Nesselrooij, J. H.; Witkamp, R. F.; van der Heijden, R.; Voshol, P. J.; Havekes, L. M.; Verpoorte, R.; van der Greef, J. Metabolomics in the Context of Systems Biology: Bridging Traditional Chinese Medicine and Molecular Pharmacology. *Phytother. Res.* **2005,** *19*(3), 173–182.

163. Wang, P. Y.; Zhu, X. L.; Lin, Z. B. Antitumor and Immunomodulatory Effects of Polysaccharides from Broken-Spore of *Ganoderma lucidum*. *Front. Pharmacol.* **2012,** *3*, 135.

164. Wang, Y.; Xu, B. Distribution of Antioxidant Activities and Total Phenolic Contents in Acetone, Ethanol, Water and Hot Water Extracts from 20 Edible Mushrooms via Sequential Extraction. *Austin J. Nutr.Food Sci.* **2014,** *2*(1), 5.

165. Wasser, S. P. Review of Medicinal Mushrooms Advances: Good News from Old Allies. *Herbal Gram* **2002,** *56*, 28–33.

166. Wasser, S. P.; Coates, P.; Blackman, M.; Cragg, G.; Levine, M.; Moss, J.; White, J. Reishi or Lingzhi (*Ganoderma lucidum*). In*Encyclopedia of Dietary Supplements;* Marcel Dekker: New York, 2005; pp 680–690.

167. Wasser, S. Medicinal Mushroom Science: Current Prospects, Advances, Evidences, and Challenges. *Biosphere* **2015,** *7*(2), 212–218.

168. Wasser, S. P.; Weis, A. L. Therapeutic Effects of Substances Occurring in Higher Basidiomycetes Mushrooms: A Modern Perspective. *Crit. Rev.Immunol.* **1999,** *19*(1).

169. Watanabe, K.; Shuto, T.; Sato, M.; Onuki, K.; Mizunoe, S.; Suzuki, S.; Sato, T.; Koga, T.; Suico, M. A.; Kai, H. Lucidenic Acids-Rich Extract from Altered form of *Ganoderma lucidum* Enhances TNFα Induction in THP-1 Monocytic Cells Possibly via its Modulation of Map Kinases p38 and JNK. *Biochem. Biophys. Res. Commun.* **2011,** *408*(1), 18–24.

170. Wen, H.; Kang, S.; Song, Y.; Song, Y.; Sung, S. H.; Park, S. Differentiation of Cultivation Sources of *Ganoderma lucidum*by NMR-Based Metabolomics Approach. *Phytochem. Anal.* **2010,** *21*(1), 73–79.

171. Weng, C. J.; Yen, G. C. The in Vitro and in Vivo Experimental Evidences Disclose the Chemopreventive Effects of *Ganoderma lucidum*on Cancer Invasion and Metastasis. *Clin. Exp. Metastasis* **2010,** *27*(5), 361–369.

172. Weng, C. J.; Chau, C. F.; Chen, K. D.; Chen, D. H.; Yen, G. C. The Anti-Invasive Effect of Lucidenic Acids Isolated from a New *Ganoderma lucidum*Strain. *Mol. Nutr. Food Res.* **2007,** *51*(12), 1472–1477.

173. Weng, C. J.; Chau, C. F.; Hsieh, Y. S.; Yang, S. F.; Yen, G. C. Lucidenic Acid Inhibits PMA-Induced Invasion of Human Hepatoma Cells Through Inactivating MAPK/ERK Signal Transduction Pathway and Reducing Binding Activities of NF-κB and AP-1. *Carcinogenesis* **2008,** *29*(1), 147–156.

174. Xia, Q.; Zhang, H.; Sun, X.; Zhao, H.; Wu, L.; Zhu, D.; Yang, G.; Shao, Y.; Zhang, X.; Mao, X. Comprehensive Review of the Structure Elucidation and Biological Activity of Triterpenoids from *Ganoderma spp. Molecules* **2014,** *19*(11), 17478–17535.

175. Xu, J. W.; Zhao, W.; Zhong, J. J. Biotechnological Production and Application of Ganoderic Acids. *Appl. Microbiol. Biotechnol.* **2010,** *87*(2), 457–466.

176. Xu, Z.; Chen, X.; Zhong, Z.; Chen, L.; Wang, Y. *Ganoderma lucidum*Polysaccharides: Immunomodulation and Potential Anti-Tumor Activities. *Am. J. Chin. Med.* **2011,** *39*(1), 15–27.

177. Yamamoto, S.; Wang, M. F.; Adjei, A. A.; Ameho, C. K. Role of Nucleosides and Nucleotides in the Immune System, Gut Reparation after Injury, and Brain Function. *Nutrition* **1997,** *13*(4), 372–374.

178. Yang, G.; Yang, L.; Zhuang, Y.; Qian, X.; Shen, Y. *Ganoderma lucidum*Polysaccharide Exerts Anti-Tumor Activity via MAPK Pathways in HL-60 Acute Leukemia Cells. *J. Recept. Signal Transduct.* **2016**, *36*(1), 6–13.

179. Yang, Z. L.; Feng, B. What is the Chinese "Lingzhi"?—A Taxonomic Mini-Review. *Mycology* **2013**, *4*(1), 1–4.

180. Yao, Y. J.; Wang, X. C.; Wang, B. Epitypification of *Ganoderma sichuanense* J. D. Zhao and X.Q. Zhang (Ganodermataceae). *Taxon* **2013**, *62*(5), 1025–1031.

181. Yildiz, O.; Can, Z.; Laghari, A. Q.; Şahin, H.; Malkoç, M. Wild Edible Mushrooms as a Natural Source of Phenolics and Antioxidants. *J. Food Biochem.* **2015**, *39*(2), 148–154.

182. Yue, Q. X.; Cao, Z. W.; Guan, S. H.; Liu, X. H.; Tao, L.; Wu, W. Y.; Li, Y. X.; Yang, P. Y.; Liu, X.; Guo, D. A. Proteomics Characterization of the Cytotoxicity Mechanism of Ganoderic Acid D and Computer-Automated Estimation of the Possible Drug Target Network. *Mol. Cell. Proteomics* **2008**, *7*(5), 949–961.

183. Yue, Q. X.; Song, X. Y.; Ma, C.; Feng, L. X.; Guan, S. H.; Wu, W. Y.; Yang, M.; Jiang, B. H.; Liu, X.; Cui, Y. J. Effects of Triterpenes from *Ganoderma lucidum*on Protein Expression Profile of HeLa Cells. *Phytomedicine* **2010**, *17*(8), 606–613.

184. Yuen, J. W.; Gohel, M. D. I., Anticancer Effects of *Ganoderma lucidum*: A Review of Scientific Evidence. *Nutr. Cancer* **2005**, *53*(1), 11–17.

185. Yuen, W. M. J.;Ed.Chemopreventive Effects of Ganoderma lucidum on Human Uroepithelial Cell Carcinoma. Ph.D. Thesis, The Hong Kong Polytechnic University, 2007, p 282.PolyU Library Call No.: [THS] LG51.H577P HTI 2007 Yuen; <http://hdl.handle.net/10397/2771>.

186. Zhang, J.; Tang, Q.; Zhou, C.; Jia, W.; Da Silva, L.; Nguyen, L. D.; Reutter, W.; Fan, H. GLIS, a Bioactive Proteoglycan Fraction from *Ganoderma lucidum*, Displays Anti-Tumor Activity by Increasing Both Humoral and Cellular Immune Response. *Life Sci.* **2010**, *87*(19), 628–637.

187. Zhang, M.; Cui, S.; Cheung, P.; Wang, Q. Antitumor Polysaccharides from Mushrooms: A Review on Their Isolation Process, Structural Characteristics and Antitumor Activity. *Trends Food Sci. Technol.* **2007**, *18*(1), 4–19.

188. Zhang, X. Q.; Ip, F. C.; Zhang, D. M.; Chen, L. X.; Zhang, W.; Li, Y. L.; Ip, N. Y.; Ye, W. C. Triterpenoids with Neurotrophic Activity from *Ganoderma lucidum. Nat. Prod. Res.* **2011**, *25*(17), 1607–1613.

189. Zhao, H. B.; Lin, S. Q.; Liu, J. H.; Lin, Z. B. Polysaccharide Extract Isolated from *Ganoderma lucidum*Protects Rat Cerebral Cortical Neurons from Hypoxia/Reoxygenation Injury. *J. Pharmacol. Sci.* **2004**, *95*(2), 294–298.

190. Zhao, H. B.; Wang, S. Z.; He, Q. H.; Yuan, L.; Chen, A. F.; Lin, Z. B. *Ganoderma* Total Sterol (GS) and GS 1 Protect Rat Cerebral Cortical Neurons from Hypoxia/Reoxygenation Injury. *Life Sci.* **2005**, *76*(9), 1027–1037.

191. Zhao, H.; Zhang, Q.; Zhao, L.; Huang, X.; Wang, J.; Kang, X. Spore Powder of *Ganoderma lucidum*Improves Cancer-Related Fatigue in Breast Cancer Patients Undergoing Endocrine Therapy: A Pilot Clinical Trial. *Evidence-Based Complementary Altern. Med.* **2012d**, *2012*, Report #809614.

192. Zhao, L.; Dong, Y.; Chen, G.; Hu, Q. Extraction, Purification, Characterization and Antitumor Activity of Polysaccharides from *Ganoderma lucidum. Carbohydr. Polym.* **2010**, *80*(3), 783–789.

193. Zhonghui, Z.; Xiaowei, Z.; Fang, F. *Ganoderma lucidum* Polysaccharides Supplementation Attenuates Exercise-Induced Oxidative Stress in Skeletal Muscle of Mice. *Saudi J.Biol. Sci.* **2014,** *21*(2), 119–123.

194. Zhou, L. W.; Cao, Y.; Wu, S. H.; Vlasák, J.; Li, D. W.; Li, M. J.; Dai, Y. C. Global Diversity of the *Ganoderma lucidum* Complex (Ganodermataceae, Polyporales) Inferred from Morphology and Multilocus Phylogeny. *Phytochemistry* **2015,** *114,* 7–15.

195. Zhou, X.; Lin, J.; Yin, Y.; Zhao, J.; Sun, X.; Tang, K. Ganodermataceae: Natural Products and Their Related Pharmacological Functions. *Am.J. Chin.Med.* **2007,** *35*(04), 559–574.

196. Zhou, Y.; Qu, Z. Q.; Zeng, Y. S.; Lin, Y. K.; Li, Y.; Chung, P.; Wong, R.; Hägg, U. Neuroprotective Effect of Pre-Administration with *Ganoderma lucidum* Spore on Rat Hippocampus. *Exp. Toxicol. Pathol.* **2012,** *64*(7), 673–680.

197. Zhou, Z. Y.; Tang, Y. P.; Xiang, J.; Wua, P.; Jin, H. M.; Wang, Z.; Mori, M.; Cai, D. F. Neuroprotective Effects of Water-Soluble *Ganoderma lucidum* Polysaccharides on Cerebral Ischemic Injury in Rats. *J. Ethnopharmacol.* **2010,** *131*(1), 154–164.

198. Zhu, X. L.; Lin, Z. B. Effects of *Ganoderma lucidum* Polysaccharides on Proliferation and Cytotoxicity of Cytokine-Induced Killer Cells. *Acta Pharmacol. Sin.* **2005,** *26*(9), 1130–1137.

199. Ziegenbein, F. C.; Hanssen, H. P.; König, W. A. Secondary Metabolites from *Ganoderma lucidum* and *Spongiporus leucomallellus*. *Phytochemistry* **2006,** *67*(2), 202–211.

200. Zu-Qin, C.; Wen-Li, H.; Xin, J.; Zong-Min, L.; Yu-Jia, H.; Ping, L.; Lin-Yong, Z. Research Progress on *Ganoderma lucidum* Intensive Processing in China. *J. Food Saf. Qual.* **2016,** *7*(2), 639–644.

CHAPTER 8

NUTRITIONAL ATTRIBUTES OF CEREAL GRAINS AND LEGUMES AS FUNCTIONAL FOOD: A REVIEW

VIKAS DADWAL, HIMANI AGRAWAL, SHRIYA BHATT, ROBIN JOSHI, and MAHESH GUPTA

CONTENTS

ABSTRACT

For hundreds of years, people have been very much aware about food nutrition and its health benefits. Food has always remained apart of tradition and rituals, which reflect its diversity from region to region. Current food science has enough tools to clarify the nutritional attributes of our present traditional food items. Whole-grain cereals and legumes are one of the major raw materials for traditional food items, which were modified as a functional food. Cereals such as wheat, maize, oats, rice, and barley are mostly employed for the preparation of conventional food and used in normal routine diet. Such functional food has an advantage to serve a physiochemical function along with providing nutrition. Different ingredients such as dietary fiber, minerals, essential amino acids, and antioxidants reflect benefits of traditional food derived from whole grains. Such properties enhance the whole-grain consumption and help manufacturers to develop more processing techniques and health care functional foods.

8.1   INTRODUCTION

A traditionally utilized whole grain is a powerful cluster of macro- and micronutrients, antioxidants, and other essential biomolecules that interact together to sustain human life.

In ancient times, the maintenance of human health and nutritional attributes of food moved in a synergistic fashion. Whole grains were consumed by grounding them into flour to make bread that symbolizes a basic need of human survival. With the significant rise in traditional knowledge, scientists, food technocrats, and industrialists had conducted a detailed learning on traditionally utilized crops. These grains are composed of starchy germ, endosperm, and bran fraction, which are similar to the intact caryopsis that maybe in its malted or sprouted form before consuming.[6] Among all whole-grain cereals, rice, wheat, corn, barley, buckwheat, sorghum, and oats are best reported for their proteins, carbohydrates, vitamins, dietary fiber, minerals, and many bioactive compounds. The total world population depends on whole grains for its basic requirement of protein and carbohydrates. The maximum part of a whole grain consists of polysaccharides, mainly starch, which is composed of many glucose units, linked together with α, 2–4 and α, 1–6 glycosidic linkages.[45] The starch which is unable to digest in small intestine and gets fermented in the colon is referred to as resistant starch.[131] This resistant starch plays a key role in the production of short-chain fatty acids which increase bacterial mass and also promote the butyrate-producing bacteria in the gut.[49] The second major bioactive compounds present in whole-grain cereals are: phytosterols, tocols, phenolic compounds, alkylresorcinols, lignans, γ-oryzanols, avenanthramides, inositols, phytic acid, ferulic acid, cinnamic acid, and betaine.[1,47,69,118,169]

Some bioactive compounds are fairly specific to certain cereals such as γ-oryzanol in rice, β-glucan in oats, avenanthramides and saponins in oats and barley, and alkylresorcinols in rye, though they are also present in cereals, for example, wheat but in fewer quantities.[113,144,150]

Phenolic acids or antioxidants are free radical scavenging agents that protect the human body from different chronic diseases and balance the oxidative stress inside the body. Oxidative stress has been associated with diseases such as cardiovascular disease (CVD), cancer, and other chronic diseases that have been reported as a major part of present deaths.[159] Phenolic acids are characterized into cinnamic and benzoic acid derivatives. Further, they are divided into two groups: hydroxycinnamic acid and hydroxybenzoic acid derivatives. Hydroxycinnamic acid derivatives comprise: caffeic, p-coumaric, sinapic, and ferulic acids, while hydroxybenzoic acid derivatives include vanillic, protocatechuic, gallic, and syringic acids.[52] Phenolic acids are present in their free, esterified, and bound forms. One among them is ferulic acid, which is present in its free, soluble, conjugated, and bound form in whole grains. Wheat bran consists of maximum amount of ferulic acid in its esterified form of arabinoxylans. Hence, it is recommended as the best source of ferulic acid.[36] The total ferulic acid content among the tested grains is in the order: corn > wheat > oats > rice.[2] As per the nutritional benefits and disease control, whole-grains meals are highly recommended. The whole-grain intake is also associated with lowered risk of type-II diabetes, CVD, and a relative reduction in weight.[164]

Legumes comprising peas, beans, and lentils are the most resourceful and nutritious foods available now. Legumes are also economically a good source of proteins, carbohydrates, rich in fiber, low fat, and have the ability to lower serum cholesterol.[9] They produce an excellent amount of primary and secondary metabolites; and the complex mixture of metabolites present in them exhibits an effective antioxidant activity. Current scientific reports claimed that grain legumes are major contributors in a balanced diet and can treat heart disease, stroke, and diabetes.[4,22] Major phenolic compounds present in legumes are: gallic, caffeic, syringic, protocatechuic, p-hydroxybenzoic, vanillic, ferulic, sinapic, p-coumaric, benzoic, ellagic and cinnamic acid, flavonoids, and other polyphenolic compounds that have antioxidative, anti-inflammatory, anti-allergic, and anticarcinogenic activities.[72] Molecular study also demonstrated that different dietary legumes such as soybeans are related with the prevention of osteoporosis.[116]

8.2 REVIEW OF LITERATURE

8.2.1 PHYTOCHEMICALS IN CEREALS

8.2.1.1 BARLEY

Barley is a major traditional cereal crop grown in temperate region. It is the annual cereal crop belonging to the grass family. It has been reported that barley has the highest antioxidant potential compared to rice and wheat.[92] Barley is always considered as a rich source of phenols and protein supplements.[24] Beside this, it is also rich in starch, dietary fiber, particularly gluten, and crude proteins, and tocols.[19] Gluten and phenolics in barley lead to the reduction in blood glucose level and cholesterol.[34] It consists of various phenols including benzoic acid, quinines, flavones, cinnamic acid derivatives, flavonols, amino-phenolic compounds, and chalcones.[21] These can be in free, bound, or esterified form.

The two major phenolic compounds of low molecular weight present in barley are: p-coumaric acid (4-hydroxycinnamic acid) and ferulic acid (4-hydroxy-3-methoxycinnamic acid).[24] Thus, natural antioxidants have paved the way for the prevention of various diseases such as atherosclerosis, cancer, and other inflammatory diseases. Even the bioactive peptides from barley have shown to be effective against diabetes and hypoglycemia.[34]

Various biochemical activities have also been reported in barley due to the presence of β-glucan as a soluble dietary fiber, which adds to its nutritional benefits.[33] Composition of β-glucan includes cellotriosyl and cellotetraosyl units joined together by 1–3 linkage.[112] The cell wall of endosperm consists of 75% β-glucan. It helps to reduce glucose absorption in the intestine by trapping bile salts due to its viscosity and thus has shown to be hypocholesterolemic and hypoglycemic agent. Owing to reduction in blood glucose level, it is also associated with decline in cardiac diseases. β-glucan has been extracted by food industries for its various applications including noncaloric thickening and stabilizing agent.[15] Moreover, β-glucan along with starch has also been studied as a prebiotic potential supporting the growth of selective organisms.[33]

8.2.1.2 BUCKWHEAT

Pseudo cereal buckwheat *(Fagopyrum esculentum* Moench*)* has gained a worldwide interest due to its nutritional and medicinal properties.[166] It belongs to *Polygonaceae* family due to its cultivation techniques. Traditionally consumed buckwheat is cultivated with the maximum production from Russia, preceded by China. Common buckwheat and tartary buckwheat are the most familiar species grown worldwide. It is well-known for its richest source of dietary fiber, vitamins, minerals, proteins, polyunsaturated essential fatty acids, and resistant starch that play a vital role in the reduction of glycemic and insulin indices.[151] It is also enriched with various flavonoids, phytosterols, inositols, which have a beneficial effect in the treatment of various chronic diseases[166] and are also effective against hypocholesterolemia, hypotension, and diabetes.[86]

Buckwheat is rich in various phenolic compounds such as syringic, vanillic, sinapic, and protocatechuic acids including quercetin and rutin. Besides many major flavonoids, rutin is reported to be present only in buckwheat.[136] The concentration of antioxidants is also higher in the form of anthocyanins and tocopherols.[13,157] Buckwheat has helped in trimming down the effect of various diseases like serum cholesterol, gallstones, tumor and restraining the growth of angiotensin 1-converting enzyme.[79] Its use has also been suggested for liver and typhoid patients due to its various bioactive compounds. The most important prospective of buckwheat is that it contains gluten, hence consumed by patients suffering from celiac disease.[13] Thus, more efforts are focused on traditionally grown buckwheat for the development of various functional foods.

8.2.1.3 CORN

Corn or maize (*Zea mays*) belongs to the family of grasses called *Poaceae*. It is grown as a staple food in many parts of world. Apart from rice and wheat, maize is ranked as the third most leading grain crops of the world.[132] It is a native crop of the Western Hemisphere and specifically originated in Mexico and then was spread all over the world. In India, it is traditionally grown during the monsoon season. This plant has been studied for its various biochemical activities including hypoglycemic, diuretic, anti-inflammatory, and antioxidant properties.[117]

Corn is a good source of carbohydrates and dietary fiber enriched with vitamins B_1, B_2, B_3, B_5, B_6, A, C, and K. It also contains selenium, folic acid, N-ferrulyl tryptamine, and N-p-coumaryl tryptamine[138] with potassium as a major nutrient.[80] β-carotene with a small concentration of selenium is also present in fair amount which makes it a good nutritional source for the improvement of thyroid gland and immune system. In comparison to other cereals, it has a higher content of proteins and fats.[80] Current reports illustrate that corn is a good source of phenolic compounds, carotenoids and phytosterols.[77,94] Basically, anthocyanins are major phenolic compounds present in it. It ranks second among the production of anthocyanins, such as peonidin-3-glucoside, pelargonidin-3-glucoside, cyanidin-3-glucoside, cyanidin-3-(3″,6″-malonylglucoside), pelargonidin-3-(6″malonylglucoside), and cyanidin-3-(3″, 6″dimalonylglucoside).[101] Oil extracted from *Zea mays* is rich in phytosterols mainly stigmasterol, sitosterol, and campesterol.

Zea mays is the richest source of macronutrients such as potassium, having diuretic properties. Corn silk fraction is also enriched with various phenols, maizenic acid, gum, and oils. Thus, it is used as a medicinal source for the treatment of nephrites, diuresis, hypoglycemia, nephrotoxicity, and as an anti-inflammatory and anti-fatigue agent. *Zea mays* is a rich source of vitamin B complex, which supports the functioning of brain, hair, heart, skin, and digestion.[138] Presence of vitamin E has been reported to circumvent the development of atherosclerosis.[108] Lecithin is a major fatty acid component in *Zea mays*; and when bound to sugar molecules it leads to the inhibition of some virus activity. Similarly, many other bioactive molecules including resistant starch in *Zea mays* helps in lowering cholesterol, atherosclerosis, and the risk of cervical cancer.[104] It has also been reported that it may possess prebiotic potential.[156] Carotenoids present in *Zea mays* may reduce melanoma cancer cells, gastric cancer cells, and leukemia cells.[62,114] Bioactive fatty acid molecules are reported to be effective against bone loss, anti-inflammatory, and antidiabetic effect.[11] Anthocyanins have been reported to be antidiabetic, anti-inflammatory, lipid-lowering, anti-atherogenic, and inhibit platelet aggregation.[138]

Apart from all nutritional benefits, it has good economic importance. Its oil fractions are used in the production of soaps, cornstarch in cosmetics and pharmaceutics as diluents, and stem fibers in the production of alcohol and paper.[80]

8.2.1.4 OATS

Mediterranean oats (*Avena sativa*) are the world's seventh ranking cereal crop with Poland ranking fifth in world in its production. It is usually grown as animal food.[30] Oats are rich in carbohydrates, minerals, proteins, fats, water-soluble β-glucan, and polyphenols. Similarly, naked oats are rich in nutritive protein as compared to other cereals. It is also enriched with micronutrients, mainly potassium, copper, magnesium, iron, thiamine, folate, and zinc.[130]

The unsaturated fats such as linoleic and oleic acid in oats are reported to be higher as compared to other cereal grains.[55,119] It is enriched with various phenolic compounds such as vanillic, p-coumaric, caffeic, hydroxybenzoic acid, and their derivatives.[31] Different oat-specific derivatives of hydroxycinnamic acid such as avenanthramides and avenalumic acids have been found specifically in oats as compared to other cereals.[95,152] Three different avenanthramides mostly found in oats are: 2c, 2p, and 2f where 2 represents 5-hydroxyanthranilic acid and c, p, and f represent caffeic, p-coumaric, and ferulic acid, respectively.[152] β-D-glucan also known as β-glucan is composed of β-(1–4)-linked glucose units, separated by β-(1–3)-linked glucose units. As a dietary fiber, it also has antioxidant properties.[23] It has also been reported for lowering cholesterol,[113] insulin, and glucose as well as further reduction in growth of heart disease and type II diabetes.[113] This cholesterol-lowering effect is due to increased viscosity of intestinal contents.[7]

8.2.1.5 RICE

For many centuries, rice (*Oryza sativa* L.) has been widely cultivated around the world, especially in Asia. It belongs to the grass family of *Poaceae* with 20 wild species among which, *Oryza sativa* (Asian rice) and *Oryza glaberrima* (African rice) are the two species that are cultivated. Rice is the staple food of almost one half of the world population with 60% consumption in Southeast Asia. Industrially consumed rice first undergoes the removal of husk which is a storehouse of a variety of bioactive compounds such as proteins, fibers, oils, and other phyto-chemicals,[106] and after which, it undergoes milling to get the white rice, causing most of the nutrients to be lost. Current reports showed that rice includes biologically active components, such as: carbohydrates, proteins, vitamins, minerals (in trace amounts), vitamin E, γ-oryzanol, and various phenolic compounds.

Lipid-soluble vitamin E includes four homologs (α, β, γ, and δ) of tocopherol and tocotrienol.[54] Rice germ contains five times higher concentration of vitamin E than the rice bran mainly α-tocopherol. Vitamin E may improve the immune system, DNA repair, and other processes.[89] It also contains vitamin B complexes (i.e., vitamin B_1, B_2, and B_6), neurotransmitter γ-amino butyric acid, and fibers which help in lowering blood glucose and sugar level.[46] γ-oryzanols also show antioxidant and cholesterol-lowering effects.[99] However, the rice by-products are rich in dietary fiber including cellulose, hemicelluloses, hydrocolloids, lignin, and pectin. The pigmented rice contains cyanidin-3-glucoside and peonidin-3-glucoside as the major anthocyanins having anticancerous, antioxidant, and anti-inflammatory activities.[29,142] It also helps in the reduction of hyperlipidemia, hyperglycemia,[63] and oxidative stress.[122] Moreover, the pigmented rice is rich in secondary metabolites such as carotenoids, flavones, flavan-3-ol, and γ-oryzanol.[106] Rice bran is enriched with fatty acids.[147] Moreover, the major polyphenols both in pigmented and nonpigmented rice include: ferulic, protocatechuic, sinapic, p-coumaric, and vanillic acid[167] having anti-inflammatory, anticarcinogenic, antioxidant, and hypoglycemic effect.[155]

8.2.1.6   WHEAT

Wheat belongs to family *Poaceae* and is the second most important staple crop around the world.[27] The major production of wheat comes from Russia followed by the United States, China, and India. India has become self-sufficient in wheat production, exporting one-third of the total production. The first type of wheat cultivated was Einkorn.[61] Whole grains are enriched with various bioactive compounds such as proteins and carbohydrates as major components and vitamins (i.e., thiamin and niacin), minerals such as selenium and manganese, lipids, terpenoids, and phenolic compounds (ferulic, p-coumaric, caffeic, syringic, gentisic, p-hydroxybenzoic, and vanillic acid with ferulic acid as the major phenolic component, carotenoids like zeaxanthin and p-cryptoxanthin), and vitamin E as minor nutrient.[103,109,139]

Wheat consists of a multilayer wheat kernel that contains endosperm, aleurone layer, pericarp, and the seed coat.[90] Endosperm alone consists of 83% of kernel weight.[79] This after milling is recovered as flour containing 80% starch and 10% proteins with a low concentration of some minerals and phytochemicals.[139] The wheat outer layer contains the pericarp, seed coats, and aleurone layer collectively called as wheat bran. In general, the content of endosperm is 83% (w/w), 14.5% (w/w) bran, and 2.5% (w/w) germ.[90] Wheat bran has been reported to have valuable components such as phenolics, soluble and insoluble dietary fiber,[56] a good amount of vitamin B complexes, and low mineral and protein content.[81] Moreover, the wheat germ has low sodium, cholesterol with substantial amount of minerals, and a good source of coenzyme Q10 and para-aminobenzoic acid; and its oil may lower total cholesterol due to the presence of policosanol.[61,81] The regular intake of traditionally grown whole-grain wheat may reduce the risk of CVD.[87]

8.2.1.7 SORGHUM

Traditional sorghum (*Sorghum bicolor*) belongs to family *Poaceae*; and is the fifth grown cereal in world after rice, wheat, barley, and maize.[133] It is the main crop grown in the semiarid and tropic areas of the world, mainly in Asia and Africa, due to its adaptation to high temperature and drought conditions, comprising 70% of the daily calorie intake.[107,38] In India, it is used as animal feed. It is rich in various bioactive compounds such as carbohydrates, proteins, vitamin E, carotenoids, dietary fiber, minerals, and various phenolics, for example, flavonoids, phenolic acid and condensed tannins.[161] Presence of these phenolic compounds has been effective in reducing the risk of diabetes, obesity, hypertension, CVD, chronic noncommunicable disease, inflammation, and cancer.[110] Because it is gluten free, it can be used by patients with celiac disease.[123]

Pericarp of sorghum grain is enriched with phenolic compounds.[100] Anthocyanins comprise major flavonoids that impart color to the grain. Anthocyanins present in sorghum are 3-deoxyanthocyanins (3-DXAs) comprising luteolinidin and apigeninidins, and their derivatives 5-methoxyluteolinidin and 7-methoxyapigeninidin. Tannins are situated between the pericarp and endosperm that constitute the high molecular weight polyphenols.[43] The main constituent of sorghum grain is starch, consisting of 70% of the total dry weight of grain.[133]The phenolic compounds reported in sorghum are: gallic, protocatechuic, p-hydroxybenzoic, gentisic, salicylic, vanillic, syringic, ferulic, caffeic, p-coumaric, cinnamic, and sinapic acid.[41] Vitamin E constitutes γ-tocopherol, α-tocopherol, and tocotrienol, which exhibit a good antioxidant property.[8]

8.2.1.8 MILLETS

Millet contains bioactive ingredients with potential therapeutic activities besides their basic nutritional value as shown in (Fig. 8.1). Millets are the richest source of various bioactive compounds like proteins, fatty acids, minerals, vitamins, dietary fibers, and phenols. Particularly, millets are the storehouse of a variety of essential amino acids, mainly sulfur-containing amino acids (methionine and cysteine). It has also shown potential prebiotic activity, thus more efforts are being put to make traditionally grown millets as a functional food.

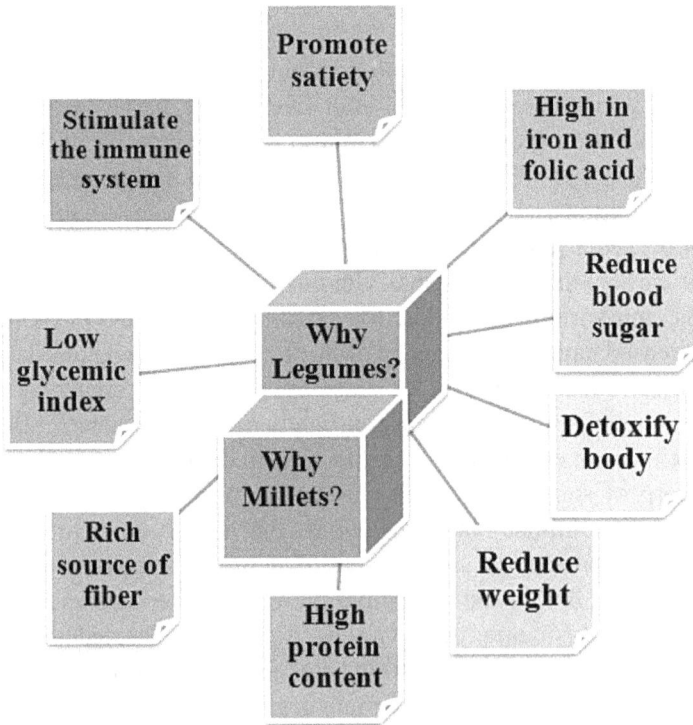

FIGURE 8.1 **(See color insert.)** Health-benefiting properties of legumes and millets.

8.2.1.9 SOYBEAN

Soybean belongs to family *Fabacea*e, genus *Glycine,* more frequently known as the leguminous family.[5] It is one of the most important seed legumes to contribute 25% of the global edible oil and two-third of protein concentrate.[69] The United States of America, Brazil, and Argentina are the largest producers of soybeans in the world, followed by China and India ranking fifth in its production. It contains 30–40% of proteins with similar amino acid composition to that of milk and nutritional value as that of animal protein.[88] It consists of a variety of bioactive compounds like lipids, proteins, phytic acids, oligosaccharides, isoflavones, saponins, and phenolic acids.[82] It also helps in reduction of the risk of obesity, CVDs, cancer, and diabetes.[74,84]

Three main isoflavones present in soybean are: daidzein, genistein, and glycitein. Conjugated aglycones comprise esterified 6″-O-malonyl-β-glycosides, 6″-O-acetyl-β-glycosides, and β-glycosides; and it includes a total of 12 isoflavones.[83] Phytoestrogens present in it exhibit antiproliferative activities and also regulate the immune system.[28,85] Despite the presence of various bioactive compounds, it has anti-nutritional effects.[127] Soybean is anti-allergic, affecting almost 0.4% of children and 0.25% of adults.[135] The Food and Agriculture Organization has listed it as one of the eight most common allergenic foods.[126] Soybean contains various allergenic glycol proteins that lead to various allergic symptoms such as eczema, atopic dermatitis, and asthma. Several soybean allergens have been classified into families such as cupin superfamily, prolamin superfamily, pathogenesis-related proteins, and profilins.[5,126,153]

8.3 RESULTS AND DISCUSSION

8.3.1 CHEMICAL COMPOSITION

Determination of moisture content is one of the commonly used parameter to understand the food properties. Every cereal is adapted to a fixed amount of water sustainability. The cost of many food products depends on the amount of water it contains. The texture, taste, appearance, and stability of foods also depend on the amount of water they contain. From the current data, it is observed that moisture content in all whole cereal grains ranges from 12.0–14.0 g/100 g. Among all cereals in this chapter, wheat, barley, oats, and millets are best suited for protein consumption. Vitamins play a very vital role in the human immune system. Hence, thiamine, riboflavin, and vitamin B_6 are present in adequate quantity in all cereal whole grains.[146]

8.3.1.1   DIETARY FIBER

Dietary fibers mainly consist of polysaccharides including soluble fibers (e.g., arabinoxylans, gum, and β-glucan) and oligosaccharides (e.g., fructo-oligosaccharides), which are mainly present in whole grains. These soluble fibers are indigestible in the human stomach, when intact with the colon, they undergo microbial fermentation.[162] Such microbial fermentation produces volatile or short chain fatty acids (acetic and butyric acids) that are metabolized within intestinal epithelial cells and contribute fuel to energy metabolisms. Oats are referred to as the best source of dietary fiber. Rice contains the lowest amount of dietary fiber (3.9%, db), while corn and oats have 12–16% db. Barley and wheat hold second position with a comparative good amount of dietary fiber.[25]

8.3.1.2   ESSENTIAL AMINO ACIDS

Amino acids act as building blocks for our body. Body cells utilize these amino acids for their repair, growth, and maintenance. Proteins are made up of long chain of amino acids, which link together and perform a specific function inside our body. Some amino acids work as neurotransmitters and establish the processing of brain cells. For the proper functioning of vitamins and minerals, free amino acids work as support molecules. Without the presence of amino acids, human body will be unable to work properly and in some severe cases, it may cause death. As the essential amino acids cannot be synthesized inside a human body, they must be derived from food. Whole grains and legumes in one meal can fulfill the actual requirement of essential amino acids. As per the current scientific reports, it was determined that wheat is the best source for the daily fulfillment of essential amino acids followed by barley, oats, rice, millets, and corn. It has been reported that glutamic acid is maximum in all the whole-grain cereals ranging between 18−29%; second, serine, leucine, and glycine are present in a comparatively good amount,[158,160] while other amino acids are present in average quantity, as shown in Table 8.1.

TABLE 8.1   Major Essential Acids, Phenolics, and Fatty Acids in Whole-Grain Cereals and Legumes.

Cereal grain	Phytochemical composition	Reference
Barley	*Essential amino acid:* Glutamic acid, proline, leucine, and isoleucine (in maximum)	[44, 121]
	Phenolics: Protocatechuic acid, p-hydroxybenzoic acid, vanillic acid, syringic acid, ferulic acid, p-coumaric acid, m-coumaric acid, o-coumaric acid, sinapic acid	[42, 93]
	Fatty acids: Palmitic acid, oleic acid, linoleic acid, linolenic acid, stearic acid	[18]
	Minerals: Potassium and phosphorus (in maximum)	[17]
Buckwheat	*Essential amino acid:* Glutamic acid, aspartic acid, and arginine (in maximum)	[44, 120]
	Phenolics: Rutin, catechin, epicatechin, caffeic acid, epicatechin galate, gallic acid, p-coumaric acid	[60]
	Fatty acids: Palmitic acid, stearic acid, arachidic acid, oleic acid, linoleic acid, linolenic acid, behenic acid, lignoceric acid	[39]
	Minerals: Copper, magnesium, and phosphorus	[17]
Corn	*Essential amino acid:* Glutamic acid, leucine (in maximum)	
	Phenolics: Protocatechuic acid, p-hydroxybenzoic acid, ferulic acid, caffeic acid, p-coumaric acid	[3, 42]
	Fatty acids: Myristic acid, palmitic acid, stearic acid, oleic acid, linoleic acid, arachidic acid, hexadecenoic acid, lignoceric acid	[17]
	Minerals: Potassium and phosphorus (in maximum)	[50]
Millets	*Essential amino acid:* Glutamic acid, leucine, analine (in maximum)	[44]
	Phenolics: Gallic acid, protocatechuic acid, p-hydroxy benzoic acid, gentisic acid, vanillic acid, syringic acid, ferulic acid, caffeic acid, p-coumaric acid	[42]
	Fatty acids: Palmitoleic acid, oleic acid, linoleic acid, linolenic acid, stearic acid, palmitic acid, arachidic acid	[128]
	Minerals: Magnesium, potassium and phosphorus (in maximum)	[17]
Oats	*Essential amino acid:* Glutamic acid, aspartic acid (in maximum)	[44]
	Phenolics: Ferulic acid, sinapic acid, p-coumaric acid, protocatechuic acid, p-hydroxy benzoic acid, vanillic acid, syringic acid, ferulic acid, caffeic acid	[102]

TABLE 8.1 *(Continued)*

Cereal grain	Phytochemical composition	Reference
	Fatty acids: Palmitic acid, oleic acid, linoleic acid, linolenic acid, and oleic acid	[3, 42]
	Minerals: Magnesium, potassium and phosphorus (in maximum)	[12, 17]
Rice	*Essential amino acid:* Glutamic acid, leucine (in maximum)	[44, 70]
	Phenolics: Ferulic acid, sinapic acid, p-coumaric acid, gallic acid, protocatechuic acid, p-hydroxybenzoic acid, vanillic acid, syringic acid, ferulic acid, sinapic acid.	[10, 42]
	Fatty acids: Myristic acid, palmitic acid, stearic acid, oleic acid, linoleic acid, arachidic acid, heneicosanoic acid, nervonic acid, margaric acid	[111]
	Minerals: Magnesium, calcium, and phosphorus (in maximum)	[17]
Sorghum	*Essential amino acid:* Glutamic acid, leucine, and proline (in maximum)	[44, 154]
	Phenolics: Ferulic acid, sinapic acid, p-coumaric acid, gallic acid, protocatechuic acid, p-hydroxy benzoic acid, gentisic acid, salicylic acid	[10]
	Fatty acids: Palmitoleic acid, oleic acid, linoleic acid, linolenic acid, stearic acid, palmitic acid	[97]
	Minerals: Potassium and phosphorus (in maximum)	[17]
Soybean	*Essential amino acid:* Glutamic acid, aspartic acid, leucine (in maximum)	[44, 102]
	Phenolics: p-Hydroxybenzoic acid, sinapic, ferulic acid, syringic acid, *p*-coumaric acid	[40]
	Fatty acids: Myristic acid, palmitic acid, stearic acid oleic acid, linoleic acid, linolenic acid	[16, 68]
	Minerals: Iron, copper, phosphorous, and magnesium	
Wheat	*Essential amino acid:* Glutamatic acid, proline, arginine, aspartic acid (in maximum)	[44, 102]
	Phenolics: Ferulic acid, apigenin, p-hydroxybenzoic acid, vanillic acid, syringic acid, p-coumaric acid	[10, 58]
	Fatty acids: Myristic acid, palmitic acid, stearic acid, arachidic acid, palmitoleic acid, oleic acid, linoleic acid, linolenic acid	[71]
	Minerals: Magnesium and phosphorus (in maximum)	[17]

8.3.1.3   MINERALS

Minerals are needed for a proper fluid balance, muscle contraction, nerve transmission, blood clotting, blood pressure regulation, immune system, healthy bones, and teeth.[17] The current study showed that traditionally utilized whole grain and legumes are the best source of macro and micro-nutrients (Table 8.1).

8.3.1.4   ANTIOXIDANTS

Phenolic compounds are characterized by the presence of one or more aromatic rings with one or more hydroxyl groups attached on either sides. Commonly known phenolic acids found in whole cereal grains include: gallic acid, vanillic acid, ferulic acid, syringic acid, caffeic acid, and p-coumaric acid (Fig. 8.2). Antioxidants are well known for their scavenging activity against free radicals and the maintenance of oxidative stress at cellular level. Whole grain and legumes are an excellent source of antioxidants. Ferulic acid is found in free, conjugated (soluble), and bound (insoluble) forms in whole cereal grains (wheat, maize, rice, and oats). Such hydroxycinnamic acid derivatives also showed activity on other oxidation models like liposome and low-density lipoprotein systems. Polyphenols are directly linked to its medicinal outcomes as shown in Table 8.1.[48]

8.4   TRADITIONAL LEGUMES, FUNCTIONAL PEPTIDES, AND THEIR HEALTH BENEFITS

Legumes have been reported to possess most peptides with potential bioactivity. This bioactivity is due to legumes that contain high-quantity and -quality proteins. Soybean (*G. max*) is a widely accepted food source; and there are a number of products that can be obtained from soybeans such as oil, flour, nuts, textured proteins, tofu, yogurt, cheese, milk, and mayonnaise. It provides an inexpensive source of vegetable protein having high nutritional value, a constituent for hundreds of chemical products, and also a potential source of bioactive peptides. Among all cereals, soybean is the most studied source of bioactive peptides and proteins. Besides being an excellent protein source, soybean also has anti-cholesterolemic, anti-hypertensive, anticancer, and antioxidative properties. Glycinin and β-conglycinin found

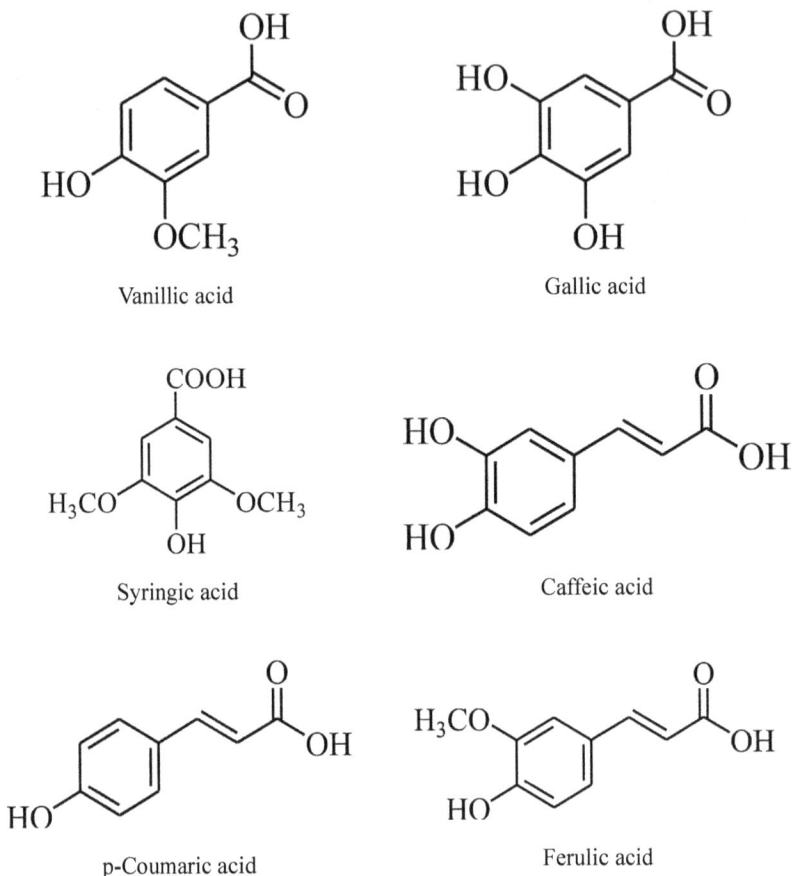

FIGURE 8.2   Phenolics as an antioxidant.

in soy protein are the precursors of most of the isolated peptides.[163] Glycinin is composed of five subunits, G1, G2, G3, G4, and G5, and conversely, β-conglycinin has three major subunits (α, α′, and β). Soy protein contains all the essential amino acids[105] that are present in animal protein along with nutritional value equivalent to high biological value animal protein. Soymilk is free of gluten, lactose, and cholesterol; hence it is suitable for vegetarians, and lactose-intolerant and milk-allergy patients.[78] There are a number of bioactive peptides that have been identified from soy milk having hypocholesterolemic, angiotensin-converting enzyme-inhibitory, anticancer, and immunomodulatory activities (Table 8.2). Aglycin, a natural bioactive peptide isolated from soybean, possesses antidiabetic potential and is stable in digestive enzymes. It has been reported that an oral dose of aglycin can prevent hyperglycemia as it was able to increase the insulin receptor in the muscles of streptozotocin-induced diabetic mice.[37] β-conglycinin and glycinin from soy protein also possess athero-protective peptides and are suggested to have preventive effects through absorption from the intestinal tract.[91]

TABLE 8.2 Peptides Identified from Soybean with Different Bioactivity.

Bioactivity	Peptides	Reference
ACE-inhibitory peptides	Val-Ala-His-Ile-Asn-Val-Gly-Lys	[129]
	Tyr-Val-Trp-Lys	
Antihypertensive	Pro-Gly-Thr-Ala-Val-Phe-Lys	[57]
	His-His-Leu	[75]
Hypotensive	Tyr-Val-Val-Phe-Lys	[141]
	Ile-Pro-Pro-Gly-Val-Pro-Try-Trp-Thr	
Anticancer	X-Met-Leu-Pro-Ser-Try-Ser-Pro-Try	[76]
Hypocholesterolemic	Leu-Pro-Tyr-Pro-Arg	
Phagocytosis-stimulatory peptide	His-Cys-Gln-Arg-Pro-Arg	[165]
	Gln-Arg-Pro-Arg	
Anticancer (Lunasin)	SKWQHQQDSCRKQKQGVNLTPCEKHIMEKI QGRGD DDDDDDDD	[73]

Generally, these bioactive peptides are inactive within the sequence of their parent protein, released by enzymatic hydrolysis, and exert a variety of physiological functions. Peptides bearing molecular mass of less than 6000 Da[149] and the chain length of 2–20 amino acids residue[98] possess higher biological activity. This activity is mainly affected by the composition and sequence of the amino acid residues present in peptide chain.[26] On the basis of this, these peptides may play a vital role, as antimicrobial,[96] immunomodulatory,[53] hypocholesterolemic,[168] mineral-binding,[32] antithrombotic,[140] antihypertensive,[67] and opiate-like[143] functions (Fig. 8.3). These bioactive peptides are also suggested as functional components of food with designed properties[51] and may be used in the development of nutraceuticals and functional foods to manage human disease conditions.

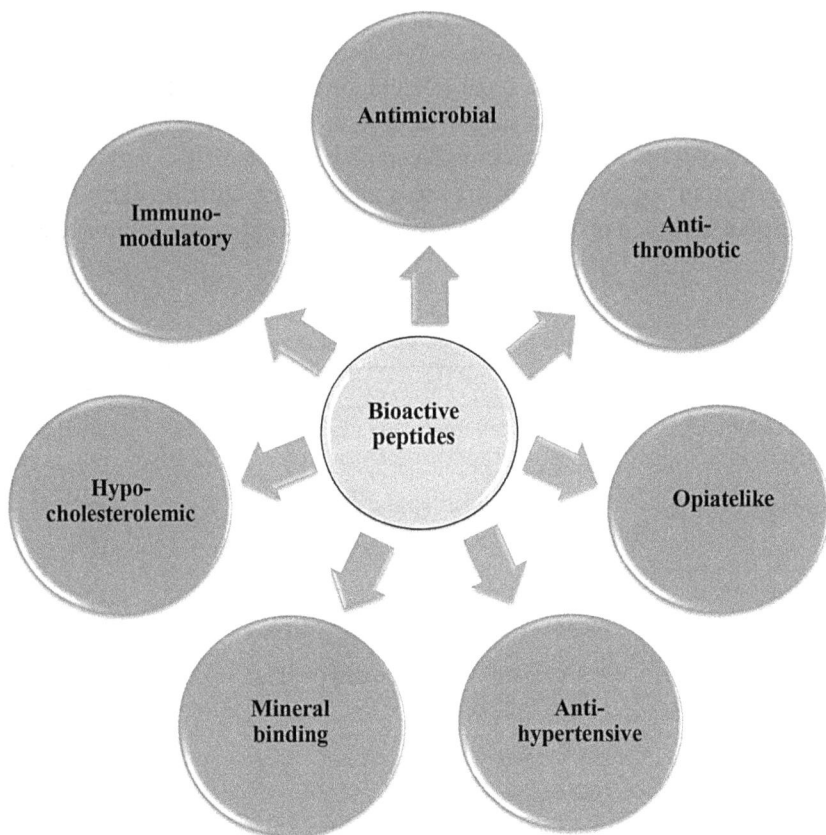

FIGURE 8.3 (See color insert.) Biofunctional activity of bioactive peptides.

Lunasin peptide containing 43 amino acids was initially found in soybean and then also identified in some pseudo cereals and cereals for instance, rice, barley, wheat, rye, and amaranth.[64–66,115] Bioactivity of the lunasin is related to its capability to arrest cell division in cancer cells, to inhibit core histone acetylation in mammalian cells, and to protect DNA from oxidative damage.[59] Bioactive peptides from legumes and cereals possess a number of physiological effects in vitro as well as in animal models. Peptide potential should not be shocking. All intercellular communications and cellular functions of the body are directed by amino acid sequences present in peptides or in protein. Proteins and peptides acquire a variety of chemical diversities, which is not offered by any other biological molecule. They are tool kits of the nature; and the more we can use native peptides or closely related analogs in order to protect health, the more we could decrease the risk of unforeseen side reactions.

8.5 WHOLE-GRAIN- AND LEGUMES-BASED FUNCTIONAL FOOD PRODUCTS

8.5.1 WHEAT AND RICE PRODUCTS USED AS FUNCTIONAL FOOD

Fermentation technology has been used for the preservation, enhancement of flavor of compounds and the production of essential amino acids and enzymes. Hence, a larger variety of food products are derived from whole-grain cereals, among them, wheat and rice are the major cereals used as functional foods in the Northern Asian region. On the basis of their production, consumption and health benefits, a detailed study has been performed from time to time (Table 8.3), showing some fermented functional food products derived from wheat and rice.[1] Such food products are referred to as heritage food that reflects the cultural attributes and traditional knowledge attached to it.

TABLE 8.3   Wheat- and Rice-Derived Functional Food Products.

Cereal grain	Food products	Reference
Wheat	*Kulcha*, *nan*, and *bhatura*	[134]
	Jalebi	[148]
	Balam	[129]
	Kurdi and *taotjo*	[20, 145]
Rice	*Adai* and *vada*	[20, 145]
	Sez	[129]
	Khaman	[137]
	Ambeli	[125]
	Bhattejaanr and *anarshe*	[6, 145]
	Dhosa	[14]
	Dhokla	[14]
	Idli	[35, 124]

8.6 SUMMARY

Nutritional value of food is the most significant parameter from ancient to modern world for the maintenance of human health. Hence with the rising importance of traditional knowledge, a detailed study has been conducted on vegetables, fruits, and cereals. They are not only well-known for their basic primary metabolites such as protein, carbohydrates, and fatty acids, but also for secondary metabolites, for example, flavonoids, anthocynins, and phenolic acids.

Antioxidants are getting popularity for their scavenging nature of free radicals in the prevention and treatment of diseases and becoming a prominent topic of interest in the development of functional foods. Whole-grain cereals have received considerable attention in the last several decades due to the presence of a unique blend of bioactive components such as phytochemicals and antioxidants; whereas pulses are an important source of macronutrients, containing almost twice the amount of protein compared to cereal grains. However, phytochemicals and antioxidants in whole grains have not received as much attention as the phytochemicals in fruits and vegetables, although the increased consumption of whole grains and whole-grain products has been associated with the reduced risk of developing chronic diseases such as CVDs, type 2 diabetes, some cancers, and all-cause mortality.

Phenolic compounds, present in legumes with their antioxidative effects on human health, include phenolic acids such as gallic acid, caffeic acid, syringic acid, flavonoids, and other polyphenolic compounds. In addition to being a source of macronutrients and minerals, pulses also contain secondary metabolites that are increasingly being recognized for their potential benefits for human health. Common processing procedures, such as soaking and cooking, may decrease the levels of these bioactive compounds and subsequent overall antioxidant activity. Millets can also fulfill the requirement of minerals like iron, calcium, and zinc in infants and young children.

Modern extraction techniques have been employed for the quantification of primary and secondary metabolites. In case of cereal grains, millets were referred to as an ancient whole grain. They are also resistant to drought, which made them most popular in Europe in 5000 Before Christ (BC) Millets have a protein structure very much similar to wheat and are gluten free, thus making this traditional food a high-end functional food as per the current food requirements. Millets also have nutraceutical assets as a source of antioxidants, which reduce the risk of CVDs, lowering blood pressure and diabetes, and in the prevention of cancer, and so forth. This chapter includes different sources of phytochemicals from cereals and pulses and the changes in these phytochemicals during the different processing methods and treatments.

KEYWORDS

- antioxidants
- balanced diet
- bioactive compounds
- bioactive peptides
- biomolecules
- cereals
- cornstarch
- dietary fiber
- essential amino acids
- fatty acids
- flavonoids
- free radicals
- functional foods
- gluten
- human health
- macronutrients
- medicinal
- micronutrients
- millets
- minerals
- nutraceutical
- nutrition
- phenolic acids
- phytochemicals
- phytosterols
- primary metabolites
- pulses
- secondary metabolites
- soy protein
- traditional foods
- whole grains

REFERENCES

1. Adams, M. R. Fermented Weaning Foods. In *Microbiology of Fermented Foods*; Wood, J. B.Ed.; Blackie Academic: London, 1998; pp 790–811.
2. Adom, K. K.; Liu, R. H. Antioxidant Activity of Grains. *J. Agric. Food Chem.* **2002a,** *50*, 6182–6187.
3. Adom, K. K.; Liu, R. H. Antioxidant Activity of Grains. *J. Agric.Food Chem.* **2002b,** *50*, 6182–6187.

4. Afshin, A.; Micha, R.; Khatibzadeh, S.; Mozaffarian, D. Consumption of Nuts and Legumes and Risk of Incident Ischemic Heart Disease, Stroke, and Diabetes: A Systematic Review and Meta-Analysis. *Am. J. Clin.Nutr.* **2014**, *1*, 278–288.

5. Agarwal, D. K.; Billore, S. D.; Sharma, A. N.; Dupare, B. U.; Srivastava, S. K. Soybean: Introduction, Improvement, and Utilization in India Problems and Prospects. *Agric. Res.* **2013**, *2*, 293–300.

6. American Association of Cereal Chemists International (AACC). Whole Grain Task Force and Definition. 2009. http://www.aaccnet.org/definitions/wholegrain.asp; Accessed on 17 March, 2014.

7. Andersson, A. M. A.; Borjesdotter, D. Effects of Environment and Variety on Content and Molecular Weight of β-Glucan in Oats. *J. Cereal Sci.* **2011**, *54*, 122–128.

8. Anunciaçao, P. C.; Cardoso, L. M.; Gomes, J. V. P.; Lucia, C. M. D.; Carvalho, C. W. P.; Galdeano, M. C.; Queirozd, V. A. V.; Alfenas, R. C. G.; Martino, H. S. D.; Ana, H. M. P. Comparing Sorghum and Wheat Whole Grain Breakfast Cereals: Sensorial Acceptance and Bioactive Compound Content. *Food Chem.* 2017, *221*, 984–989.

9. Apata, D. F.; Ologhobo, A. D. Biochemical Evaluation of Some Nigerian Legume Seeds. *Food chem.* **1994**, *49*, 333–338.

10. Awika, J. M.; Rooney, L. W. Sorghum Phytochemicals and Their Potential Impact on Human Health. *Phytochemistry* **2004**, *65*, 1199–1221.

11. Balasubashini, M. S.; Rukkumani, R.; Viswanathan, P.; Menon, V. P. Ferulic Acid Alleviates Lipid Peroxidation in Diabetic Rats. *J. Phytother. Res.* **2004**, *18*, 210–214.

12. Banas, A.; Debski, H.; Banas, W.; Heneen, W. K.; Dahlqvist, A.; Bafor, M.; Gummeson, P.; Marttila, S.; Ekman, A.; Carlsson, A. S.; Stymne, S. Lipids in Grain Tissues of Oat (*Avena sativa*): Differences in Content, Time of Deposition, and Fatty Acid Composition. *J. Exp.Bot.* **2007**, *58*, 2463–2470.

13. Bastida, J. A. G.; Piskuła, M.; Ski, Z. H. Recent Advances in Development of Gluten-Free Buckwheat Products. *Trends Food Sci.Technol.* **2015**, *44*, 58–65.

14. Battacharya. S.; Bhat, K. K. Steady Shear Rheology of Rice Blakgram Suspensions and Suitability of Rheological Models. *J. Food Eng.* **1997**, *32*, 241–250.

15. Bayer, V. L.; Francisco, A. D.; Chan, A.; Oro, T.; Ogliari, P. J.; Barreto, P. L. M. β-Glucans Extraction and Partial Characterization. *Food Chem.* **2014**, *154*, 84–89.

16. Bellaloui, N.; Bruns, H. A.; Gillen, A. M.; Abbas, H. K.; Zablotowicz, R. M.; Mengistu, A.; Paris, R. L. Soybean Seed Protein, Oil, Fatty Acids, and Mineral Composition as Influenced by Soybean-Corn Rotation. *Agric. Sci.* **2010**, *1*, 102–109.

17. Beloshapka, A. N.; Buff, P. R.; Fahey, G. C. J.; Swanson, K. S. Compositional Analysis of Whole Grains, Processed Grains, Grain Co-products, and Other Carbohydrate Sources with Applicability to Pet Animal Nutrition. *Foods* **2016**, *5*, 23.

18. Bhatty, R. S.; Rossnagel, B. G. Lipid and Fatty Acid Composition of Riso 1508 and Normal Barley. *Cereal Chem.* **1980**, *57*, 382–386.

19. Biel, W.; Jacyno, E. Chemical Composition and Nutritive Value of Spring Hulled Barley Varieties. *Bulg. J. Agric. Sci.* **2004**, *52*, 5195–5200.

20. Blandinob, A.; Aseeria, A. M. E.; Pandiellaa, S. S.; Canterob, D.; Webba, C. Review: Cereal-Based Fermented Foods and Beverages. *Food Res. Int.* **2003**, *36*, 527–543.

21. Bonoli, M.; Verardo, V.; Marconi, E.; Caboni, M. F. Antioxidant Phenols in Barley (*Hordeum vulgare l.*) Flour: Comparative Spectrophotometric Study Among Extraction Methods of Free and Bound Phenolic Compounds. *J. Agric. Food Chem.* **2004,** *52,* 5195–5200.

22. Bouchenak, M.; Senhadji, M. L. Nutritional Quality of Legumes, and Their Role in Cardiometabolic Risk Prevention: A Review. *J. Med. Food* **2013,** *16,* 1–14.

23. Brindzova, L.; Certik, M.; Rapta, P.; Zalibera, M.; Mikulajova, A.; Takacsova, M. Antioxidant Activity, β-glucan and Lipid Contents of Oat Varieties. *Czech J. Food Sci.* **2008,** *26,* 163–173.

24. Carvalho, D. O.; Curto, A. F.; Guido, L. F. Determination of Phenolic Content in Different Barley Varieties and Corresponding Malts by Liquid Chromatography-Diode Array Detection-Electrospray Ionization Tandem Mass Spectrometry. *J. Antioxid* **2015,** *4,* 563–576.

25. Charalampopoulos, D.; Wang, R.; Pandiella, S. S.; Webb, C. Application of Cereals and Cereal Components in Functional Foods: A Review. *Int. J. Food Microbiol.* **2002,** *79,* 131–141.

26. Chen, H. M.; Muramoto, K.; Yamauchi, F.; Fujimoto, K.; Nokihara, K. Antioxidative Properties of Histidine-Containing Peptides Designed from Peptide Fragments Found in the Digests of a Soybean Protein. *J. Agric. Food Chem.* **1998,** *46,* 49–53.

27. Chen, C. Y. O.; Kamil, A.; Blumberg, J. B. Phytochemical Composition and Antioxidant Capacity of Whole Wheat Products. *Int.J. Food Sci. Nutr.* **2015,** *66,* 63–70.

28. Cho, K. M.; Hong, S. Y.; Math, R. K.; Lee, J. H.; Kambiranda, D. M.; Kim, J. M.; Islam, S. M. A.; Yun, M. G.; Cho, J. J.; Lim, W. J.; Yun, H. D. Biotransformation of Phenolics (Isoflavones, Flavonols and Phenolic Acids) during the Fermentation of Cheonggukjang by *Bacillus pumilus* HY1. *Food Chem.* **2009,** *114,* 413–419.

29. Choi, S. P.; Kim, S. P.; Friedman, M. Antitumor Effects of Dietary Black and Brown Rice Brans in Tumor-Bearing Mice: Relationship to Composition. *Mol.Nutr.Food Res.* **2012,** *57,* 390–400.

30. Choubey, R. N.; Roy, A. K. Forage Oat Breeding in India –Achievements and Prospects, Chapter5; 20/6/2005; http://www.fao.org/ag/AGP/AGPC/doc/Proceedings/nepal2005/chapter5.pdf Accessed on May 5, 2017.

31. Cova, M. K.; Malinová, E. Ferulic and Coumaric Acids, Total Phenolic Compounds and their Correlation in Selected Oat Genotypes. *Czech J.Food Sci.* **2007,** *25,* 325–332.

32. Cross, K. J.; Huq, N. L.; Palamara, J. E.; Perich, J. W.; Reynolds, E. C. Physico-chemical Characterization of Casein Phosphopeptideamorphous Calcium Phosphate Nanocomplexes. *J. Biol.Chem.* **2005,** *280,* 15, 362–15, 369.

33. Das, M.; Kaur, S. Status of Barley as a Dietary Component for Human. *Res. Rev.: J. Food Dairy Technol.* **2016,** *S1,* 25–30.

34. Datt, M. A.; Ereifej, K.; Zaiton, A. A.; Alrababah, M.; Almajwal, A.; Rababah, T.; Yang, W. Anti-Oxidant, Anti-Diabetic, and Anti- Hypertensive Effects of Extracted Phenolics and Hydrolyzed Peptides from Barley Protein Fractions. *Int. J. Food Sci.* **2012,** *15,* 781–795.

35. Desikachar, H. S. R.; Radhakrishnamurthy, R.; Rao, R. G.; Kol, K. S. B.; Srinivasan, M.; Subrahmanyan, V. Studies on Idli Fermentation: Part1. Some Accompanying Changes in the Batter.*J. Sci.Ind.Res.* **1960,** *19,* 168–172.

36. Dewanto, V.; Wu, X.; Liu, R. H. Processing Sweet Corn has Higher Antioxidant Activity. *J. Agric.Food Chem.* **2002**, *50*, 4959–496.

37. Dhananjay, S.; Kulkarni, S.; Kapanoor, K. G.; Naganagouda, V. K.; Veerappa, H. M. Reduction of Flatus-Inducing Factors in Soymilk by Immobilized α-Galactosidase. *Biotechnol. Appl. Biochem.* **2006**, *45*, 51–57.

38. Dicko, M. H.; Gruppen, H.; Traore, A. S.; Berkel, V. W. J. H.; Voragen, A. G. J. Review: Sorghum Grain as Human Food in Africa: Relevance of Content of Starch and Amylase Activities. *Afr. J. Biotechnol.* **2006**, *5*, 384–395.

39. Dorrell, D. G. Fatty Acid Composition of Buckwheat Seed. *J. Am. Oil Chem.Soc.* **1971**, *48*, 693–696.

40. Dueñas, M.; Hernández, T.; Robredo, S.; Lamparski, G.; Estrella, I.; Muñoz, R. Bioactive Phenolic Compounds of Soybean (*Glycine max cv. Merit*): Modifications by Different Microbiological Fermentations. *Pol. J. Food Nutr. Sci.* **2012**, *62*, 241–250.

41. Dykes, L.; Rooney, L. W. Review: Sorghum and Millet Phenols and Antioxidants. *J. Cereal Sci.* **2006**, *44*, 236–251.

42. Dykes, L.; Rooney, L. W. Phenolic Compounds in Cereal Grains and Their Health Benefits. *Cereal Foods World* **2007**, *52*, 105–111.

43. Earp, C. F.; McDonough, C. M.; Rooney, L. W. Microscopy of Pericarp Development in the Caryopsis of *Sorghum bicolor (L.) Moench.J. Cereal Sci.* **2004**, *39*, 21–27.

44. Ejeta, G.; Hassen, M. M.; Mertz, E. T. *In vitro* Digestibility and Amino Acid Composition of Pearl Millet (*Pennisetum typhoides*) and Other Cereals. *Appl. Biol.* **1987**, *84*, 6016–6019.

45. Englyst, H. N.; Kingman, S. M.; Cummings, J. H. Classification and Measurement of Nutritionally Important Starch Fractions. *Eur. J. Clin. Nutr.* **1992**, *46*, 33–50.

46. Esa, N. M.; Ling, T. B.; Esa, L. S. P. By-Products of Rice Processing: An Overview of Health Benefits and Applications. *J. Rice Res.* **2013**, *1*, 1–11.

47. Fardet, A. New Hypotheses for the Health-Protective Mechanisms of Whole-Grain Cereals: What is Beyond Fiber? *Nutr. Res. Rev.* **2010**, *23*, 65–134.

48. Fardet, A.; Rock, E.; Rémésy, C. Is the In Vitro Antioxidant Potential of Whole-Grain Cereals and Cereal Products Well Reflected In Vivo? *J. Cereal Sci.* **2008**, *48*, 258–276.

49. Fred, B.; Bernd, K.; Eva, A.Resistant Starch and the Butyrate Revolution. *Trends Food Sci. Technol.* **2002**, *13*, 251–261.

50. Fredric, J.; Brown, J. B. The Fatty Acids of Corn Oil. *J. Am. Chem. Soc.* **1945**, *67*, 1899–1900.

51. Galvez, A. F.; Revilleza, M. J. R.; Lumen, B. O. ANovel Methionine-Rich Protein from Soybean Cotyledon: Cloning and Characterization of cDNA. *Plant Physiol.* **1997**, *114*, 1567–1569.

52. Gani, A.; Wani, S. M.; Masoodi, F. A.; Hameed, G. Whole-Grain Cereal Bioactive Compounds and their Health Benefits: A Review. *J. Food Process.Technol.* **2012**, *3*, 146.

53. Gauthier, S. F.; Pouliot, Y.; Saint-Sauveur, D. Immunomodulatory Peptides Obtained by the Enzymatic Hydrolysis of Whey Proteins. *Int. Dairy J.* **2006**, *16*, 1315–1323.

54. Ham, H.; Oh, S. K.; Lee, J. S.; Choi, I. S.; Jeong, H. S.; Kim, I. H.; Lee, J.; Yoon, S. W. Antioxidant Activities and Contents of Phytochemicals in Methanolic Extracts of Specialty Rice Cultivars in Korea. *Food Sci. Biotechnol.* **2013**, *22*, 631–637.

55. Hareland, G. A.; Manthey F. A. Oats. In*Encyclopedia of Food Sciences and Nutrition*; 1stEd; Elsevier Academic Press: Oxford, 2003; pp 4213–4220.

56. Hemery, Y.; Rouau, X.; Pellerin, L. V.; Barron, C.; Abecassis, J. Dry Processes to Develop Wheat Fractions and Products with Enhanced Nutritional Quality. *J. Cereal Sci.* **2007**, *46*, 327–347.

57. Hernandez, L. B.; Amigo, L.; Ramos, M.; Recio, I. Angiotensin Converting Enzyme Inhibitory Activity in Commercial Fermented Products. Formation of Peptides under Simulated Gastrointestinal Digestion.*J. Agric.Food Chem.* **2004**, *52*, 1504–1510.

58. Hernández, L.; Afonso, D.; Rodríguez, E. M.; Díaz, C. Phenolic Compounds in Wheat Grain Cultivars. *Plant Foods Hum. Nutr.* **2011**, *66*, 408–415.

59. Hernandez-Ledesma, B.; Hsieh, C. C.; de Lumen, B. O. Chemopreventive Properties of Peptide Lunasin: A Review. *Protein Pept. Lett* **2013**, *20*, 424–432.

60. Inglett, G. E.; Chen, D.; Berhow, M.; Lee, S. Antioxidant Activity of Commercial Buckwheat Flours and Their Free and Bound Phenolic Compositions. *Food Chem.* **2011**, *125*, 923–929.

61. Irmak, S.; Dunford, N. T. Policosanol Contents and Compositions of Wheat Varieties. *J. Agric. Food Chem.* **2005**, *53*, 5583 – 5586.

62. Jang, S. H.; Lim, J. W.; Kim, H. Mechanism of β-Carotene-Induced Apoptosis of Gastric Cancer Cells: Involvement of Ataxia-Telangiectasia-Mutated. *Ann. N. Y. Acad. Sci.* **2009**, *1171*, 156–162.

63. Jang, H. H.; Park, M. Y.; Kim, H. W.; Lee, Y. M.; Hwang, K. A.; Park, J. H.; Park, D. S.; Kwon, O. Black Rice (*Oryza sativa L.*) Extract Attenuates Hepatic Steatosis in C57BL/6 J Mice Fed a High-Fat Diet Via Fatty Acid Oxidation. *Nutr.Metab.* **2012**, *9*, 27.

64. Jeong, H. J.; Jeong, J. B.; Kim, D. S.; Park, J. H.; Lee, J. B.; Kweon, D. H.; Chung, G. Y.; Seo, E. W.; de Lumen, B. O. The Cancer Preventive Peptide Lunasin from Wheat Inhibits Core Histone Acetylation. *Cancer Lett.* **2007**, *255,* 42–48.

65. Jeong, H. J.; Lee, J. R.; Jeong, J. B.; Park J. H.; Cheong, Y. K.; de Lumen, B. O. The Preventive Seed Peptide Lunasin from Rye is Bioavailable and Bioactive. *Nutr. Cancer* **2009**, *61*, 680–686.

66. Jeong, H. J.; Jeong, J. B.; Hsieh, C. C.; Ledesma, H. B.; de Lumen, B. O. Lunasin is Prevalent in Barley and is Bioavailable and Bioactive In Vivo and In Vitro Studies. *Nutr. Cancer* **2010**, *62*, 1113–1119.

67. Jia, J.; Maa, H.; Zhao, W.; Wang, Z.; Tian, W.; Luo, L.; He, R. The Use of Ultrasound for Enzymatic Preparation of ACE-Inhibitory Peptides from Wheat Germ Protein. *Food Chem.* **2010**, *119*, 336–342.

68. Jokić, S.; Sudar, R.; Svilović, S.; Vidović, S.; Bilić, M.; Velić, D.; Jurković, V. Fatty Acid Composition of Oil Obtained from Soybeans by Extraction with Supercritical Carbon Dioxide. *Czech J. Food Sci.* **2013**, *31*, 116–125.

69. Jones, J. M.; Reicks, M.; Adams, J.; Fulcher, G.; Marquart, L. Becoming Proactive with the Whole-Grains Message. *Nutr. Today* **2004**, *39*, 10–17.

70. Kalman, D. S. Amino Acid Composition of Organic Brown Rice Protein Concentrate and Isolate Compared to Soy and Whey Concentrates and Isolates. *Food* **2014**, *3*, 394–402.

71. Kan, A. Characterization of the Fatty Acid and Mineral Composition of Selected Cereal Cultivars from Turkey. *Rec.Nat. Prod.* **2015**, *9*, 124–134.

72. Khang, D. T.; Dung, T. N.; Elzaawely, A. A.; Xuan, T. D. Phenolic Profiles and Antioxidant Activity of Germinated Legumes. *J. Foods* **2016**, *5*, 27.

73. Kim, H. J.; Bae, I. Y.; Ahn, C. W.; Lee, S.; Lee, H. G. Purification and Identification of Adipogenesis Inhibitory Peptide from Black Soybean Protein Hydrolysate. *Peptides* **2007**, *28*, 2098–2103.

74. Kim, H. G.; Kim, G. W.; Oh, H.; Yoo, S. Y.; Kim, Y. O.; Oh, M. S. Influence of Roasting on the Antioxidant Activity of Small Black Soybean (*Glycine max L. Merrill*). LWT. *Food Sci. Technol* **2011**, *44*, 992–998.

75. Kitts, D. D.; Weiler, K. Bioactive Proteins and Peptides from Food Sources. Applications of Bioprocesses used in Isolation and Recovery. *Curr.Pharm.Des.* **2003**, *9*, 1309–1323.

76. Kodera, T.; Nio, N. Angiotensin converting enzyme inhibitors. EP1352911; PCT/JP2002/000194; 2003; A1.

77. Kopsell, D. A.; Armel, G. R.; Mueller, T. C.; Sams, C. C.; Deyton, D. E.; McElroy, J. S.; Kopsell, D. E. Increase in Nutritionally Important Sweet Corn Kernel Carotenoids Following Mesotrione and Atrazine Applications. *J. Agric.Food Chem.* **2009**, *57*, 6362–6368.

78. Korhonen, H.; Leppälä, P. A. Milk Protein-Derived Bioactive Peptides—Novel Opportunities for Health Promotion. *Bull.Int.Dairy Fed.* **2001**, *363*, 17–26.

79. Koyama, M.; Nakamura, C.; Nakamura, K. Changes in Phenols Contents from Buckwheat Sprouts During Growth Stage. *J. Food Sci. Technol.* **2013**, *50*, 86–93.

80. Kumar, D.; Jhariya, N. A. Nutritional, Medicinal and Economical Importance of Corn: AMini Review. *J. Pharm. Sci.* **2013**, *2*, 7–8.

81. Kumar, P.; Yadava, R. K.; Gollen, B.; Kumar, S.; Verma, R. K.; Yadav, S. Nutritional Contents and Medicinal Properties of Wheat: A Review. *Life Sci. Med. Res.* **2011**, *22*, 1–10.

82. Kwak, C. S.; Lee, M. S.; Park, S. C. Higher Antioxidant Properties of Chungkookjang, a Fermented Soybean Paste, May Be Due to Increased Aglycone and Malonyl Glycosideiso Flavone During Fermentation. *Nutr. Res.* **2007**, *27*, 719–727.

83. Lee, J. H.; Choung, M. G. Determination of Optimal Acid Hydrolysis Time of Soybean Isoflavones Using Drying Oven and Microwave Assisted Methods. *Food Chem.* **2011**, *129*, 577–582.

84. Lee, J. H.; Cho, K. M. Changes Occurring in Compositional Components of Black Soybeans Maintained at Room Temperature for Different Storage Periods. *Food Chem.* **2012**, *131*, 161–169.

85. Lee, J. H.; Seo, W. T.; Lim, W. J.; Cho, K. M. Phenolic Contents and Antioxidant Activities from Different Tissues of Baekseohyang (*Daphne kiusiana*). *Food Sci. Biotechnol.* **2011**, *20*, 695–702.

86. Lee, C.; Shen, S.; Lai, Y.; Wu, S. Rutin and Quercetin, Bioactive Compounds from Tartary Buckwheat, Prevent Liver Inflammatory Injury. *Food Funct.* **2013**, *4*, 794–802.

87. Leoncini, E.; Prata, E.; Malaguti, M.; Marotti, I.; Carretero, A. S.; Catizone, P.; Dinelli, G.; Hrelia, S. Phytochemical Profile and Nutraceutical Value of Old and Modern Common Wheat Cultivars. PLOS one **2012**, *7*, e45997. https://doi.org/10.1371/journal.pone.0045, 997

88. Li, L.; Wang, C.; Qiang, S.; Zhao, J.; Song, S.; Jin, W.; Wang, B.; Huang, Y. Z. L.; Wang, Z. Mass Spectrometric Analysis of n-Glycoforms of Soybean Allergenic Glycoproteins Separated by SDS-PAGE. *J. Agric. Food chem.* **2016**, *64*, 7367−7376.

89. Liu, H. Whole Grain Phytochemicals and Health. *J. Cereal Sci.* **2007**, *46*, 207–219.

90. Liu, Y.; Ng, P. K. W. Relationship Between Bran Characteristics and Bran Starch of Selected Soft Wheat Grown in Michigan. *Food Chem.* **2016**, *197*, 427–435.

91. Lu, J.; Zeng, Y.; Hou, W.; Zhang, S.; Li, L.; Luo, X.; Xi, W.; Chen, Z.; Xiang, M. The Soybean Peptide Aglycin Regulates Glucose Homeostasis in Type II Diabetic Mice Via IR/IRS1 Pathway. *J. NutrBiochem.* **2012**, *23*, 1449–1457.

92. Madhujith, T.; Izydorczyk, M.; Shahidi, F. Antioxidant Activity of Pearled Barley Fractions. *J. Agric.Food Chem.* **2006**, *54*, 3283–3289.

93. Maillard, M. N.; Berset, C. Evolution of Antioxidant Activity during Kilning-Role of Insoluble Bound Phenolic-Acids of Barley and Malt. *J.Agric.Food Chem.* **1995**, *43*, 1789–1793.

94. Martinez, L. X.; Ros, R. M. O.; Alfaro, V. G.; Lee, C. H.; Parkin, K. L.; Garcia, H. S. Antioxidant Activity, Phenolic Compounds and Anthocyanins Content of Eighteen Strains of Mexican Maize. *Food Sci. Technol.* **2009**, *42*, 1187–1192.

95. Mattila, P.; Pihlava, J. M.; Hellström, J. Contents of Phenolic Acids, Alkyl- and Alkenylresorcinols and Avenanthramides in Commercial Grain Products. *J. Agric.Food Chem.* **2005**, *53*, 8290–8295.

96. McCann, K. B.; Shiell, B. J.; Michalski, W. P.; Lee, A.; Wan, J.; Roginski, H.; Coventry, M. J. Isolation and Characterisation of a Novel Antibacterial Peptide from Bovine aS1-Casein. *Int. Dairy J.* **2006**, *16*, 316–323.

97. Mehmood, S.;Orhan, I.; Ahsan, Z.; Aslan, S.; Gulfraz, M. Fatty Acid Composition of Seed Oil of Different *Sorghum bicolor* Varieties. *Food Chem,* **2008**, *109*, 855–959.

98. Meisel, H.; FitzGerald, R. J. Biofunctional Peptides from Milk Proteins: Mineral Binding and Cytomodulatory Effects. *Curr.Pharm.Des.* **2003**, *9*, 1289–1295.

99. Miller, A.; Engel, K. H. Content of γ-Oryzanol and Composition of Steryl Ferulates in Brown Rice (*Oryza sativa L.*) of European Origin. *J. Agric. Food Chem.* **2006**, *54*, 8127–8133.

100. Moraes, E. A.; Marinelli, R. S.; Lenquiste, S. A.; Steel, C. J.; Menezes, C. B.; Queiroz, V. A. V. Sorghum Flour Fractions: Correlations among Polysaccharides, Phenolic Compounds, Antioxidant Activity and Glycemic Index. *Food Chem.* **2015**, *180,* 116–123.

101. Moreno, Y. S.; Sanchez, G. S.; Hernandez, D. R.; Lobato, N. R. Characterization of Anthocyanin Extracts from Maize Kernels. *J. Chromatogr. Sci.* **2005**, *43*, 483–487.

102. Morey, D. D.; Evans, J. J. Amino Acid Composition of Six Grains and Wheat Forage. *Cereal Chem.* **1983**, *60*, 461–464.

103. Mpofu, A.; Sapirstein, H. D.; Beta, T. Genotype and Environmental Variation in Phenolic Content, Phenolic Acid Composition, and Antioxidant Activity of Hard Spring Wheat. *J. Agric. Food Chem.* **2006**, *54*, 1265−1270.

104. Murphy, M. M.; Douglass, J. S.; Birkett, A. Resistant Starch Intakes in the United States. *J. Am. Diet. Assoc.* **2008**, *108*, 67–78.

105. Nagarajan, S.; Burris, R. L.; Stewart, B. W.; Wilkerson, J. E.; Badger, T. M. Dietary Soy Protein Isolate Ameliorates Atherosclerotic Lesions in Apolipoprotein E-Deficient Mice Potentially by Inhibiting Monocyte Chemoattractant Protein-1 Expression. *J. Nutr.* **2008**, *138*, 332–337.

106. Nantiyakul, N.; Furse, S., Fisk, I.; Foster, T. J.; Tucker, G.; Gray, D. A. Phytochemical Composition of *Oryza sativa* (Rice) Branoil Bodies in Crude and Purified Isolates. *J. Am. Chem. Soc.* **2012**, *89*, 1867–1872.

107. Needham, S. A.; Beck, E. J.; Johnson, S. K.; Tapsell, L. C. Sorghum: An Underutilized Cereal Whole Grain with the Potential to Assist in the Prevention of Chronic Disease. *Food Rev. Int.* **2015**, *31*, 401–437.

108. Norojarvi, M. L.; Eldin, A. K.; Appelqvist, L.; Dimberg, H. L.; Ohrvall, M.; Vessby, B. Corn and Sesame Oils Increase Serum γ-Tocopherol Concentrations in Healthy Swedish Women. *J. Nutr.* **2001**, *131*, 1195–1201.

109. Okarter, N.; Liu, C. S.; Sorrells, M. B.; Liu, R. H. Phytochemical Content and Antioxidant Activity of Six Diverse Varieties of Whole Wheat. *Food Chem.* **2010**, *119*, 249–257.

110. Oliveira, K. G.; Queiroz, V. A. V.; Carlos, L. A.; Cardoso, L. M.; Ana, H. M. P.; Anunciaçao, P. C.; Menezes, C. B.; Silva, E. C.; Barros, F. Effect of the Storage Time and Temperature on Phenolic Compounds of Sorghum Grain and Flour. *Food Chem.* **2017**, *216*, 390–398.

111. Oluremi, O. I.; Solomon, A. O.; Saheed, A. A. Fatty Acids, Metal Composition and Physico-Chemical Parameters of Igbemo Ekiti Rice Bran Oil. *J. Environ. Chem. Ecotoxicol.* **2013**, *5*, 39–46.

112. Oscarsson, M.; Andersson, R.; Salomonsson, A. C.; Man, P. A. Chemical Composition of Barley Samples Focusing on Dietary Fiber Components. *J. Cereal Sci.* **1996**, *24*, 161–170.

113. Othman, R. A.; Moghadasian, M. H.; Jones, P. J. H. Cholesterol-Lowering Effects of Oat β-Glucan. *Nutr.n Rev.* **2011**, *69*, 299–309.

114. Palozza, P.; Serini, S.; Torsello, A.; Nicuolo, F. D.; Maggiano, N.; Ranelletti, F. O.; Calviello, G. Mechanism of Activation of Caspase Cascade during β-Carotene-Induced Apoptosis in Human Tumor Cells. *Nutr.Cancer* **2003**, *47*, 76–87.

115. Park, J. H.; Jeong, H. J.; de Lumen, B. O. Contents and Bioactivities of Lunasin, Bowman-Birk Inhibitor, and Isoflavones in Soybean Seed. *J. Agric. Food Chem.* **2005**, *53*, 7686–7690.

116. Park, Y.; Moon, H. J.; Paik, D. J.; Kim, D. Y. Effect of Dietary Legumes on Bone-Specific Gene Expression in Ovariectomized Rats. *Nutr. Res. Pract.* **2013**, *7*, 185–191.

117. Parle, M.; Dhamija, I. Zea maize: A Modern Craze. *Int.Res. J. Pharm.* **2013**, *4*, 39–43.

118. Patel, M.; Naik, S. K. Gamma Oryzanol from Rice Bran Oil—A Review. *J. Sci. Ind. Res.* **2004**, *63*, 569–578.

119. Petkov, K.; Biel, W.; Kowieska, A.; Jaskowska, I. The Composition and Nutritive Value of Naked Oat Grain (*Avena sativa* var. nuda). *J.Animal Feed Sci.* **2001**, *10*, 303–307.

120. Pomeranz, Y.; Robbins, G. S. Amino Acid Composition of Buckwheat. *J. Agric. Food Chem.* **1972**, *20*, 271–274.

121. Pomeranz, Y.; Robbins, G. S.; Smith, R. T.; Craddock, J. C.; Gilbertson, J. T.; Moseman, J. G. Protein Content and Amino Acid Composition of Barleys from the World Collection. *Cereal Chem.* **1976**, *53*, 497–504.

122. Posuwan, J.; Prangthip, P.; Leardkamolkarn, V.; Yamborisut, U.; Surasiang, R.; Charoensiri, R. Long-Term Supplementation of High Pigmented Rice Bran Oil (*Oryza sativaL.*) on Amelioration of Oxidative Stress and Histological Changes in Streptozotocin-Induced Diabetic Rats Fed a High Fat Diet; Riceberry Bran Oil. *Food Chem.* **2013**, *138*, 501−508.

123. Queiroz, V. A. V.; Silva, C. S. D.; Menezes, C. B. D.; Schaffert, R. E.; Guimaraes, F. F. M.; Guimaraes, L. J.; Guimaraes, P. E. O.; Tardin, F. D. Nutritional Composition of Sorghum [*Sorghum bicolor* (L.) Moench] Genotypes Cultivated without and with Water Stress.*J. Cereal Sci.* **2015**, *65*, 103–111.

124. Radhakrishnamurthy, R.; Desikachar, H. S. R.; Srinivasan, M.; Subrahmanyan, V. Studies on Idli Fermentation II. Relative Participation of Blackgram Flour and Rice Semolina in the Fermentation.*J. Sci. Ind.Res.* **1961**, *20*, 342–344.

125. Ramakrishnan, C. V. The Studies on Indian Fermented Food.*Baroda J Nutr.* **1979**, *6*, 1–57.

126. Riascos, J. J.; Weissinger, A. K.; Weissinger, S. M.; Burks, A. W. Hypoallergenic Legume Crops and Food Allergy: Factors Affecting Feasibility and Risk. *J. Agric. Food Chem.* **2010**, *58*, 20−27.

127. Ritt, A. B. B.; Mulinari, F.; Vasconcelos, I. M.; Carlini, C. R. Antinutritional and/or Toxic Factors in Soybean (*Glycine max* (L) Merril) Seeds: Comparison of Different Cultivars Adapted to the Southern Region of Brazil. *J. Sci.Food Agric.* **2004**, *84*, 263–270.

128. Rooney, I. W. Sorghum and Pearl Millet Lipids. *Cereal Chem.* **1978**, *55*, 584–590.

129. Roy, B.; Kala, C. P.; Nehal, F. A.; Majila, B. S. Indigenous Fermented Food and Beverages: A Potential for Economic Development of the High Altitude Societies in Uttaranchal. *J. Hum.Ecol.* **2004**, *15*, 45–49.

130. Ruxton, C.; Cobbs, R. The Role of Oats and Oat Products in the UK Diet.*Comp.Nutr.* **2015**, *14*, 55–57.

131. Sajilata, M. G.; Singhal, R. S.; Kulkarni, P. R. Resistant Starch. *Compr. Rev. Food Sci.Food Saf.* **2006**, *5*, 1–17.

132. Sandhu, K. S.; Singh, N.; Malhi, N. S. Some Properties of Corn Grains and their Flours: Physicochemical, Functional and Chapatti-Making Properties of Flours. *Food Chem.* **2007**, *101*, 938–946.

133. Sang, Y.; Bean, S.; Seib, P. A.; Pedersen, J.; Shi, Y. C. Structure and Functional Properties of Sorghum Starches Differing in Amylose Content. *J. Agric.Food Chem.* **2008**, *56*, 6680–6685.

134. Sanjeev, K. S.; Sandhu, K. D. Indian Fermented Foods; Microbiological and Biochemical Aspects. *Ind. J. Microbiol.* **1990**, *30*, 135–157.

135. Savage, J. H.; Kaeding, A. J.; Matsui, E. C.; Wood, R. A. The Natural History of Soy Allergy. *J. Allergy Clin.Immunol.* **2010**, *125*, 683−686.

136. Sedej, I.; Sakac, M.; Mandic, A.; Misan, A.; Tumbas, V.; Brunet, J. C. A. Buckwheat (*Fagopyrum esculentum* Moench) Grain and Fractions: Antioxidant Compounds and Activities. *Food Sci.* **2012**, *77*, 954–959.

137. Sekar, S.; Mariappan, S. Usage of Traditional Fermented Products by Indian Rural Folks and IPR. *Ind.J.Tradit.Knowl.* **2007**, *6*, 111–120.

138. Shah, T. F.; Prasad K.; Kumar, P. Maize—APotential Source of Human Nutrition and Health: AReview. *Cogent FoodAgric.* **2016**, *2*(1), e-article: 1, 166, 995

139. Shewry, P. R.; Hawkesford, M. J.; Piironen, V.; Lampi, A. M.; Gebruers, K.; Boros, D.; Andersson, A. A. M.; Aman, P.; Rakszegi, M.; Bedo, Z.; Ward, J. L. Natural Variation in Grain Composition of Wheat and Related Cereals. *J. Agric.Food Chem.* **2013**, *61*, 8295–8303.

140. Shimizu, M.; Sawashita, N.; Morimatsu, F.; Ichikawa, J.; Taguchi, Y.; Ijiri, Y.; Yamamoto, J. Antithrombotic Papain-Hydrolyzed Peptides Isolated from Pork Meat. *Thromb.Res.* **2008**, *123*, 753–757.

141. Shin, Z. I.; Yu, R.; Park, S. A.; Chung, D. K.; Ahn, C. W.; Nam, H. S.; Kim, K. S.; Lee, H. J.; Leu, H. H. An Angiotensin I Converting Enzyme Inhibitory Peptide Derived from Korean Soybean Paste, Exerts Antihypertensive Activity In Vivo. *J. Agric.Food Chem.* **2001**, *49*, 3004–3009.

142. Shipp, J.; Aal, E. S. A. Food Applications and Physiological Effects of Anthocyanins as Functional Food Ingredients. *Open Food Sci. J.* **2010**, *4*, 7–22.

143. Sienkiewicz-Szlapka, E.; Jarmolowska, B.; Krawczuk, S.; Kostyra, E.; Kostyra, H.; Iwan, M. Contents of Agonistic and Antagonistic Opioid Peptides in Different Cheese Varieties. *Int.Dairy J.* **2009**, *19*, 258–263.

144. Slavin, J. Why Whole Grains are Protective: Biological Mechanisms. *Proc.Nutr.Soc.* **2003**, *62*(1), 129–134.

145. Soni, S. K.; Sandhu, D. K. Indian Fermented Foods: Microbiological and Biochemical Aspects. *Ind. J. Microbiol.* **1990**, *30*, 135–157.

146. Souci, S. W.; Fachmann, W.; Kraut, H. (Eds.) *Food composition and nutrition tables. 7th Ed, CRC press Taylor & Francis Group*; 2008; ISBN 9, 780, 849, 341, 410; p 1300.

147. Sriseadka, T.; Wongpornchai S.; Rayanakorn M. Quantification of Flavonoids in Black Rice by Liquid Chromatography-Negative Electrospray Ionization Tandem Mass Spectrometry. *J. Agric. Food Chem.* **2012**, *60*, 11, 723–11, 732.

148. *Handbook of Indigenous Fermented Foods.* Steinkraus, K. H. (Ed.) 2nd Ed; Elsevier: Great Britain; 1999; p 131.

149. Sun, J.; He, H.; Xie, B. J. Novel Antioxidant Peptides from Fermented Mushroom *Ganoderma lucidum. J. Agric. Food Chem.* **2004**, *52*, 6646–6652.

150. Sur, R.; Nigam, A.; Grote, D.; Liebel, F.; Southall, M. D. Avenanthramides, Polyphenols from Oats, Exhibit Anti-Inflammatory and Anti-Itch Activity. *Arch. Dermatol. Res.* **2008**, *10*, 569–74.

151. Terpinc, P.; Cigic, B.; Polak, T.; Haribar, J.; Pozrl, T. LC-MS Analysis of Phenolic Compounds and Antioxidant Activity of Buckwheat at Different Stages of Malting. *Food Chem.* **2016**, *210*, 9–17.

152. Verardo, V.; Serea, C.; Segal, R.; Cabo F. M. Free and Bound Minor Polar Compounds in Oats: Different Extraction Methods and Analytical Determinations. *J. Cereal Sci.* **2011**, *54*, 211–217.

153. Verma, A. K.; Kumar, S.; Das, M.; Dwivedi, P. D. A Comprehensive Review of Legume Allergy. *J. Allergy Clin.Immunol.* **2013**, *45*, 30–46.

154. Virupaksha, T. K.; Sastry, L. V. S. Studies on the Protein Content and Amino Acid Composition of Some Varieties of Grain Sorghum. *J. Agric. Food Chem.* **1968**, *16,* 199–203.

155. Walter, M.; Marchesan, E.; Fabrício, P.; Massoni, S.; Silva, L. P. D.; Meneghetti, G.; Sartori, S.; Ferreira, R. B. Antioxidant Properties of Rice Grains with Light Brown, Red and Black Pericarp Colors and the Effect of Processing. *Food Res. Int.* **2013,** *50,* 698–703.

156. Wang, X.; Brown, I. L.; Khaled, D.; Mahoney, M. C.; Evans, A. J.; Conway, P. L. Manipulation of Colonic Bacteria and Volatile Fatty Acid Production by Dietary High Amylose Maize (Amylomaize) Starch Granules. *J. Appl. Microbiol.* **2002,** *93,* 390–397.

157. Watanabe, M. An Anthocyanin Compound in Buckwheat Sprouts and its Contribution to Antioxidant Capacity. *Biosci., Biotechnol., Biochem.* **2007,** *71,* 579–582.

158. Wieser, H.; Seilmeier, W.; Eggert, M.; Belitz, H. D. Tryptophan Content of Cereal Proteins. *Eur. Food Res.Technol.* **1983,** *177,* 457–460.

159. Willcox, J. K.; Ash, S. L.; Catignani, G. L. Antioxidants and Prevention of Chronic Disease. *Crit. Rev. Food Sci. Nutr.* **2004,** *44,* 275–95.

160. Wood, P. J. Cereal β-Glucans in Diet and Health. *J. Cereal Sci.* **2007,** *46,* 230–238.

161. Wua, G.; Johnson, S. K.; Bornman, J. F.; Bennett, S. J.; Fang, Z. Changes in Whole Grain Polyphenols and Antioxidant Activity of Six Sorghum Genotypes under Different Irrigation Treatments. *Food Chem.* **2017,** *214,* 199–207.

162. Wursch, P. F.; Pi, S. X. The Role of Viscous Soluble Fiber in the Metabolic Control of Diabetes. *Diabetes Care* **1997,** *20,* 1774–1780.

163. Wynstra, R. J. *Expanding the Use of Soybeans.* College of Agriculture, University of Illinois; Urbana—Champaign; 1986; p20.

164. Ye, Q. E.; Sara A.; Chacko; Chou, E. L.; Kugizaki, M.; Liu, S. Greater Whole-Grain Intake is Associated with Lower Risk of Type 2 Diabetes, Cardiovascular Disease, and Weight Gain. *J. Nutr.* **2014,** *142,* 1304–1313.

165. Yoshikawa, M.; Fujita, H.; Matoba, N.; Takenaka, Y.; Yamamoto, T.; Yamauchi, R.; Tsuruki, H.; Takahata, K. Bioactive Peptides Derived from Food Proteins Preventing Lifestyle-Related Diseases. *Bio Factors* **2000,** *12,* 143–146.

166. Zhang, Z. L.; Zhou, M. L.; Tang, Y.; Li, F. L.; Tang, Y. X.; Shao, J. R.; Xue, W. T. Bioactive Compounds in Functional Buckwheat Food. *Food Res.Int.* **2012,** *49,* 389–395.

167. Zhang, H.; Shao, Y.; Bao, J.; Beta, T. Phenolic Compounds and Antioxidant Properties of Breeding Lines between the White and Black Rice. *Food Chem.* **2015,** *172,* 630–639.

168. Zhong, F.; Liu, J.; Ma, J.; Shoemaker, C. F. Preparation of Hypocholesterol Peptides from Soy Protein and their Hypocholesterolemic Effect in Mice. *Food Res. Int.* **2007,** *40,* 661–667.

169. Zieli, A.; Ki, H.; Koz, A.; Owska, H. Antioxidant Activity and Total Phenolics in Selected Cereal Grains and their Different Morphological Fractions. *J. Agri. Food Chem.* **2000,** *48,* 2008–2016.

PART IV
Innovative Use of Medicinal Plants

CHAPTER 9

THERAPEUTIC IMPLICATIONS OF *RHODIOLA* SP. FOR HIGH ALTITUDE MALADIES: A REVIEW

KALPANA KUMARI BARHWAL, KUSHAL KUMAR,
SURYANARAYAN BISWAL, and SUNIL KUMAR HOTA

CONTENTS

ABSTRACT

Both traditional medicine and the recent research findings indicate toward the adaptogenic potential of *Rhodiola* for extreme high-altitude environment. Considering the increase in mountain warfare in recent years, *Rhodiola* surely has a potential to be used as a prophylactic to combat stress as well as an adaptogen for extreme environmental conditions. A concerted effort for the conservation, propagation, sustainable utilization, and large-scale adoption of *Rhodiola*-based nutraceuticals by the combatants is, however, required to optimally utilize this *Sanjeevani* of the trans-Himalayas in combat operations.

9.1   INTRODUCTION

The snow-clad mountains have always poised a formidable challenge to the adventurous souls, who derive inspiration, solace, self-esteem, or confidence by scaling these heights. Although mountaineering was an integral part of warfare training in Central Asia since ages, it was limited to crossing the high passes to invade Northern India. There has been a constant progress in the nature of mountaineering expeditions in recent times both in terms of technique and technology. Though mountaineering is considered an adventure sport in several countries, the strategic location of Himalayas as a natural border of India has made mountaineering and survival in extreme conditions of high altitude, a defense requirement.

The term high altitude refers to terrestrial elevations over 1500 m and has been categorized as high altitude (1500–3500 m), very high altitude (3500–5500 m), and extreme altitude (5500–8500 m).[19] Ascent to high altitude is associated with fall in partial pressure of oxygen, resulting in decreased oxygen saturation of arterial blood and tissue hypoxia. This condition is often referred to as hypobaric hypoxia since hypobaric or decreased atmospheric pressure at high altitude results in hypoxia, that is, reduced oxygen supply to the tissues.[26] Nearly, 80% of individuals experience headache within hours of arrival at high altitude, which is the most common and initial presentation of hypobaric hypoxia-induced illness.[30,75] Subsequent symptoms like nausea, vomiting, and dyspnea (shortness of breath)—which occur between 6 and 96 h after arrival at high altitude—are collectively termed as acute mountain sickness.[54] These symptoms, if ignored, may in turn result in life-threatening conditions like high altitude cerebral edema and pulmonary edema.[36] On the contrary, the body tries to adjust to the hypoxic conditions through a series of physiological changes that include faster blood circulation, increase in the hematocrit, and so forth, through a process commonly referred to as acclimatization.[9]

Recent reports suggest that despite acclimatization to high altitude, prolonged exposure to hypobaric hypoxia results in several adverse effects on the human physiology.[63] Hypophagia, alterations in gastric motility, cognitive impairment, and sleep apnea are some of the adverse physiological effects that persist even after acclimatization at high altitude.[58,59] Research by authors shows that prolonged stay at high altitude resulted in changes in autonomic rhythm, a progressive increase in circulatory homocysteine, which is an important cardiac risk factor and domain-specific temporal decline in cognitive performance.[61] There have been reports on altered functioning of mitochondrial transport chain and the production of reactive oxygen species causing consequent damage to lipids, proteins, and deoxyribonucleic acid on prolonged exposure to hypobaric hypoxia.[26,67] This increased physiological stress may be one of the reasons for compromised physical performance at high altitude.[48] Chronic exposure to hypobaric hypoxia also results in increased level of various inflammatory markers: tumor necrosis factor alpha, interleukins-6, and so forth, that may cause inflammation in various types of tissues.[42,56] In addition to this, long-term stay at high altitude leads to cognition dysfunctions, such as impairment in attention, memory, judgment, and emotion.[26]

Apart from hypobaric hypoxia, the high altitude environment is also characterized by extremely low temperatures and high ultraviolet radiations. For every 1000 m increase in altitude, the level of ultraviolet radiations increases by about 12%.[68] Since ultraviolet radiations possess high energy, they may cause damage to biological tissues causing sunburn, skin ageing, blemishes, immunosuppression, and photo-carcinogenesis.[65] On the other hand, the extreme cold at high altitude may cause cold injuries like joint pains and frostbite at extremities, which sometimes may even require amputation.[20,21] Hence, both acute and chronic exposure to extreme altitude poses a challenge to the survival and performance of humans (Fig. 9.1).

Despite several years of research, the treatment for high altitude-related maladies is symptomatic and de-induction is still considered as the most effective measure. The therapeutic compounds commonly used for the treatment of hypobaric hypoxia-related health problems like acute mountain sickness and edema are limited to acetazolamide, nifedipine, and dexamethasone and the development of alternative and more effective prophylactics for ameliorating high-altitude illness still remains a challenge for researchers worldwide.[23,62]

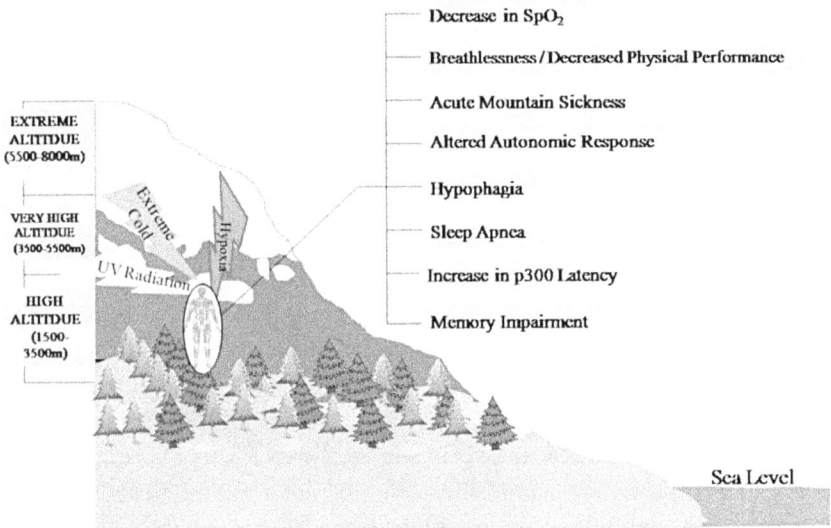

FIGURE 9.1 **(See color insert.)** Adverse effects of high altitude on human health.

Traditional knowledge-based information and plants indigenous to these extreme altitude conditions that have been used by the local population since ages and find mention in traditional literature are, therefore, being explored by researchers for their ability to improve endurance to high altitude stress. One such plant that has gained tremendous attention due to its medicinal properties is *Rhodiola*. Considering the indispensable requirement of the Indian army to deploy its combatants at extreme altitude locations, *Rhodiola* plant has been extensively studied by scientists for its adaptogenic or adaptogens, radioprotectant, antioxidant, and immunomodulatory properties, thereby propelling it into the league of Sanjeevani.

This chapter reviews the therapeutic potential and implications of *Rhodiola* sp. for high altitude maladies.

9.2 USES OF *RHODIOLA* IN TRADITIONAL MEDICINE

The genus *Rhodiola* L. (Crassulaceae) consists of approximately 100 species, distributed in the mountainous regions of Southwest China and the Himalayas, where the environment is vulnerable to unfavorable conditions.[6,41,78] Use of *Rhodiola* in the traditional system of medicine can be traced back to as early as 300 *anno Domini* (A. D.: in the year of Lord), based upon its description in the Handbook of Traditional Tibetan Drugs among 175 most important Tibetan drugs.[64] *Rhodiola* was used to treat lung diseases, particularly lung-heat disorders. It is mentioned in the Handbook of Traditional Tibetan Drugs in 10 formulations out of which nine are for the treatment of lung diseases.[40,64] One of the species of *Rhodiola*, namely *Rhodiola rosea* (commonly called as golden root, rose root, roseroot, western roseroot, Aaron's rod, Arctic root, king's crown, *lignum rhodium*, orpin rose) has also been included in the Ayurveda (traditional Indian system of medicine) as an adaptogen.[57] In addition to this, *Rhodiola* has also been used in traditional medicine in European countries like Russia.[13]

Out of 100 species of *Rhodiola*, *R. rosea* and *Rhodiola imbricata* have been extensively reported in recent literature for their adaptogenic,[18,33] immunomodulatory,[46] memory-enhancing,[51] and radioprotective properties.[2,11] Approximately, 140 compounds have been isolated from the root and rhizome of *R. rosea*.[51] The isolated compounds are of different classes that include essential oils, trans-cinnamic alcohol glycosides, flavonoids, organic acids, fats, phenolics including tannins and proteins.[34] Salidroside is a principal phenolic compound isolated from *Rhodiola*, other than rosavin, triandrin, and tyrosol.[37]

9.2.1 ADAPTOGENIC AND PERFORMANCE-ENHANCING PROPERTIES

An adaptogen may be defined as the compound that increases the power of resistance to multiple stressors and has a normalizing influence, irrespective of the direction of change from normal physiological conditions that is caused by the stressors, and is innocuous without influencing the normal body functions more than required.[33,57] Chronic exposure to hypobaric hypoxia has been widely considered to be a powerful stressor even for acclimatized dwellers and results in compromised physical performance at high altitude.[73,74] Altered physical performance at high altitude has been attributed to the reduction in fiber size and mass of muscles.[43] Reduction in oxygen availability at high altitude also results in alterations in the mitochondrial transport chain, leading to the production of reactive oxygen species[10] and the depletion of intracellular energy stores like phosphorcreatinine and adenosine triphosphate in muscles.[24]

Rhodiola has been widely reported for its adaptogenic and performance-enhancing properties.[69] It has also been shown that *Rhodiola* increases the muscle strength and improves the physical performance in humans.[15] Oral treatment with extracts from *R. rosea* and *Rhodiola crenulata* roots significantly prolonged the exhaustive swimming in rats and increased adenosine triphosphate synthesis in skeletal muscle mitochondria.[1] ADAPT -32, a fixed-dose combination of *Eleutherococcus senticosus* root extract, *Schisandra chinensis* berry extract, and *R. rosea* root extract has been reported to stimulate the release of stress hormones in systemic circulation, which augments innate defense responses against mild stressors and acts as a herbal adaptogen.[52] Clinical studies reveal that acute oral treatment with *R. rosea* significantly decreases the heart rate equivalent to what we experience during submaximal exercise and appears to improve the physical performance.[50] Salidroside, an active ingredient found in the root of *R. rosea,* has been widely studied for its adaptogenic and physical performance-enhancing properties.[3,52] It has been reported to protect the myocardium of rats in an acute exhaustive injury, possibly, by reducing the content of malondialdehyde, increasing the content of superoxide dismutase, and increasing phosphorylated extracellular-related kinase, and decreasing phosphorylated-p38 protein expressions in rat myocardium.[71] Salidroside has also been reported to increase mitochondrial biogenesis and protect the endothelium of rats during hydrogen peroxide-induced endothelial dysfunctions.[76]

9.2.2 RHODIOLA AS HERBAL IMMUNOMODULATOR

Immunomodulators or biologic response modifiers are compounds that are capable of interacting with the immune system to upregulate or downregulate specific aspects of the host response.[66] Acute as well as chronic exposure to high altitude affects the immune system as an outcome of stressor-induced altered physiological responses as well as altered metabolic functions.[44,45] Altered physiologic and metabolic functions during hypoxia promote several transcription factors like nuclear factor-kappa B and hypoxia-inducible factor-1 alpha.[39] Nuclear factor-kappa B, a hypoxia-inducible transcription factor, plays a central role in the stimulation of pro-inflammatory cytokines tumor necrosis factor-α and interleukin-6.[39] Activation of the nuclear factor-kappa B at high altitudes is caused due to hypobaric hypoxia-induced increase in the expression of toll-like receptors.[42]

Rhodiola sp. has been demonstrated to possess immunomodulatory properties.[46,81] The extract prepared from *R. crenulata* significantly reduced the level of hypoxia-inducible transcription factor and attenuated pulmonary edema in rats, during acute exposure to hypoxia.[38] Rosin and salidroside isolated from *R. rosea* have been reported to reduce the levels of pro-inflammatory cytokines, namely tumor necrosis factor-alpha and interleukin-1 beta in lipopolysaccharide-induced toxicity in murine microglial BV2 cells.[38] In addition to this, salidroside has been shown to attenuate ovalbumin-induced inflammation in lungs of mice through the modulation of p38 mitogen-activated protein kinase and nuclear factor-kappa B.[77]

The aqueous extract of *R. imbricata,* activates pro-inflammatory mediators through phosphorylated inhibitory kappa B and transcription factor nuclear factor-kappa B in human peripheral blood mononuclear cells, induces the expression of toll-like receptor-4 and intracellular granzyme-B in splenocytes, and results in stimulation of the immune system.[47,46] In addition to this, a homogeneous polysaccharide prepared from *R. rosea* has been reported to increase the production of pro-inflammatory cytokines in S-180 tumor cells, thereby providing a novel therapeutic approach to treat cancer.[8]

9.2.3   COGNITIVE-ENHANCING ABILITIES OF RHODIOLA SP.

The human brain accounts for 1–2% of the total body mass but is responsible for 20% of the body's total oxygen consumption. A large amount of energy is required to maintain normal neuronal activity in the central nervous system, thus making it most susceptible to hypoxic insult. There is burgeoning evidence of hypobaric hypoxia-induced impairment of cognition, vigilance, and memory in humans at different altitudes and durations.[4,5,29,36,49,61] This hypobaric hypoxia-induced cognitive impairment has been associated with decreased acetylcholine, oxidative stress, and neurodegeneration in different regions of the brain.[26] Research studies by authors reveal that glutamate excitotoxicity, oxidative stress, and mitochondrial dysfunction[28,25,27] are the primary hallmarks of hypobaric hypoxia-induced neurodegeneration and subsequent memory impairment (Fig. 9.2).

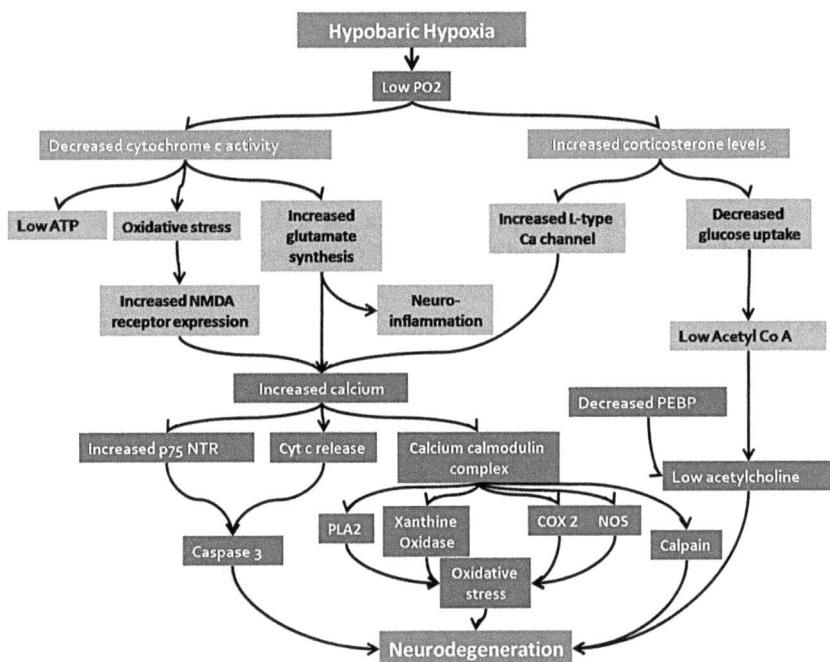

FIGURE 9.2 (See color insert.) Molecular mechanisms leading to neurodegeneration in hypobaric hypoxia.

R. rosea has been traditionally used in Russia for improving memory and learning functions.[22] Alcohol aqueous extract of *R. rosea* has been shown to improve both short- and long-term memory in animal models[53] and inhibits acetylcholinesterase activity in vitro, indicating its potential as a therapeutic agent for neurological disorders with an altered neurotransmitter.[22] It has been suggested that the presence of phenolic compounds like salidroside and rosin in *Rhodiola* sp. may contribute toward its neuroprotective effect.[38]Salidroside, isolated from *Rhodiola sachalinensis* has been reported to protect the pheochromocytoma (PC12) cells against amyloid beta-induced neuronal damage possibly due to its antioxidant activity.[31]

Studies conducted on animal models show that salidroside attenuates amyloid beta-induced learning and memory deficit due to its antioxidant potential.[80] Extract from *Rhodiola* sp. has been reported to attenuate intra-cerebro-ventricularly injected streptozotocin-induced neurotoxicity[55] and scopolamine-induced learning and memory impairment in rats.[16]

Protection of neurons by salidroside may be due to its potential to induce expression of thioredoxin, heme oxygenase-1, and peroxiredoxin-I as well as ability to downregulate proapoptotic genes like Bax and the upregulation of antiapoptotic genes like Bcl-2 and Bcl-X (L).[79] Salidroside also increases phosphorylation of Akt and thus provides neuroprotection through PI3K/Akt-dependent pathways.[12]

9.2.4   RHODIOLA AS RADIOPROTECTIVE AGENT

Extreme high altitude regions are characterized by high intensity of ultra-violet A and B. Ultraviolet C radiations may also be detected in some extremely high altitude locations. Ultraviolet radiation is a very potent initiator of photochemical reactions through the excitation of electrons and this can result in energy transfer or chemical modification of the exposed molecule. Singlet oxygen, hydrogen peroxide, and hydroxyl free radicals are generated on the ultraviolet exposure that can cause damage to cellular proteins, lipids, and saccharides and produce structural damage to the deoxyribonucleic acid, impair the immune system, and lead to cancer.[65,72] On the other hand, herbs confer radioprotective properties due to the presence of antioxidants such as vitamins (vitamin C and E), flavonoids, and phenolic acids.[35] *R. imbricata* has been studied extensively for its radio-protective efficacy.[2] Studies on the hydroalcoholic extract of the rhizome of *R. imbricata*, both in vivo and in vitro, reveal that *Rhodiola* provides radioprotection in gamma radiation-induced damage to the tissues.[2] The Hydroalcoholic extract was found to be more effective in terms of radio-protective efficiency when compared to aqueous extract both at lower as well as high doses.[17] Schriner et al.[60] show that *R. rosea* supplementa-tion could protect cultured cells against ultraviolet light.[60] The ultraviolet protective effect of *Rhodiola* could be attributed to its antioxidant activity and the presence of natural antioxidants and free radical quenchers.[60]

9.2.5 OTHER MEDICINAL PROPERTIES OF RHODIOLA

In addition to its use as an adaptogen, cognitive enhancer, immunomodulator, and a radioprotectant, *Rhodiola* has been reported to have immense potential for the treatment of several health ailments. *R. rosea* has been shown to alleviate renal damage induced by unilateral ureter obstruction in rats. Recently, *R. crenulata* has been shown to attenuate acute hyobaric hypoxia-induced pulmonary edema in animals.[38]

Flavonoids isolated from *R. rosea,* such as gossypetin and kaempferol—have been shown to inhibit the activity of neuraminidases in vitro.[32] Salidroside has also been reported as an antiviral agent, both in vivo and in vitro, against coxsackievirus B3 possibly by modulation of mRNA expression of interferon-gamma, interleukin-10, tumor necrosis factor-alpha, and interleukin-2.[70] Clinically, *R. rosea* has shown to significantly decrease the mean Hamilton anxiety rating scale and reducing the symptoms of generalized anxiety disorder.[7] Standardized extract SHR-5 of rhizomes of *R rosea* has shown antidepressant potency in patients with mild to moderate depression during phase-III clinical studies.[14]

9.3 SUMMARY

Rhodiola has emerged as a potential performance enhancer because of the burgeoning evidence on its medicinal properties and its implications on health benefits. Both traditional medicine and the recent research findings indicate toward the adaptogenic potential of *Rhodiola* for extreme high altitude environment. Apart from this, several clinical studies conducted worldwide do not show any adverse effects of *Rhodiola* supplementation on human health. Considering the increase in mountain warfare in recent years, not only in India, but also in the global scenario that includes combat operations in Afghanistan, Waziristan, Northern Himalayas, Eastern Himalayas, and elsewhere, *Rhodiola* surely has a potential to be used as a prophylactic to combat stress as well as an adaptogen for extreme environmental conditions.

In the Indian scenario, the plant has been given its due importance through the constant research and development efforts of Defense Research and Development Organization. The Defense Institute of High Altitude Research at Ladakh has developed several herbal nutraceuticals from *Rhodiola* that have demonstrated to improve physical and cognitive efficiency of the troops. Considering the fact that every technology being used in modern day warfare is driven by the man behind the machine, this wonder herb can surely improve the lethality, precision, and efficiency of defense technologies in the combat scenario, which is actually human driven. A concerted effort for the conservation, propagation, sustainable utilization, and large-scale adoption of *Rhodiola*-based nutraceuticals by the combatants is however required to optimally utilize this Sanjeevani of the trans-Himalayas in combat operations.

KEYWORDS

- acute mountain sickness
- anxiety
- autonomic responses
- cognition
- fatigue
- free radicals
- glutamate excitotoxicity
- herbal adaptogen
- herbal immunomodulator
- high altitude
- himalayas
- hypobaric hypoxia
- hypophagia
- inflammation
- interleukin-I
- learning and memory
- neuroprotection
- nutraceuticals
- oxidative stress
- pharmacological activity
- phytochemistry
- pro-inflammatory cytokines
- radiation exposure
- radioprotection
- *Rhodiola*
- rosavin
- salidroside
- sleep disturbance
- traditional medicine
- triandrin
- tyrosol
- ultraviolet radiations

REFERENCES

1. Abidov, M.; Crendal, F.; Grachev, S.; Seifulla, R.; Ziegenfuss, T. Effect of Extracts from *Rhodiola rosea* and *Rhodiola crenulata* (Crassulaceae) Roots on ATP Content in Mitochondria of Skeletal Muscles. *Bull. Exp. Biol. Med.* **2003,** *136*(6), 585–587.

2. Arora, R.; Chawla, R.; Sagar, R.; Prasad, J.; Singh, S.; Kumar, R.; Sharma, A.; Singh, S.; Sharma, R. K. Evaluation of Radio-Protective Activities *Rhodiola imbricate Edgew*-A High Altitude Plant. *Mol. Cell. Biochem.* **2005,** *273*(1–2), 209–223.

3. Asea, A.; Kaur, P.; Panossian, A.; Wikman, K. G. Evaluation of Molecular Chaperons Hsp72 and Neuropeptide Y As Characteristic Markers of AdaptogenicActivity of Plant Extracts. *Phytomedicine* **2013,** *20* (14), 1323–1329.

4. Bartholomew, C. J.; Jensen, W.; Petros, T. V.; Ferraro, F. R.; Fire, K. M.; Biberdorf, D.; Fraley, E.; Schalk, J.; Blumkin, D. The Effect of Moderate Levels of Simulated Altitude on Sustained Cognitive Performance. *Int. J. Aviat. Psychol.* **1999,** *9*, 351–359.

5. Bolmont, B.; Thullier, F.; Abraini, J. H. Relationships between Mood States and Performances in Reaction Time, Psychomotor Ability, and Mental Efficiency during a 31-Day Gradual Decompression in a Hypobaric Chamber from Sea Level to 8848 m Equivalent Altitude. *Physiol. Behav.* **2000,** *71*, 469–476.

6. Brown, R. P.; Gerbarg, P. L.; Ramazanov, Z. *Rhodiola rosea*: A Phytomedicinal Overview. *Herbal Gram* **2002,** *56*, 40–52.

7. Bystritsky, A.; Kerwin, L.; Feusner, J. D. A Pilot Study of *Rhodiola rosea* (Rhodax) for Generalized Anxiety Disorder (GAD). *J. Altern. Complement. Med.* **2008,** *14*(2), 175–180.

8. Cai, Z.; Li, W.; Wang, H.; Yan, W.; Zhou, Y.; Wang, G.; Cui, J.; Wang, F. Antitumor Effects of a Purified Polysaccharide from *Rhodiola rosea* and its Action Mechanism. *Carbohydr. Polym.* **2012,** *90*(1), 296–300.

9. Calbet, J. A. Chronic Hypoxia Increases Blood Pressure and Noradrenaline Spillover in Healthy Humans. *J. Physiol.* **2003,** *551*(Pt 1), 379–386.

10. Chandel, N. S.; Maltepe, E.; Goldwasser, E.; Mathieu, C. E.; Simon, M. C.; Schumacker, P. T. Mitochondrial Reactive Oxygen Species Trigger Hypoxia-Induced Transcription. *Proc. Natl. Acad. Sci. U. S. A.* **1998,** *95*(20), 11715–117120.

11. Chawla, R.; Jaiswal, S.; Kumar, R.; Arora, R.; Sharma, R. K. Himalayan Bioresource *Rhodiola imbricate* As a Promising Radioprotector For Nuclear and Radiological Emergencies. *J. Pharm. Bio. Allied Sci.* **2010,** *2*(3), 213–219.

12. Chen, S. F.; Tsai, H. J.; Hung, T. H.; Chen, C. C.; Lee, C. Y.; Wu, C. H.; Wang, P. Y.; Liao, N. C. Salidroside Improves Behavioral and Histological Outcomes and Reduces Apoptosis Via PI3K/Akt Signaling after Experimental Traumatic Brain Injury. *PLoS One* **2012,** *7*(9), e45763.

13. Darbinyan, V.; Kteyan, A.; Panossian, A.;Gabrielian, E.; Wikman, G,; Wagner, H. *Rhodiola rosea* in Stress Induced Fatigue—A Double Blind Cross-Over Study of a Standardized Extract SHR-5 with a Repeated Low-Dose Regimen on the Mental Performance of Healthy Physicians during Night Duty. *Phytomedicine* **2000,** *7*(5), 365–37.

14. Darbinyan, V.; Aslanyan, G.; Amroyan, E.; Gabrielyan, E.; Malmström, C.; Panossian, A. Clinical Trial of *Rhodiola rosea* L. Extract SHR-5 in the Treatment of Mild to Moderate Depression. *Nord. J. Psychiatry* **2007,** *61*(5), 343–348.

15. De Bock, K.;Eijnde, B. O.;Ramaekers, M.;Hespel, P. Acute *Rhodiola rosea*Intake can Improve Endurance Exercise Performance. *Int. J. Sport Nutr. Exercise Metab.* **2004,** *14*(3), 298–307.

16. Getova, D. P.; Mihaylova, A. S. Effects of *Rhodiola rosea* Extract on Passive Avoidance Tests in Rats. *Cent. Eur. J. Med.* **2013,** *8*(2), 176–181.

17. Goel, H. C.;Bala, M.; Prasad, J.; Singh, S.;Agrawala, P. K.;Swahney R. C. Radio-protection by *Rhodiola imbricata* in Mice against Whole-Body Lethal Irradiation. *J. Med. Food* **2006,** *9*(2), 154–160.

18. Gupta, V.; Saggu, S.; Tulsawani, R. K.; Sawhney, R. C.; Kumar, R. A Dose Dependent Adaptogenic and Safety Evaluation of *Rhodiola imbricate Edgew*, a High Altitude Rhizome. *Food Chem.Toxicol*. **2008,** *46*(5), 1645–1652.

19. Hackett, P. H.; Roach, R. C. High-Altitude Medicine. In *Auerbach Wilderness medicine*; P.S. Ed.; Mosby: Philadelphia, 2001; pp 2–43.

20. Hashmi, M. A.; Rashid, M.; Haleem, A.;Bokhari, S. A.; Hussain, T. Frostbite: Epidemiology at High Altitude in the Karakoram Mountains. *Ann. R. Coll. Surg. Engl*. **1998,** *80*(2), 91–95.

21. He, B.; Wang, J.; Qian, G.; Hu, M.; Qu, X.; Wei, Z.; Li, J.; Chen, Y.; Chen, H.; Zhou, Q.; Wang, G. Analysis of High-Altitude De-Acclimatization Syndrome after Exposure to High Altitudes: A Cluster-Randomized Controlled Trial. *PLoS One* **2013,** *8*(5), e62072.

22. Hillhouse, B. J.; French, C. J.; Neil Tower, G. H. Acetylcholine Esterase Inhibitors in *Rhodiola rosea*. *Pharm. Biol*. **2004,** *42*(1), 68–72.

23. Hohenhaus, E.; Niroomand, F.; Goerre, S.; Vock, P.; Oelz, O.; Bärtsch, P. Nifedipine Does Not Prevent Acute Mountain Sickness. *Am. J.Respir. Crit. Care Med*. **1994,** *150*(3), 857–60.

24. Holloway, C. J.; Montgomery, H. E.; Murray, A. J.;Cochlin, L. E.; Codreanu, I.; Hopwood, N.; Johnson, A. W.; Rider, O. J.;Levett, D. Z.; Tyler, D. J.; Francis, J. M.;Neubauer, S.;Grocott, M. P.; Clarke, K. *Caudwell extreme Everest Research Group*: Cardiac Response to Hypobaric Hypoxia: Persistent Changes in Cardiac Mass, Function, and Energy Metabolism after a Trek to Mt. Everest Base Camp. *FASEB J*. **2011,** *25*(2), 792–796.

25. Hota, S. K.; Barhwal, K.; Ray, K.; Singh, S. B.;Ilavazhagan, G. Ceftriaxone Rescue Hippocampal Neurons from Excitotoxicity and Enhances Memory Retrieval in Chronic Hypobaric Hypoxia. *Neurobiol. Learn. Mem*. **2008,** *89*, 522–532.

26. Hota, S. K.; Barhwal, K.; Singh, S. B.; Sairam, M.; Ilavazhagan, G. NR1 and GluR2 Expression Mediates Excitotoxicity in Chronic Hypobaric Hypoxia. *J. Neurosci. Res*. **2008,** *86*, 1142–1152.

27. Hota, S. K.; Hota, K. B.; Prasad, D.; Ilavazhagan, G.; Singh, S. B. Oxidative-Stress-Induced Alterations in Sp Factors Mediate Transcriptional Regulation of the NR1 Subunit in Hippocampus during Hypoxia. *Free Radical Biol. Med*. **2010,** *49*(2), 178–191.

28. Hota, K. B.; Hota, S. K.;Chaurasia, O. P.; Singh, S. B. Acetyl-L-carnitine-Mediated Neuroprotection during Hypoxia is Attributed to ERK1/2-Nrf2-Regulated Mitochondrial Biosynthesis. *Hippocampus* **2012,** *22*(4), 723–736.

29. Hota, S. K.; Sharma, V. K.; Hota, K.; Das, S.; Dhar, P.; Mahapatra, B. B.; Srivastava, R. B.; Singh, S. B. Multi-Domain Cognitive Screening Test for Neuropsychological Assessment of Cognitive Decline in Acclimatized Lowlanders Staying at High Altitude. *Indian J. Med. Res*. **2012,** *136*(3), 411–420.

30. Jafarian, S.; Abolfazli, R.; Gorouhi, F.; Rezaie, S.; Lotfi, J. Gabapentin for Prevention of Hypobaric Hypoxia-Induced Headache: Randomized Double-Blind Clinical Trial. *J. Neurol. Neurosurg. Psychiatry*. **2008,** *79*(3), 321–333.

31. Jang, S. I.; Pae, H. O.; Choi, B. M.; Oh, G. S.; Jeong, S.; Lee, H. J.; Kim, H. Y.; Kang, K. J.; Yun, Y. G.; Kim, Y. C.; Chung, H. T. Salidroside from *Rhodiola sachalinens* is Protects Neuronal PC12 Cells against Cytotoxicity Induced by Amyloid-Beta. *Immunopharmacol. Immunotoxicol.* **2003**, *25*(3), 295–304.

32. Jeong, H. J.; Ryu, Y. B.; Park, S. J.; Kim, J. H.; Kwon, H. J.; Kim, J. H.; Park, K. H.; Rho, M. C.; Lee, W. S. Neuraminidase Inhibitory Activities of FlavonolsIsolated from *Rhodiola rosea* Roots and Their In Vitro Anti-Influenza Viral Activities. *Bioorg. Med. Chem.* **2009**, *17*(19), 6816–6823.

33. Kelly, G. S. *Rhodiola rosea*: A Possible Plant Adaptogen. *Altern. Med. Rev.* **2001**, *6*(3), 293–302.

34. Kołodziej, B.; Sugier, D. Influence of Plants Age on the Chemical Composition of Roseroot (*Rhodiola rosea* L.). *Acta Sci. Pol. Hortorum Cultus* **2013**, *12*(3), 147–160.

35. Korać, R. R.; Khambholja, K. M. Potential of Herbs in Skin Protection from Ultraviolet Radiation. *Pharmacogn. Rev.* **2011**, *5*(10), 164–173.

36. Kumar, K.; Sharma, S.; Vashishtha, V.; Kumar, A.; Barhwal, K.; Hota, S. K.; Malairaman, U.; Singh, B. *Terminaliaarjuna*Bark Extract Improves Diuresis and Attenuates Acute Hypobaric Hypoxia Induced Cerebral Vascular Leakage. *J. Ethnopharmacol.* **2016**, *180*, 43–53.

37. Kurkin, V. A.; Zapesochnaya, G. G. Chemical Composition and Pharmacological Properties of *Rhodiola rosea* L. Khim. *Farm. Zurnal* **1986**, *20*, 1231–1244.

38. Lee, S. Y.; Li, M. H.; Shi, L. S.; Chu, H.; Ho, C. W.; Chang, T. C. *Rhodiola crenulata* Extract Alleviates Hypoxic Pulmonary Edema in Rats. *J. Evidence-Based Complementary Altern. Med.* **2013**, *2013*, 718739.

39. LemosVde, A.; dos Santos, R. V.; Lira, F. S.; Rodrigues, B.; Tufik, S.; de Mello, M. T. Can High Altitude Influence Cytokines and Sleep? *Mediators Inflamm.* **2013**, *2013*, 279365.

40. Li, Y. The Characteristics of Tibetan Medicine Preparation in *SibuYidian* (Tibetan Medical Dictionary). *Chin. J. Ethnomed. Ethnopharm.* **1995**, *14*, 16–19.

41. Liu, Z.; Liu, Y.; Liu, C.; Song, Z.; Li, Q.; Zha, Q.; Lu, C.; Wang, C.; Ning, Z.; Zhang, Y.; Tian, C.; Lu, A. The Chemotaxonomic Classification of *Rhodiola* Plants and its Correlation with Morphological Characteristics and Genetic Taxonomy. *Chem. Cent. J.* **2013**, *7*(1), 118.

42. Luo, H.; Guo, P.; Zhou, Q. Role of TLR4/NF-κB in Damage to Intestinal Mucosa Barrier Function and Bacterial Translocation in Rats Exposed to Hypoxia. *PLoS One* **2012**, *7*(10), e46291.

43. Mathieu-Costello, O. Muscle Adaptation to Altitude: Tissue Capillarity and Capacity for Aerobic Metabolism. *High Alt. Med. Biol.* **2001**, *2*(3), 413–425.

44. Mazzeo, R. S.; Wolfel, E. E.; Butterfield, G. E.; Reeves, J. T. Sympathetic Responses during 21 Days at High Altitude (4,300 m) As Determined by Urinary and Arterial Catecholamines. *Metabolism* **1994**, *43*, 1226–1232.

45. Mazzeo, R. S.; Brooks, G. A.; Butterfield, G. E.;Podolin, D. A.;Wolfel, E. E.; Reeves J. T. Acclimatization to High Altitude Increases Muscle Sympathetic Activity Both at Rest and during Exercise. *Am. J. Physiol.* **1995**, *269*, 201–207.

46. Mishra, K. P.; Padwad, Y. S; Jain, M.; Karan, D.; Ganju, L.; Sawhney, R. C. Aqueous Extract of *Rhodiola imbricata* Rhizome Stimulates Pro-Inflammatory Mediators Via

Phosphorylated I-kappa-B and Transcription Factor Nuclear Factor-Kappa-B. *Immunopharmacol. Immunotoxicol.* **2006,** *28*(2), 201–212.

47. Mishra, K. P.; Ganju, L.; Singh S. B. Anti-Cellular and Immunomodulatory Potential of Aqueous Extract of *Rhodiola imbricate* Rhizome. *Immunopharmacol. Immunotoxicol.* **2012,** *34*(3), 513–518.

48. Nakanishi, K.; Tajima, F.; Nakamura, A.; Yagura, S.; Ookawara, T.; Yamashita, H.; Suzuki, K.; Taniguchi, N.; Ohno, H. Effects of Hypobaric Hypoxia on Antioxidant Enzymes in Rats. *J. Physiol.* **1995,** *489*, 869–876.

49. Neuhaus, C.; Hinkelbein, J. Cognitive Responses to Hypobaric Hypoxia: Implications for Aviation Training. *Psychol. Res.Behav.Manag.* **2014,** *7*, 297–302.

50. Noreen, E. E.; Buckley, J. G.; Lewis, S. L.; Brandauer, J.; Stuempfle, K. J. The Effects of an Acute Dose of *Rhodiola rosea* on Endurance Exercise Performance. *J. Strength Cond. Res.* **2013,** *27*(3), 839–847.

51. Panossian, A.;Wikman, G.; Sarris, J. Rosenroot (*Rhodiola rosea*): Traditional Use, Chemical Composition, Pharmacology and Clinical Efficacy. *Phytomedicine* **2010,** *17*(7), 481–493.

52. Panossian, A.;Wikman, G.; Kaur, P.; Asea, A. Adaptogens Stimulate Neuropeptide y and hsp72 Expression and Release in Neuroglia Cells. *Front.Neurosci.* **2012,** *6*, 6.

53. Petkov, V. D.; Yonkov, D.; Mosharoff, A.; Kambourova, T.; Alova, L.; Petkov, V. V.; Todorov, I. Effects of Alcohol Aqueous Extract from *Rhodiola rosea* L. Roots on Learning and Memory. *Acta Physiol. Pharmacol. Bulg.* **1986,** *12*(1), 3–16.

54. Pigman, C. E. Acute Mountain Sickness. *Sports Med.* **1991,** *2*(2), 71–79.

55. Qu, Z. Q.; Zhou, Y.; Zeng, Y. S.; Lin, Y. K.; Li, Y.; Zhong, Z. Q.; Chan, W. Y. Protective Effects of a *Rhodiola crenulata* Extract and Salidroside on Hippocampal Neurogenesis Against Streptozotocin-Induced Neural Injury in the Rat. *PLoS One* **2012,** *7*(1), e29641.

56. Rashid, M.; Fahim, M.; Kotwani, A. Efficacy of *Tadalafil* in Chronic Hypobaric Hypoxia-Induced Pulmonary Hypertension: Possible Mechanisms. *Fundam. Clin. Pharmacol.* **2013,** *27*(3), 271–278.

57. Rege, N. N.; Thatte, U. M.;Dahanukar, S. A. Adaptogenic Properties of Six RasayanaHerbs Used in AyurvedicMedicine. *Phytother. Res.* **1999,** *13*, 275–291.

58. Riepl, R. L.; Fischer, R.;Hautmann, H.; Hartmann, G.; Müller, T. D.; Tschöp, M.; Toepfer, M.; Otto, B. Influence of Acute Exposure to High Altitude on Basal and Postprandial Plasma Levels of Gastroentero Pancreatic Peptides. *PLoS One* **2012,** *7*(9), e44445.

59. San, T.; Polat, S.; Cingi, C.; Eskiizmir, G.; Oghan, F.; Cakir, B. Effects of High Altitude on Sleep and Respiratory System and theirs Adaptations. *Sci. World J.* **2013,** *10*, 241569.

60. Schriner, S. E.; Avanesian, A.; Liu, Y.; Luesch, H.; Jafari, M. Protection of Human Cultured Cells against Oxidative Stress by *Rhodiola rosea*Without Activation of Antioxidant Defenses. *Free Radic. Biol. Med.* **2009,** *47*(5), 577–584.

61. Sharma, V. K.; Das, S. K.; Dhar, P.; Hota, K. B.; Mahapatra, B. B.; Vashishtha, V.; Kumar, A.; Hota, S. K.; Norboo, T.; Srivastava, R, B. Domain Specific Changes in Cognition at High Altitude and its Correlation with Hyperhomocysteinemia. *PLoS One* **2014,** *9*(7), e101448.

62. Sikri, G.; Srinivasa, A. B.; Grewal, R. S. Is Concurrent Prophylactic Use of Acetazolamide and Dexamethasone Superior to Acetazolamide Alone in Un-Acclimatized Lowlanders on Ascent to High Altitude? *Indian J. Physiol. Pharmacol.* **2014**, *58*(1), 87–91.

63. Subudhi, A. W.; Bourdillon, N.; Bucher, J.; Davis, C.; Elliott, J. E.; Eutermoster, M.; Evero, O.; Fan, J. L.; Jameson-Van Houten, S.; Julian, C. G.; Kark, J.; Kark, S.; Kayser, B.; Kern, J. P.; Kim, S. E.; Lathan, C.; Laurie, S. S.; Lovering, A. T; Paterson, R.; Polaner, D. M.; Ryan, B. J.; Spira, J. L.; Tsao, J. W.; Wachsmuth, N. B.; Roach, R. C. AltitudeOmics: The Integrative Physiology of Human Acclimatization to Hypobaric Hypoxia and its Retention upon Reascent. *PLoS One* **2014**, *9*(3), e9219.

64. Tsarong, T. J. *Handbook of Traditional Tibetan Drugs: Their Nomenclature, Composition, Use, and Dosage.* Tibetan Medical Publications; Kalimpong, India, 1986; p 101.

65. Tuchinda, C.; Srivannaboon, S.; Lim, H. W. Photoprotection by Window Glass, Automobile Glass, and Sunglasses. *J. Am. Acad.Dermatol.* **2006**, *54*(5), 845–854.

66. Tzianabos, A. O. Polysaccharide Immunomodulators as Therapeutic Agents: Structural Aspects and Biologic Function. *Clin. Microbiol. Rev.* **2000**, *13*(4), 523–533.

67. Uttara, B.; Singh, A. V.; Zamboni, P.; Mahajan, R. Oxidative Stress and Neurodegenerative Diseases: A Review of Upstream and Downstream Antioxidant Therapeutic Options. *Curr. Neuropharmacol.* **2009**, *7*(1), 65–74.

68. Vincent, W. F. Solar Ultraviolet-B Radiation and Aquatic Primary Production: Damage, Protection, and Recovery. *Environ. Rev.* **1993**, *1*, 1–11.

69. Walker, T. B.; Robergs, R. A. Does *Rhodiola rosea* Possess Ergogenic Properties? *Int. J. Sport Nutr. Exerc. Metab.* **2006**, *16*(3), 305–315.

70. Wang, H.; Ding, Y.; Zhou, J.; Sun, X.; Wang, S. The In Vitro and In Vivo Antiviral Effects of Salidroside from *Rhodiola rosea* L. against Coxsackie Virus B3. *Phytomedicine* **2009**, *16*(2–3), 146–155.

71. Wang, Y.; Xu, P.; Wang, Y.; Liu, H.; Zhou, Y.; Cao, X. The Protection of Salidroside Of the Heart against Acute Exhaustive Injury and Molecular Mechanism in Rat. *Oxid. Med. Cell. Longev.* **2013**, *2013*, 507832.

72. Weiss, J. F.;Landauer, M. R. Radioprotection by Antioxidants. *Ann. N. Y. Acad. Sci.* **2000**, *899*, 44–60.

73. West, J. B. Physiology of Extreme Altitude. In *Handbook of Physiology, Section 4: Environmental Physiology*; *Blatteis, C., Eds.;*Oxford University Press: New York, 1996; pp 1307–1325.

74. Wicklera, S. J.; Greenea, H. M. High Altitude Acclimatization and Athletic Performance in Horses. *Equine Comp. Exerc. Physiol.* **2004**, *1*(3), 167–170.

75. Wilson, M. H.; Newman, S.; Imray, C. H. The Cerebral Effects of Ascent to High Altitudes. *Lancet Neurol.* **2009**, *8*, 175–91.

76. Xing, S.; Yang, X.; Li, W.;Bian, F.; Wu, D.; Chi, J.; Xu, G.; Zhang, Y.; Jin, S. Salidroside Stimulates Mitochondrial Biogenesis and Protects against H_2O_2-Induced Endothelial Dysfunction. *Oxid. Med. Cell. Longev.* **2014**, *2014*, 904834.

77. Yan, G. H.; Choi, Y. H. Salidroside Attenuates Allergic Airway Inflammation Through Negative Regulation of Nuclear Factor-Kappa B and p38 Mitogen-Activated Protein Kinase. *J. Pharmacol. Sci.* **2014**, *126*(2), 126–135.

78. You, J.; Liu, W.; Zhao, Y.; Zhu, Y.; Zhang, W.; Wang, Y.; Lu, F.; Song, Z. Microsatellite Markers in *Rhodiola* (Crassulaceae), a Medicinal Herb Genus Widely Used in Traditional Chinese Medicine. *Appl. Plant Sci.* **2013,** *1*(3), 15.

79. Zhang, L.; Yu, H.; Sun, Y.; Lin, X.; Chen, B.; Tan, C.; Cao, G.; Wang, Z. Protective Effects of Salidroside on Hydrogen Peroxide-Induced Apoptosis in SH-SY5Y Human Neuroblastoma Cells. *Eur. J. Pharmacol.* **2007,** *564*(1–3), 18–25.

80. Zhang, J.; Zhen, Y. F.; Pu-Bu-Ci-Ren.; Song, L. G.; Kong W. N.; Shao, T. M.; Li, X.; Chai, X. Q. Salidroside Attenuates Beta Amyloid-Induced Cognitive Deficits Via Modulating Oxidative Stress and Inflammatory Mediators in Rat Hippocampus. *Behav. Brain Res.* **2013,** *244*, 70–81.

81. Zhu, C.; Guan, F.; Wang, C.; Jin, L. H. The Protective Effects of *Rhodiola crenulata* Extracts on *Drosophila melanogaster* Gut Immunity Induced by Bacteria and SDS Toxicity. *Phytother. Res.* **2014,** *28*(12), 1861–1866.

CHAPTER 10

HERBAL PLANTS AS POTENTIAL BIOAVAILABILITY ENHANCERS

JOSLINE Y. SALIB and SAYED A. EL-TOUMY

CONTENTS

ABSTRACT

Although flavonoids have shown promising health properties under experimental conditions, yet low bioavailability of some flavonoids needs to be enhanced for full exploitation of their therapeutic benefits in prevention and treatment of diseases. The available scientific research on herbal bioenhancers has shown to produce more efficacious and safe medicine with significant enhancing effect on bioavailability when coadministered or pretreated with many drugs and nutraceuticals. Among these natural extracts/compounds such as *Zingiber officinale*, *Allium sativum*, *Carum carvi*, *Cuminum cyminum*, quercetin, naringin, genistein, curcumin, glycyrrhizin, and capsaicin are included and discussed in this chapter.

10.1   INTRODUCTION

The history of medicinal plant use for treating diseases and ailments probably dates back to human civilization. Medicinal plants are an important part of our natural wealth. They serve as important therapeutic agents as well as valuable raw materials for manufacturing numerous traditional and modern medicines. Ancestors were compelled to use any natural substance that they could find to ease their sufferings caused by acute and chronic illnesses, physical discomforts, wounds and injuries, and even terminal illnesses. Since ancient times, plants with therapeutic properties have secured an important place in the healing practices and treatment of diseases.[31,50,66]

Recently, the World Health Organization (WHO) estimated that 80% of people worldwide rely on herbal medicines partially for their primary health care. During the past three decades, the demand and utilization of medicinal plants have increased globally. Now, there is a consensus regarding the importance of medicinal plants and traditional health systems in solving the health care problems, efficacy, and safety of medicinal plants in curing various diseases. Because of this growing awareness, the international trade in medicinal plants is growing phenomenally.[4,71]

The term "Alternative Medicine" has become very common in Western culture, as it focuses on the idea of using the plants for medicinal purpose. Even so, most of these pills and capsules we take and use during our daily life came from plants. Medicinal plants have been frequently used as raw materials for extraction of active ingredients, which are used in the synthesis of different drugs. Similarly, laxatives, blood thinners, antibiotics, and anti-malaria medications, contain ingredients from plants. Moreover, the active ingredients of Taxol, vincristine, and morphine have been isolated from foxglove, periwinkle, yew, and opium poppy, respectively.

Although the high potential of medicinal plant extracts has been revealed in various pharmacological activities, yet the results are much less promising or disappear in vivo investigations. This is because most of the biologically active constituents of plants are polar or water-soluble molecules[16] and these water-soluble phytoconstituents (like flavonoids, tannins, glycosidicaglycones, and so forth.) are poorly absorbed either due to their large molecular size; and therefore, cannot be absorbed from the intestine into the blood by simple diffusion,[75] or due to their poor lipid solubility; and often fail to pass through the small intestine because of its lipoidal nature, resulting poor bioavailability.[48] Thus, there is a great interest and medical need for the improvement of bioavailability of a large number of herbal drugs and plant extracts.

Bioavailability is the rate and extent to which a substance enters the systemic circulation and becomes available at the required site of the action.[10] Maximum bioavailability is attained by drugs administered through intravenous route, whereas drugs administered orally are poorly bioavailable as they readily undergoes first pass metabolism and incomplete absorption. Such unutilized drug in the body may lead to adverse effects and also drug resistance.

It has been observed that the isolation and purification of the constituents of an extract may lead to a partial or total loss of specific biological activity for the purified constituent. The natural constituent synergy becomes lost probably due to the removal of chemical-related substances contributing the synergistic effect of the active principle (s).[7] Very often the chemical complexity of the crude or partially purified extract seems to be an essential for the bioavailability of the active constituents. Extracts when taken orally, some constituents may get destroyed in the gastric environment. As standardized extracts are established, poor bioavailability often limits their clinical utility due to these reasons.

Therefore, there is a great need of the molecules that themselves have no same therapeutic activity but when combined with other drugs/

molecules enhance their bioavailability, decrease the time of administration and prevent their toxic effects. Many natural compounds from
medicinal plants have the capacity to augment the bioavailability when
coadministered with another drug.

The application of herbal bioenhancer in improving drug delivery has
been of special interest to scientists as they increase the bioavailability
and absorption of the coadministered drugs. Many approaches have been
developed to improve the oral bioavailability, such as inclusion by solubility and bioavailability enhancers, structural modification, and entrap-
ment with the lipophilic carriers.[45,67,75]

This chapter presents an overview on the natural herbal bioenhancers,
their suggested mechanisms of action and the role of some natural extracts
and/or compounds.

10.2   HERBAL BIOENHANCER CONCEPT

Bioenhancers lead to the enhancement of pharmacologic effect of the
drug combined with an active drug. Such formulations have been found to
increase the bioavailability/bioefficacy of a number of drugs even when
reduced the doses of drugs are present in such formulations. They reduce
the dose, and shorten the treatment period thus reducing drug-resistance
problems. Thus, the treatment is cost-effective, minimizes drug toxicity
and adverse reactions. When used in combination with a number of drug
classes (such as antibiotics, antituberculosis, antiviral, antifungal, and
anticancerous drugs) they are quite effective. Oral absorption of vitamins,
minerals, herbal extracts, amino acids, and other nutrients is improved by
them. They act through several mechanisms, which may affect mainly the
absorption process, drug metabolism, or action on drug target. They can
be classified based on their natural origin as well as based on the various
mechanisms elicited by them when in combination with drugs to improve
their bioavailability.

Thus, bioenhancers are chemical entities, which promote and augment
the bioavailability of the drugs that are mixed with them and do not exhibit
a synergistic effect with the drug.[22,57]

Moreover, bioenhancer should have novel properties such as nontoxic
to humans or animals, effective at a very low concentration in a combination, should be easy to formulate, and enhance uptake/absorption and
activity of the drug molecules.[23]

10.3   MECHANISM OF ACTION OF HERBAL BIOENHANCERS

There are several mechanisms of action through which herbal bioenhancers act. Different herbal bioenhancers may have same or different mechanism of action. Nutritional bioenhancers are enhancing absorption by acting on the gastrointestinal tract (GIT). Antimicrobial bioenhancers mostly act on drug metabolism process. Among the various mechanisms of action postulated for herbal bioenhancers, some are as follows:

a) By modulating the active transporters located in various locations, for example, P-glycoprotein (P-gp) is an efflux pump that pumps out drugs and prevents it from reaching the target site. Bioenhancers in such case act by inhibiting the P-gp.

b) Decreasing the elimination process thereby extending the sojourn of drug in the body by;

   • Inhibiting the drug metabolizing enzymes like CYP 3A4, CYP1A1, CYP1B2,and CYP2E1, in the liver, gut, lungs, and various other locations. In addition, this helps to overcome the first pass effect of administered drugs.

   • Inhibiting the renal clearance by preventing glomerular filtration, active tubular secretion by inhibiting P-gp, and facilitating passive tubular reabsorption. Sometimes biliary clearance is also affected by inhibiting the uridine diphosphate glucose glucuronyltransferase enzyme, which conjugates and inactivates the drug.[49]

c) Enhancing the absorption of orally administered drugs from the GIT by reduction in hydrochloric acid secretion and increase in gastrointestinal blood supply.

d) Inhibition of gastrointestinal transit, gastric emptying time, and intestinal motility.[3]

e) Modifications in GIT epithelial cell membrane permeability.[37]

f) Suppression of first pass metabolism and inhibition of drug metabolizing enzyme[5] and stimulation of gamma glutamyltranspeptidase activity which enhances uptake of amino acids.[36]

10.4 CLASSIFICATION OF BIOENHANCERS

TABLE 10.1 Classification of Bioenhancers.

Classification based on origin	
Plant origin	**Animal origin**
Allicin ginger	Cow urine distillate
*Aloe vera*g lycyrrhizin	
Black cumin lysergol	
Carumcarvi Naringin	
Curcumin peppermint oil	
Genistein Sinomenine	
Classification based on Mechanism of action	
Mechanism type	Examples
Inhibiting the P-gp efflux pump and any other pumps	*Carumcarvi*, sinomenin, *Cuminum cyminum*, naringin, genistein, and quercetin.
Suppressor of CYP-450 enzyme and isoenzyme	Naringin, quercetin, gallic acid and its esters
Modifications in GIT epithelial cell membrane permeability	*Zingiber officinale,* Aloe vera, *Drumstick pods* and Glycyrrhizin

10.5 MEDICINAL PLANTS AND THEIR BIOACTIVE COMPOUNDS AS DRUG BIOAVAILABILITY ENHANCERS

10.5.1 GINGER

Figure 10.1 indicates various types of bioenhancers. Ginger (*Zingiber officinale*) is a rhizome, which contains potent active constituents known as gingerols. The major pungent principle of ginger is[6] gingerol which increases the motility of the GIT in laboratory animals, and have analgesic, sedative, antipyretic and antibacterial properties.[54] Ginger extracts possess activities like antiulcer activity,[76,77] antithrombotic activity,[72] antimicrobial activity,[35] antifungal activity,[26] anti-inflammatory activity,[30,40,55,72] and anticancer activity.[69]

Ginger has a powerful effect on GIT mucous membrane. It regulates the intestinal function × 10–30 mg/kg body weight as bioenhancer. The bioavailability of different antibiotics like azithromycin (85%), erythromycin (105%), cephalexin (85%), cefadroxil (65%), amoxicillin (90%), and cloxacillin (90%) are increased by it.[59] The composition containing *Z. officinale* alone provides bioavailability/bioenhancing activity in the range of 30–75%, the dosage of bioenhancer from *Z. officinale* as a bioactive fraction is in the range of 5–15 mg/kg body weight, preferably 30 mg/kg body weight.

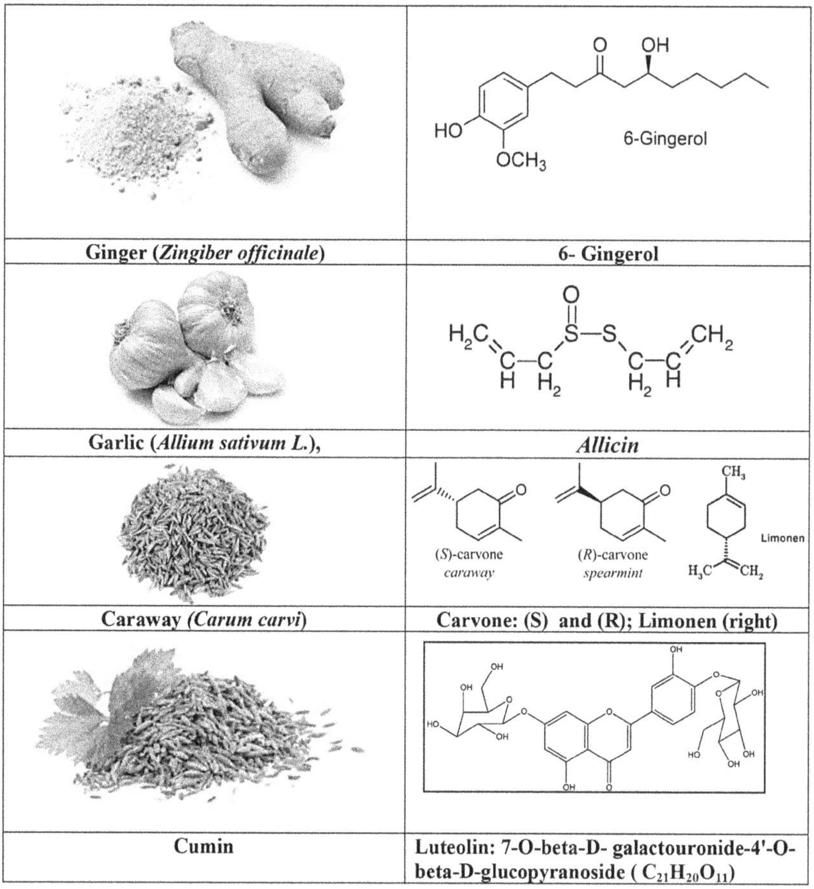
6-Gingerol
Ginger (Zingiber officinale)
6- Gingerol
Garlic (Allium sativum L.),
Allicin
(S)-carvone
caraway
(R)-carvone
spearmint
Limonen
Caraway (Carum carvi)
Carvone: (S) and (R); Limonen (right)
Cumin
Luteolin: 7-O-beta-D- galactouronide-4'-O-beta-D-glucopyranoside ($C_{21}H_{20}O_{11}$)

FIGURE 10.1 (See color insert.) Different types of bioenhancers.

The extracts or its fractions have been found to be highly selective in their bioavailability enhancing activity varying from almost nearly significant (20%) to highly significant (200%).[59,60]

10.5.2 GARLIC

Garlic (*Allium sativum* L.) has an exquisite defense system, composed of as many different components as the human immune system. Garlic has been used worldwide since ancient times, not only as a food but also as a medicine. As early as 3000 B. C., in ancient civilizations, including Egyptian, Phoenicians, Greek, Indian, Roman, Babylonian, Viking, and Chinese, garlic was used for the treatment of heart conditions, arthritis, pulmonary complaints, abdominal growths (particularly uterine), respiratory infections, skin disease, symptoms of aging, diarrhea, headache, bites, worms, wounds, ulcers, and tumors.[6,27,62]

Allicin, the active bioenhancer phytomolecule in garlic, enhances the fungicidal activity of amphotericin B (AmB) against pathogenic fungi such as *Candida albicans*, *Aspergillus fumigatus*, and yeast *Saccharomyces cerevisiae*.[22]Ogitaet al.[53] reported that allicin enhances AmB-induced vacuole membrane damage by inhibiting ergosterol tracking from the plasma membrane to the vacuole membrane.

10.5.3 CARAWAY

Caraway *(Carumcarvi)* contains caraway oil obtained from dried and crushed seeds. Carvone and limonene are the chief constituents of the oil and their odor and flavor are mainly attributed to them.[73]*C. carvi* (Meridian fennel and Persian cumin) exhibits activities like antiulcer effects,[39] hypoglycemic effect,[24] diuretic activity,[42] antioxidant activity,[51] and antiaflatoxigenic activity.[1]

The effective dose of the bioenhancer extract is in the range of 5–100 mg/kg body weight and the dose bioactive fraction of bioenhancer is in the range of 1–55 mg/kg body weight. It has been reported that *C. carvi* enhance bioavailability of antibiotics, antifungal, antiviral, and anticancerous drug. The extract orits fractions were found to be 20–110% more active; and when used in combination with *Z. officinale,* it was found to be more effective, in the range of 10–150 mg/kg body weight. *C.carvi*in different combinations showed prominent activity ranging from 25 to 95%.[59,61]

10.5.4 CUMIN

The main component of Cumin (*Cuminum cyminum*) is 3',5-dihydroxy-flavone-7-O-β-D-galactouronide-4'-O-β-D-glucopyranoside (Luteolin 7-O-beta-D-galactouronide-4'-O-beta-D-glucopyranoside). *C. cyminum* exhibits activities like estrogenic activity,[47] hypolipidaemic activity,[19] antinociceptive and anti-inflammatory activity,[65] anticonvulsant effect,[64] anticancer activity,[29] antimicrobial activity,[18,28] antitussive effect,[8] anti-oxidant activity,[25] and antifungal activity.[18,56]

The doses of its fractions responsible for the bioavailability enhancement activity ranged from 0.5 to 25 mg/kg body weight. Percentage enhancement of bioavailability for rifampicin is 250%, for cycloserine is 89%, for ethionamide is 78%. Sachin et al. studied the enhancement of rifampicin levels in rat plasma by 3',5-dihydroxyflavone-7-O-β-D-galactouronide-4'-O-β-D-glucopyranoside. The results obtained revealed that the C_{max} of rifampicin was enhanced by 35% and the area under plasma concentration curve (AUC) was enhanced by 53%.[63]

Apart from the above bioenhancing effects, black cumin also enhances the bioavailability of antibiotics (cefadroxil: 90% and cloxacillin: 94%), antifungal (fluconazole: 170%), antiviral (zidovudine: 330%) and anti-cancer (5-fluorouracil: 335%) drugs.

10.5.5 QUERCETIN

The citrus fruits and vegetables contain quercetin that has bioactivities for human health.[52] Studies[68] on the pharmacokinetic profile of some drugs alone and with quercetin have shown that it leads to increase in the absorption rate constant, peak concentration and the AUC – time curve. Umathe et al.[74] found that pioglitazone is readily metabolized by cytochrome P-450 (CYP 3A4).

For the more detailed study, the reader is referred to, "Ajazudin et al. Role of Herbal Bioactives as a Potential Bioavailability Enhancer for Active Pharmaceutical Ingredients. *Fitoterapia*. **2014**, 97, 1–14"; and refs 12,13,15,32,52,68,74 at the end of this chapter.

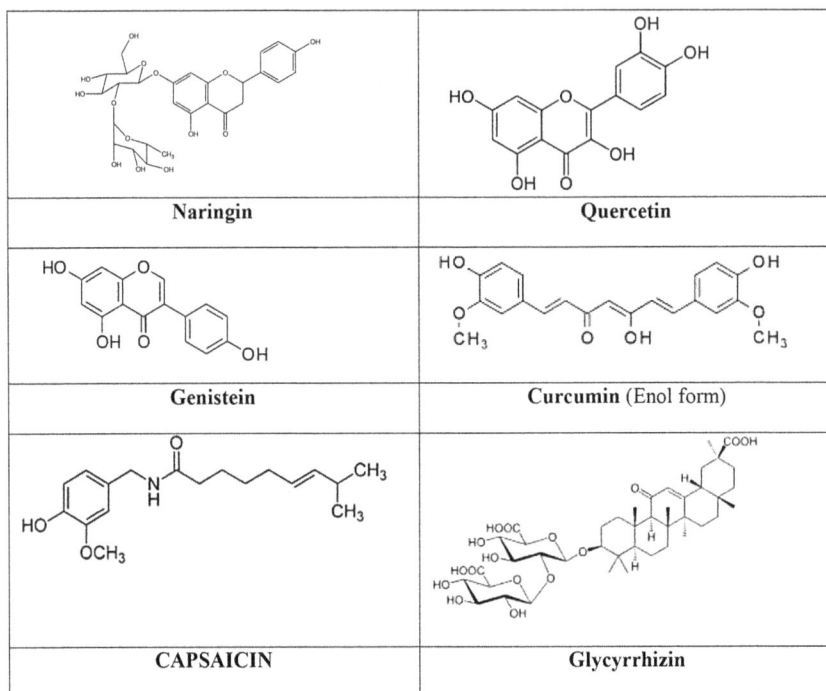

FIGURE 10.2 Bioactive compounds.

10.5.6 NARINGIN

Grapefruit, apples, onions, and tea are source of Naringin[20] that has bioactivities for human health.[52] In experiments by Choi et al.[12] there was a significant enhancement in relative bioavailability (from 100 to 207%) and absolute bioavailability (from $6.09 \pm 1.10\%$ to $12.6 \pm 1.61\%$) of diltiazem. Hence, it can be concluded that naringin acts through inhibition of P-gp and intestinal metabolism of diltiazem, thereby, reducing the dose and adverse effect of such potent drug.[12]

For more information, the reader is referred to, "Ajazudin et al. Role of Herbal Bioactives as a Potential Bioavailability Enhancer for Active Pharmaceutical Ingredients. *Fitoterapia*.2014, 97, 1–14"; and refs 12,14,44,52, 68,80 at the end of this chapter.

10.5.7 GENISTEIN

Certain bioactive compounds in medicinal plants are shown in Figure 10.2. For more detailed research, the reader is referred to, "Ajazudin et al. Role of Herbal Bioactives as a Potential Bioavailability Enhancer for Active Pharmaceutical Ingredients. *Fitoterapia.* 2014, 97, 1–14"; and refers[21,33,41,43] at the end of this chapter confirm beneficial bioactivities for human health.

10.5.8 CURCUMIN

Turmeric is a principle source of Curcumin. It is one of the most commonly used in India. For more detailed research, the reader is referred to, "Ajazudin et al. Role of Herbal Bioactives as a Potential Bioavailability Enhancer for Active Pharmaceutical Ingredients. *Fitoterapia.* 2014, 97, 1–14"; and research by various investigators[2,58,78,79] confirm beneficial bioactivities for human health.

10.5.9 GLYCYRRHIZIN

For more detailed research, the reader is referred to, "Ajazudin et al. Role of Herbal Bioactives as a Potential Bioavailability Enhancer for Active Pharmaceutical Ingredients. *Fitoterapia.*2014, 97, 1–14"; and research by various investigators[11,34,38] confirm beneficial bioactivities for a cough suppressant.

10.5.10 CAPSAICIN

Capsaicin (8-methyl-N-vanillyl-6-nonenamide) is the active component of chili peppers (*Capsicum annuum*). It is an irritant for mammals, including humans, and produces a sensation of burning in any tissue with which it comes into contact.

Cruz et al.[17] reported that *C. annuum* reduces the bioavailability of aspirin after oral administration in rats. L´opez et al.[46] have shown that capsaicin has little or no impact on the bioavailability of ciprofloxacin.

Absorption and bioavailability of theophylline from a sustained release gelatin capsule were investigated in 10-male rabbits after oral administration (20 mg/kg), with and without a ground capsicum fruit suspension. Comparison of pharmacokinetic parameters showed that the concomitant absorption of capsicum increases AUC from 86.06 ± 9.78 μg/ml·h to 138.32 ± 17.27 μg/ml·h, $P < 0.001$ and the maximum serum concentration (C_{max}) from 6.65 ± 0.76–8.78 ± 0.98 μg/ml, $P < 0.01$. A second administration of the capsicum suspension produced a new rise of theophylline plasma levels in every rabbit after 11-h dosing. This indicated that bioavailability of theophylline was enhanced due to the action of capsaicin.[9]

10.6 SUMMARY

Many scientists and numerous pharmaceutical industries are emphasizing on the improvement of bioavailability of a large number of potent drugs, which are poorly bioavailable. The utilization of natural bioavailability enhancers provides an innovative concept for reducing the dose of a drug and makes the cost of treatment economical and available to the broader section of the society. The natural bioenhancers also reduce the development of drug resistance by microbes, which pose a major problem to human beings.

The chapter presents the bioavailability enhancement techniques for some herbal extracts and/or natural compounds that may increase the solubility and/or permeability of different drugs for enhancing the drug bioavailability.

KEYWORDS

- absolute bioavailability
- abundant compound
- active pharmaceutical ingredient
- *Allium sativum*
- alternative medicine
- bioavailability
- bioavailability mechanism
- bioefficacy
- capsaicin
- *Carumcarvi*
- classification
- *Cuminum cyminum*
- curcumin
- derivatives
- drug delivery system
- essential oils
- flavonoids
- formulation

- genistein
- glycyrrhizin
- herbal bioenhancers
- herbal drugs
- medicinal plants
- naringin

- natural enhancer
- pharmacokinetic parameters
- phenolic acids
- quercetin
- *Zingiber officinale*

REFERENCES

1. Abyaneh, M. R.; Ghahfarokhi, M. S.; Rezaee, M. B.; Jaimand, K.; Alinezhad, S.; Saberi, R.; Yoshinari, T. Chemical Composition and Antiaflatoxigenic Activity of *Carumcarvi* L., *Thymus vulgar is* and *Citrus aurantifolia* Essential Oils. *Food Control.* **2009,** *20*(11), 1018–1024.

2. Ahshawat, M. S.; Saraf, S.; Saraf, S. Preparation and Characterization of Herbal Creams for Improvement of Skin Viscoelastic Properties. *Int. J. Cosmet. Sci.* **2008,** *30,* 183–193.

3. Bajad, S.; Bedi, K. L.; Singla, A. K.; Johri, R. K. Piperine Inhibits Gastric Emptying and Gastrointestinal Transit in Rats and Mice. *Planta Med.* **2001,** 67, 176–179.

4. Bannerman, R. H. Traditional Medicine and Healthcare Coverage. WHO, Geneva, 1983; pages 342.

5. Bhardwaj, R. K.; Glaeser, H.; Becquemont, L.; Klotz, U.; Gupta, S. K.; Fromm, M. F. Piperine, a major constituent of black pepper, inhibits human P-glycoprotein and CYP3A4. *J. Pharmacol. Exp. Ther.* **2002,** *302*(2), 645–650.

6. Block, E. The Chemistry of Garlic and Onions. *Sci. Am.* **1985,** *252,* 114–119.

7. Bombardelli, E.; Curri, S. B.; Loggia Della, R.; Del, N. P.; Tubaro, A.; Gariboldi, P. Complexes between Phospholipids and Vegetal Derivatives of Biological Interest. *Fitoterapia* **1989,** *60,* 1–9.

8. Boskabady, M. H.; Kiani, S.; Azizi, H.; Khatami, T. Antitussive effect of *Cuminumcyminum* Linn. in guinea pigs. *Indian J. Nat. Prod. Resour.* **2006,** *5,* 266–269.

9. Bouraoui, A. Toumi, H. Ben Mustapha, H.; Brazier, J. L. Effects of CapsicumFruit on Theophylline Absorption and Bioavailability in Rabbits. *Drug-Nutr. Interact.* **1988,** *5*(4), 345–350.

10. Brahmankar, D. B.; Jaiswal, S. Eds., Biopharmaceutics and Pharmacokinetics: A Treatise, 1st Edition; Vallabh Prakashan, 1995; pp 24–26.

11. Chen, L.; Yang, J.; Davey, A. K.; Chen, Y. X.; Wang, J. P.; Liu, X. Q. Effects of Diammonium Glycyrrhizinate on the Pharmacokinetics of Aconitine in Rats and the Potential Mechanism. *Xenobiotica* **2009,** *39,* 955–963.

12. Choi, J. S.; Han, H. K. Enhanced Oral Exposure of Diltiazem by the Concomitant use of Naringin in Rats. *Int. J. Pharm.* **2005,** *305*(1–2), 122–128.

13. Choi, J. S.; Li, X. Enhanced Diltiazem Bioavailability after Oral Administration of Diltiazem with Quercetin to Rabbits. *Int. J. Pharm.* **2005,** *297*(1–2), 1–8.

14. Choi, J. S.; Shin, S. C. Enhanced Paclitaxel Bioavailability after Oral Coadministration of Paclitaxel Prodrug with Naringin to Rats. *Int. J. Pharm.* **2005**, *292*, 149–156.

15. Choi, J. S.; Piao, Y. J.; Kang, K. W. Effects of Quercetin on the Bioavailability of Doxorubicin in Rats: Role of CYP3A4 and P-gp Inhibition by Quercetin. *Arch. Pharmacal Res.* **2011**, *34*, 607–613.

16. Cosa, P.; Vlietinck, A. J.; Berghe, D. V.; Maes, L. Anti-infective Potential of Natural Products: How to Develop a Stronger In Vitro 'Proof-of-Concept'. *J. Ethnopharmacol.* **2006**, *106*, 290–302.

17. Cruz, L.; Castañeda-Hernández, G.; Navarrete, A. Ingestion of Chilli Pepper (*Capsicum annuum*) Reduces Salicylate Bioavailability after Oral Aspirin Administration in the Rat. *Can. J. Physiol. Pharmacol.* **1999**, *77*(6), 441–446.

18. De Martino, L.; De Feo, V.; Fratianni, F.; Nazzaro, F. Chemistry, Antioxidant, Antibacterial and Antifungal Activities of Volatile Oils and Their Components. *Nat. Prod. Commun.* **2009**, *4*(12), 1741–1750.

19. Dhandapani, S.; Subramanian, V. R.; Rajagopal, S.; Namasivayam, N. Hypolipidemic Effect of *Cuminumcyminum* L. on Alloxan-Induced Diabetic Rats. Pharmacological Research, 2002, 46 (3), 251–255.

20. Dixon, R. A.; Steele, C. L. Flavonoids and Isoflavonoids—a Gold Mine for Metabolic Engineering. *Trends Plant Sci.* **1999**, *4*, 394–400.

21. Doyle, L.; Ross, D. D. Multidrug Resistance Mediated by the Breast Cancer Resistance Protein BCRP (ABCG2). *Oncogene* **2003**, *22*, 7340–7358.

22. Drabu, S.; Khatri, S.; Babu, S.; Lohani, P. Use of Herbal Bioenhancers to Increase the Bioavailability of Drugs. *Res. J. Pharm., Biol. Chem. Sci.* **2011**, *2*(4), 108–119.

23. Dudhatra, G. B.; Modi, S. K.; Awale, M. M.; Patel, H. B.; Modi, C. M.; Kumar, A.; Kamani, D. R.; Chauhan B. N. A Comprehensive Review on Pharmacotherapeutics of Herbal Bioenhancers. *Sci. World J.* **2012**, 1–33.

24. Eddouks, M.; Lemhadri, A.; Michel, J. B. Caraway and Caper: Potential Anti Hyperglycaemic Plants in Diabetic Rats. *J. Ethnopharmacol.* **2004**, *94*(1), 143–148.

25. El-Ghorab, A. H.; Nauman, M.; Anjum, F. M.; Hussain, S.; Nadeem, M. A Comparative Study on Chemical Composition and Antioxidant Activity of Ginger (*Zingiberofficinale*) and Cumin (*Cuminumcyminum*). *J. Agric. Food Chem.* **2010**, *58*(14), 8231–8237.

26. Ficker, C.; Smith, M. L.; Akpagana, K.; Gbeassor, M.; Zhang, J.; Durst, T.; Assabgui, R.;Arnason, J. T. Bioassay-Guided Isolation and Identification of Antifungal Compounds from Ginger. *Phytother. Res.* **2003**, *17*, 897–902.

27. Freeman, F.; Kodera, Y. Garlic Chemistry: Stability of S-(2-Propenyl) 2-Propene-1-Sulfinothioate (Allicin) in Blood, Solvents, and Simulated Physiological Fluids. *J. Agric. Food Chem.* **1995**, *43*, 2332–2338.

28. Gachkar, L.; Yadegari, D.; Rezaei, M. B.; Taghizadeh, M.; Astaneh, S. A.; Rasooli, I. Chemical and Biological Characteristics of *Cuminumcyminum* and *Rosmarinus officinalis* Essential Oils. *Food Chem.* **2007**, *102*(3), 898–904.

29. Gagandeep, S.; Dhanalakshmi, S.; M´endiz,E.; Rao,A. R.; Kale,R. K. Chemopreventive Effects of *Cuminumcyminum* in Chemically Induced Forestomach and Uterine Cervix Tumors in Murine Model Systems. *Nutr. Cancer* **2003**, *47*(2), 171–180.

30. Grzanna, R.; Lindmark, L.; Frondoza, C. G. Ginger—an Herbal Medicinal Product with Broad Anti-Inflammatory Actions. *J. Med. Food* **2005**,*8*, 125–32.

31. Hong, F. F. History of Medicine in China. *McGill J. Med.* **2004**, *8*(1), 7984.

32. Hsiu, S. L.; Hou, Y. C.; Wang, Y. H.; Tsao, C. W.; Su, S. F.; Chao, P. D. Quercetin Significantly Decreased Cyclosporin Oral Bioavailability in Pigs and Rats. *Life Sci.* **2002,** *72,* 227–235.

33. Huisman, M. T.; Chhatta, A. A.; van Tellingen, O.; Beijnen, J. H.; Schinkel, A. H. MRP2 (ABCC2) Transports Taxanes and Confers Paclitaxel Resistance and Both Processes are Stimulated by Probenecid. *Int. J. Cancer* **2005,** *116,* 824–829.

34. Imai, T.; Sakai, M.; Ohtake, H.; Azuma, H.; Otagiri, M. Absorption-Enhancing Effect of Glycyrrhizin Induced in the Presence of Capric Acid. *Int. J. Pharm.* **2005,** *294,* 11–21.

35. Jagetia, G. C.; Baliga, M. S.; Venkatesh, P.; Ulloor, J. N. Influence of Ginger Rhizome (*Zingiber officinale* Rosc) on Survival, Glutathione and Lipid Peroxidation in Mice After Whole-Body Exposure to Gamma Radiation. *Radiat. Res.* **2003,** *160,* 584–92.

36. Kesarwani, K.; Gupta, R. Bioavailability enhancers of herbal origin: An Overview. *Asian Pac. J. Trop. Biomed.* **2013,** *3*(4), 253–266.

37. Khajuria, A.; Thusu, N.; Zutshi, U. Piperine Modulates Permeability Characteristics of Intestine by Inducing Alterations in Membrane Dynamics: Influence on Brush Border Membrane Fluidity, Ultrastructure and Enzyme Kinetics. *Phytomedicine* **2002,** *9*(3): 224–231.

38. Khanuja, S.; Arya, J.; Srivastava, S.; Shasany, A.; Kumar, T. S.; Darokar M.; Kumar, S. Antibiotic Pharmaceutical Composition with Lysergol as Bio-enhancer and Method of Treatment. 2007, United States Patent Number, US20070060604 A1.

39. Khayyal, M. T.; El-Ghazaly, M. A.; Kenawy, S. A.; Seif-el-Nasr, M.; Mahran, L. G.; Kafafi, Y. A.; Okpanyi, S. N. Antiulcerogenic Effect of Some Gastrointestinally Acting Plant Extracts and their Combination. *Arzneim. Forsch.* **2001,** *51*(7), 545–553.

40. Kim, J. K.; Kim, Y.; Na, K. M.; Surh, Y. J.; Kim, T. Y. [6]-Gingerol Prevents UVB Induced ROS Production and COX-2 Expression In Vitro and In Vivo. *Free Radical Res.* **2007,** *41,* 603–14.

41. Kurzer, M. S.; Xu, X. Dietary Phytoestrogens. *Annu. Rev. Nutr.* **1997,** *17,* 353–381.

42. Lahlou, S.; Tahraoui, A.; Israili, Z.; Lyoussi, B. Diuretic Activity of the Aqueous Extracts of *Carumcarvi* and *Tanacetum vulgare* in Normal Rats. *J. Ethnopharmacol.* **2007,** *110*(3), 458–463.

43. Li, X.; Choi, J. S. Effect of Genistein on the Pharmacokinetics of Paclitaxel Administered Orally or Intravenously in Rats. *Int. J. Pharm.* **2007,** *337,* 188–193.

44. Lim, S. C.; Choi, J. S. Effects of Naringin on the Pharmacokinetics of Intravenous Paclitaxel in Rats. *Biopharm. Drug Dispos.* **2006,** *27,* 443–447.

45. Longer, M. A.; Ching, H. S.; Robinson, J. R. Oral Delivery of Chlorthiazide using a Bioadhesive Polymer. *J. Pharm. Sci.* **1985,** *74,* 406–411.

46. López, H. S.; Olvera, L. G.; Jiméenez, R. A.; Olvera, C. G.; Gómez, F. J. Administration of Ciprofloxacin and Capsaicin in Rats to Achieve Higher Maximal Serum Concentrations. *Arzneim. Forsch.* **2007,** *57*(5), 286–290.

47. Malini, T.; Vanithakumari, G. Estrogenic Activity of *Cuminumcyminum* in Rats. *Indian J. Exp. Biol.* **1987,** *25*(7), 442–444.

48. Manach, C.; Scalbert, A.; Morand, C. Polyphenols: Food Sources and Bioavailability. *Am. J. Clin. Nutr.* **2004,** *79,* 727–747.

49. Mekala, P.; Arivuchelvan, A. Bioenhancer for Animal Health and Production: A Review. *Agriculture* **2012,** 1–6.

50. Motaleb, M. A. Approaches to Conservation of Medicinal Plants and Traditional Knowledge; A focus on the Chittagong hill tracts. Firoz, R.; Adrika, A.; Khan, N. A. Eds.; IUCN Bangladesh Country Office, Dhaka, Bangladesh, 2010; pp viii+30.

51. Najda, A.; Dyduch, J.; Brzozowski, N. Flavonoid Content and Antioxidant Activity of Caraway Roots (*Carumcarvi* L). *Veg. Crops Res. Bull.* **2008,** *68,* 127–133.

52. Nijveldt, R. J.; van Nood, E.; van Hoorn, D. E.; Boelens, P. G.; van Norren, K.; van Leeuwen, P. A. Flavonoids: AReview of Probable Mechanisms of Action and Potential Applications. *Am. J. Clin. Nutr.* **2001,** *74,* 418–425.

53. Ogita, A.; Fujita, K. I.; Tanaka, T. Enhancement of the Fungicidal Activity of Amphotericin B by Allicin: Effects on Intracellular Ergosterol Trafficking, *Planta Med.* **2009,** *75*(3), 222–226.

54. O'Hara, M.; Kiefer, D.; Farrell, K.; Kemper, K. A Review of 12 Commonly used Medicinal Herbs. *Arch. Fam. Med.* **1998,** *7*(6), 523–536.

55. Ojewole, J. A. Analgesic, Antiinflammatory and Hypoglycaemic Effects of Ethanol Extract of *Zingiber officinale* (Roscoe) Rhizomes (*Zingiberaceae*) in Mice and Rats. *Phytother. Res.* **2006,** *20,* 764–72.

56. Pai, M. B. H.; Prashant, G. M.; Murlikrishna, K. S.; Shivakumar, K. M.; Chandu, G. N. Antifungal Efficacy of *Punicagranatum, Acacia nilotica, Cuminumcyminum* and *Foeniculum vulgare* on *Candida albicans*: An In Vitro Study. *Indian J. Dent. Res.* **2010,** *21*(3), 334–336.

57. Patil, U. M.; Singh, A.; Chakraborty, A. K. Role of Piperine as a Bioavailability Enhancer. *Int. J. Rec. Adv Pharm. Res.* **2011,** *1*(4), 16–23.

58. Pavithra, B. H.; Prakash, N.; Jayakumar, K. Modification of Pharmacokinetics of Norfloxacin Following Oral Administration of Curcumin in Rabbits. *J. Vet. Sci.* **2009,** *10,* 293–297.

59. Qazi G. N.; Bedi K. L.; Johri R. K.; Tikoo M. K.; Tikoo A. K.; Sharma S. C.; Abdullah, T.; Suri, O.; Gupta, B.; Suri, K.; Satti, N.; Khajuria, A. Bioavailability Enhancing Activity of *Carumcarvi* Extracts and Fractions Thereof. United States Patent Number, US20030228381A1, 2003a.

60. Qazi, G. N.; Tikoo, L. C.; Gupta, A. K.; Ganjoo, K. S.; Gupta, D. K.; Jaggi, B. S.; Singh, R. P.; Singh, R. P.; Singh, G.; Chandan, K. B.; Suri, K. A.; Satti K. N.; Gupta V. N.; Bakshi S. K.; Bedi K. L.; Suri O. P.; Puri, S. C.; Somal, P.; Singh, S.; Khajuria, A. Bioavailability Enhancing Activity of *Zingiberofficinale* and its Extracts/Fractions Thereof. World Intellectual Property Organization, International Publication Number, WO03049753A1, European Patent, Number EP 1,465,646, 2003b.

61. Qazi G. N.; Bedi K. L.; Johri R. K.; Tikoo M. K.; Tikoo A. K.; Sharma S. C. Bioavailability Enhancing Activity of *Carumcarvi* Extracts and Fractions Thereof. United States Patent Number, US20070020347A1, 2007.

62. Rivlin, R. S. Historical Perspective on the Use of Garlic. *J. Nutr.* **2001,** *131,* 951S–954S.

63. Sachin, B. S.; Sharma, S. C.; Sethi, S.; Tasduq, S. A. Herbal Modulation of Drug Bioavailability: Enhancement of Rifampicin Levels in Plasma by Herbal Products and a Flavonoid Glycoside Derived from *Cuminumcyminum*. *Phytother. Res.* **2007,** *21,* 157.

64. Sayyah, M.; Mahboubi, A.; Kamalinejad, M. Anticonvulsant Effect of the Fruit Essential Oil of *Cuminumcyminum* in Mice. *Pharm. Biol.* **2002a,** *40*(6), 478–480.
65. Sayyah, M.; Peirovi, A.; Kamalinejad, M. Anti-Nociceptive Effect of the Fruit Essential Oil of *Cuminumcyminum* L. in Rat. *Iran. Biomed. J.* **2002b,** *6*(4), 141–145.
66. Sazada, S.; Ruchi, Y.; Kavita, Y.; Feroze, A. W.; Mukesh, K. M.; Sudarshana, S.; Farah, J. Allelopathic Potentialities of Different Concentration of Aqueous Leaf Extracts of Some Arable Trees on Germination and Radicle Growth of *Cicerarietinum* Var.-C-235. *Global J. Mol. Sci.* **2009,** *4*(2), 91–95.
67. Sharma, S.; Sikarwar, M.Phytosome: A Review. PlantaIndica, 2005, 1, 1–3.
68. Shin, S. C.; Choi, J. S.; Li, X. Enhanced Bioavailability of Tamoxifen after Oral Administration of Tamoxifen with Quercetin in Rats. *Int. J. Pharm.* **2006,** *313*, 144–149.
69. Shukla, Y.; Singh, M. Cancer Preventive Properties of Ginger: A Brief Review. *Food Chem. Toxicol.* **2007,** *45*, 683–90.
70. Sparreboom, A.; van Asperen, J.; Mayer, U.; Schinkel, A. H.; Smit, J. W.; Meijer, D. K.; Borst, P.; Nooijen, W. J.; Beijnen, J. H.; van Tellingen, O. Limited Oral Bioavailability and Active Epithelial Excretion of Paclitaxel (Taxol) Caused by P-Glycoprotein in the Intestine. *Proc. Natl. Acad. Sci. U. S. A.*, **1997,** *94*, 2031–2035.
71. New WHO guidelines to promote proper use of alternative medicines. The World Health Organization; January 2004, p 1.
72. Thomson, M.; Al-Qattan, K. K.; Al-Sawan, S. M.; Alnaqeeb, M. A.; Khan, I.; Ali, M. The Use of Ginger (*Zingiber officinale* Rosc.) As a Potential Antiinflammatory and Antithrombotic Agent. *Prostaglandins Leukot. Essent. Fatty Acids* **2002,** *67*, 475–8.
73. Toxopeus, H.; Bouwmeester, H. J. Improvement of Caraway Essential Oil and Carvone Production in The Netherlands. *Ind. Crops Prod.* **1992,** *1*(2–4), 295–301.
74. Umathe, S. N.; Dixit, P. V.; Kumar, V.; Bansod, K. U.; Wanjari, M. M. Quercetin Pretreatment Increases the Bioavailability of Pioglitazone in Rats: Involvement of CYP3A Inhibition. *Biochem. Pharmacol.* **2008,** *75,* 1670–1676.
75. Venkatesan, N.; Babu, B. S.; Vyas, S. P. Protected Particulate Drug Carriers for Prolonged Systemic Circulation–A Review. *Indian J. Pharm. Sci.* **2000,** *62*, 327–333.
76. Wu, H.; Ye, D.; Bai, Y.; Zhao, Y. Effect of Dry Ginger and Roasted Ginger on Experimental Gastric Ulcers in Rats. *Zhongguo Zhong Yao Za Zhi.* **1990,** *15*, 278–280.
77. Yamahara, J.; Mochizuki, M.; Rong, H. Q.; Matsuda, H.; Fujimura, H. The Anti-Ulcer Effect in Rats of Ginger Constituents. *J. Ethnopharmacol.* **1988,** *23*, 299–304.
78. Yan, Y. D.; Kim, D. H.; Sung, J. H.; Yong, C. S.; Choi, H. G. Enhanced Oral Bioavailability of Docetaxel in Rats by Four Consecutive Days of Pretreatment with Curcumin. *Int. J. Pharm.* **2010,** *399*, 116–120.
79. Zhang, W.; Lim, L. Y. Effects of Spice Constituents on P-Glycoprotein Mediated Transport and CYP3A4-Mediated Metabolism In Vitro. *Int. J. Cancer* **2008,** *36,* 1283–1290.
80. Zhang, H.; Wong, C. W.; Coville, P. F.; Wanwimolruk, S. Effect of the Grapefruit Flavonoid Naringin on Pharmacokinetics of Quinine in Rats. *Drug Metab. Drug Interact.* **2000,** *17*, 351–363.
81. Zhang, W.; Tan, T. M.; Lim, L. Y. Impact of Curcumin-Induced Changes in P Glycoprotein and CYP3A Expression on the Pharmacokinetics of Peroral Celiprolol and Midazolam in Rats. *Int. J. Cancer* **2007,** *35*, 110–5.

CHAPTER 11

FORMULATED NATURAL SELECTIVE ESTROGEN RECEPTOR MODULATORS: A KEY TO RESTORING WOMEN'S HEALTH

A. ANITA MARGRET, S. AISHWARYA, and J. THEBORAL

CONTENTS

ABSTRACT

Phytoestrogens are considered as an effective treatment against the annoyance of women's ailments such as osteoporosis, breast cancer stress, and cardiovascular diseases (CVDs). Small amounts of phytoestrogen appear in many common food sources and there are innumerable ways to include such compounds in the diet. Soybeans, soy products, alfalfa fodder, and flaxseed are the foodstuffs that can be taken directly to increase the amount of phytoestrogens in the body. Some forms of phytoestrogen are being made into specific products, which solely help to introduce the phytoestrogen into the body. Presently, phytoestrogenic compounds are available commercially and are more commonly sold in capsules, pills, and in the form of external applicant such as ointments and cream.

11.1   INTRODUCTION

The health care of women is an essential component and there is a substantial need to uphold their welfare as their issues vary profoundly compared with that of men.[43,109] A multitude of health problems faced by women is a significant contributor to well-being and improving the health outcomes can contribute to economic gain through the creation of quality human capital.[2] Presently, women encounter various types of health issues, where maternal and reproductive illness is a foremost concern along all other crises such as cancer, cardiovascular disease (CVD), lung disease depression, dementia, osteoporosis, and anemia. Reproductive health is a massive facet that needs to be monitored cautiously and endocrine (hormonal) health includes menstruation, birth control, and menopause.[56] Cancer intrudes their normal subsistence. On the other hand, bone health is an indispensable component of women health care. Osteoporosis affects one in two women over the age of 50 but it can affect all ages. Women, with no symptoms of premenopausal and who are treated for (breast) cancer, may have an increased risk of developing osteoporosis.[109] Treatments such as chemotherapy and ovarian ablation and suppression can cause an early menopause and therefore a rapid and significant reduction in bone density. Therefore, it is essential to investigate this synchronized issue and needs attention to the healthy women.

There is a need of modulators (selective estrogen receptor modulators, SERMs) that block estrogen's action in certain cells and can activate in other cells, such as bone, liver, and uterine cells. Cells in other tissues in the body, such as bones and the uterus, also have estrogen receptors (ERs). But each ER has a slightly different structure, depending on the kind of cell it is in. Breast cell ERs are different from bone cell ERs and both of those ERs are different from uterine ERs. SERMs are "selective" implying that a SERM can activate estrogen's action in other cells, such as bone, liver, and uterine cells. For premenopausal women diagnosed with hormone-receptor-positive breast cancer, the SERM is the standard hormonal therapy treatment. Concerns regarding safety of hormone replacement therapy (HRT) have led to an interest in the use of phytoestrogens for a variety of menopause-related health complaints in women.

This chapter focuses on the various medicinal plants, which can elevate the estrogen levels and strengthen the well-being standards of women.

11.2   ESTROGEN RECEPTORS (ERS) AND ITS MODULATORS

11.2.1   ESTROGEN RECEPTORS: THE TARGET-SPECIFIC REGULATORS

Estrogen (American English) or oestrogen (British English) is a significant sex hormone to regulate the growth, development, and physiology of the human reproductive system.[18,74] Estrogens also affect the activity of neuroendocrine, skeletal, adipogenesis, and cardiovascular systems.[96] The biological functions of estrogen are mediated by binding to the ERs: estrogen receptor alpha (ERα) and estrogen receptor beta (ERβ). Estrogen signaling is selectively stimulated or inhibited depending upon a balance between ERα and ERβ activities in target organs. Further, these receptors are transcriptional factors to regulate the expression of specific genes in different tissues in aligned-dependent manner. Estrogens are small, carbon-rich molecules built from cholesterol.

This is quite different than insulin and growth hormone, which are sensed by receptors on the cell surface. When estrogen enters the nucleus, it binds to the ER, causing it to pair up and form a dimer. This dimer then binds to several dozen specific sites in the DNA, strategically placed next to the genes that need to be activated. Then, the DNA-bound receptor activates the DNA-reading machinery and starts the production of messenger RNA. They are encoded by distinct genes located on different chromosomes. The human ERα gene is located on chromosome-6, whereas the ERβ gene is on chromosome-14[49] The full-length human ERα protein has 595 amino acids and a molecular size of 66 kDa, whereas the full-length human ERβ protein has 530 amino acids and a molecular size of 54 kDa. Similar to other nuclear receptors (NRs), ERs have five domains with distinct functions.[94] The N-terminal of the A/B domains of ERs consists of activation function-1 (AF1), which contributes to the transcriptional activity of ERs and is an essential domain for interaction with co-regulators.

AF1 is the least conserved region with only 30% identity between ERα and ERβ. Functional studies have shown that ERβ has low levels of AF1 activity. The A/B domains also contain amino acids that are targets of posttranscriptional modifications, including splicing to stimulate AF1 activity.[8] The C domain encodes a centrally located DNA-binding domain essential for sequence-specific binging of ERs to DNA and regulating the expression of target genes.[29] The D domain, a hinge region, includes amino acid sequences that stimulate nuclear localization signaling and facilitate posttranslational modification of ERs, resulting in the activation of ER signaling in cells. Finally, the E/F domain, located in the C-terminal region, contains a ligand-binding domain (LBD) that serves as an interaction site with co-regulators and ligand-dependent activation function-2 (AF2). ER/co-regulator complexes act specifically on target genes in particular organs according to extracellular stimuli.[113] Analysis of ERα and ERβ tissue distribution suggests that ERs have high specificity on the target tissue.[112] ERα is highly expressed in the uterus, prostate stroma, ovarian theca cells, Leydig cells in testes, epididymis, breast, and liver.[50] ERβ is highly expressed in prostate epithelium, testes, ovarian granulosa cells, bone marrow, and brain.[107] As mentioned above, ERα and ERβ have different downstream transcriptional activities, resulting in their tissue-specific biological actions.[37]

11.3   ER PATHWAY

Estrogens and SERMs activate estrogen genes by a series of events that occur after their binding to the ER. The interaction of the hormone with the naive receptor induces conformational changes of the ligand–receptor binding to nuclear proteins, adaptor proteins, or co-regulators[63,70,89] that induces the dissociation of heat-shock proteins associated with the inactive receptor. This results in receptor activation and an interaction with DNA.[15,61] This ligand–receptor complex binds to DNA response elements, called estrogen response elements, located in the promoter region of the estrogen target genes, initiating the transcription process and mRNA synthesis. The agonistic and antagonistic feature is based on either inducing or inhibiting gene transcription by the ligand–ER dimer that depends on the type of cell and the presence of co-regulator proteins along with the relevant gene promoters. Rapid-acting, non-genomic pathways are also activated by estrogens and SERMs. The pathway that activates SERMs has been presented by other investigators.[62,90,91,106] The ER contains a LBD, which includes a series of amino acids called (AF2), essential for the activation of genes that mediate the estrogen effect in reproductive tissues as the breast or uterus.

Therefore, the different ligands can induce different gene transcription processes. For example, the union of the LBD with one type of SERM results in a partial agonistic effect in the uterus, whereas this same interaction is fully antagonistic in the breast. In contrast, when binding to estradiol, the conformation of the ligand–receptor complex permits an interaction with a coactivator that results in a fully agonistic effect in breast tissue.[62] Conformational shape variation contributes the antagonist effect in which the interaction with the co-regulatory protein is not feasible and the transcription process cannot be produced. In general, because of these differences in the three-dimensional conformation of the ligand–receptor complex, there is a wide range of subsequent actions from full activation in the case of estradiol to complete antagonism in the case of the pure antiestrogens. The different SERMs exhibit intermediate properties because they induce transitional conformations closer to one or the other boundary.[45]

11.4   AGONIST AND ANTAGONIST ATTRIBUTES OF SELECTIVE ESTROGEN RECEPTOR MODULATORS (SERMS)

SERMs are synthetic nonsteroidal agents that bind to the ER and produce a change in the biological activity of the receptor depending on the tissue type. The primary target site for SERMs, the ER, is a NR. SERM binds to the ER and causes a change in the shape of the ER that allows recruitment of coactivators, which elicit an estrogenic response, and designates itself as corepressors if its response is antiestrogenic. The binding of the co-regulatory molecules leads to the activation of the promoter sequence of the estrogenic-responsive gene.[40] This process is also controlled by the degradation and disassembly of complexes at the gene promoter site, which causes renewed activation of the signal to initiate RNA synthesis and consequently can specifically modulate the estrogen responsiveness of a target tissue. SERMs are agents that bind to ERs but act either as agonists or antagonists in different tissues. For example, some SERMs act as agonists on the bone and uterus ERs, and antagonists on the breast ERs. The growth of some forms of breast cancers is dependent on estrogen. SERMs that act as antagonists on breast tissue are used in the treatment of breast cancer. Estrogen is important in maintaining bone structure in women, and therefore SERMs can also be useful in preventing postmenopausal osteoporosis. SERMs have increased our understanding of hormone-receptor regulatory mechanisms. Their development has permitted a targeted efficacy profile avoiding some of the side effects of the hormone therapy. Their clinical utility relies today mostly on the effects on breast cancer and bone.

11.5 CATEGORIZATION OF SYNTHETIC SERMS

There are several modules of chemical moieties, which are classified as SERMs, whereas there exist more than 70 molecules that belong to them.[64] Figure 11.1 illustrates the five significant chemical groups (triphenylethylenes, benzothiophenes, tetrahydronaphtylenes, indoles, and benzopyrans) that are considered as non-hormonal compounds, which are capable of activating the ER.[14] These chemical SERMs block the effects of estrogen in the targeted tissue.

Tamoxifene is widely used for this indication and is a reference compound for prevention and treatment.[25,41,108] Toremifene has been also marketed for breast cancer treatment.[16] Tamoxifene induces positive effects on bone density, whereas toremifene use is accompanied by slight reductions in bone density.[59] Both toremifene and tamoxifene have an estrogen-agonistic effect on the endometrium.[71,87] Raloxifene is the chief molecule of benzothiophenes and is widely used for osteoporosis treatment and prevention. This class of SERMs is antiresorptive on bone.

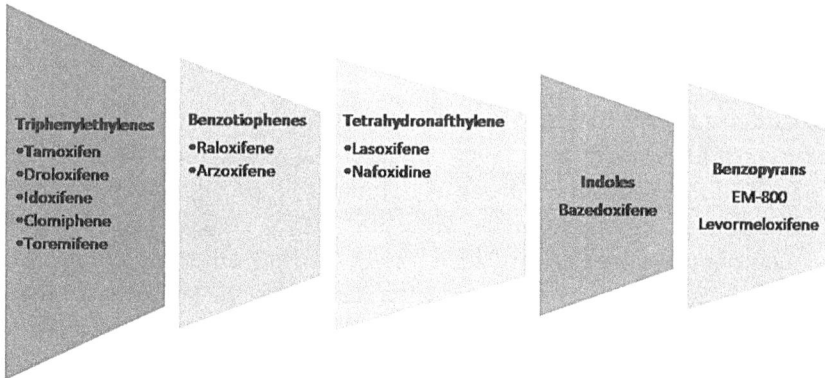

FIGURE 11.1   (See color insert.) Chemical classification of selective estrogen receptor modulators (SERMs).

11.6   ADVERSE EFFECTS OF SERMS

SERMs may cause some serious side effects, including blood clots, stroke, and endometrial cancer. Some chemical SERMs may cause specific side effects as listed below:

- Abnormal vaginal bleeding or discharge
- Chest pain
- Dizziness
- Leg swelling or tenderness
- Pain or pressure in the pelvis
- Severe headache
- Shortness of breath
- Sudden difficulty in vision
- Weakness, tingling, or numbness in your face, arm, or leg

The most common side effects of SERMs are:

- Fatigue
- Hot flashes
- Mood swings
- Night sweats
- Vaginal discharge

11.7 PHYTOESTROGENS: THE REDEEMERS TO CONTEST UNDESIRABLE COROLLARY

Diets rich in plant-derived products may supply a variety of phytoestrogens capable of producing a range of pharmacological effects in the human body. Phytoestrogens are a group of chemicals found in plants that can act similar to the hormone estrogen. More than 300 foods have been shown to contain phytoestrogens. Most food phytoestrogens are from one of three chemical classes: the isoflavonoids, the lignans, or the coumestans. They are actively being researched for beneficial effects of menopausal stress and ailments.[35] Phytoestrogens are compounds found in many plants, such as herbs, grains, and fruits. The presence of vital phytonutrients, such as vitamins, omega fatty acids, and lignans (potent cancer fighters), provides an excellent source of protein and fiber that balances the nutrient level during menopausal phase.

- **Effect on hormone levels**: Phytoestrogen with a high concentration of lignans is a great choice for women of all ages, as a natural way to normalize the menstrual cycle, manage menopause, and lower the risk of osteoporosis, cancer, and heart disease by balancing the hormone level.
- **Anticancer effects:** Extensive studies on both breast and colon cancer indicate that phytoestrogen may play an important role in cancer treatment as well as prevention.
- **Protection against bone loss**: The phytoestrogen supplements protection against bone loss may increase bone density and reduces the risk of osteoporosis.
- **Reduce the risk of CVD**: Plant source generally decreases the saturated fatty acid level and increases the good cholesterol. Hence, blood pressure is decreased, and the development of atherosclerosis and inflammation is suppressed, thereby enhancing blood vessel tone.

Contemporarily, there has been an amplified emphasis on phytoestrogens to be an enhanced elucidation toward HRT.[83] The schematic classification of dietary source of estrogen derivatives has been illustrated by Cos et al.[77]

11.8   MECHANISMS OF PHYTOESTROGENS TARGETING WOMEN AILMENTS

The secretion of estrogen production stimulated by the sex organs can be hindered by various reasons to cause discomfort in the physical and mental status of women. The chief source of decline in the level of estrogen is menopause, whereas numerous women in their premenopausal phase suffer from its tedious ailments.

Low level of estrogen can induce extreme distress such as sleep disturbances and extreme fatigue that distort the normal existence of women. This physiological condition can cause other apprehensions such as:

- Bladder and vaginal infection
- Depression
- Dryness in epithelial regions
- Headaches
- Joint pain

A decline in this natural hormone may be a dynamic factor to increase heart diseases among postmenopausal women. Estrogen is alleged to have an optimistic effect on the inner layer of artery wall, helping to keep blood vessels flexible. Hence, the hormone induces the contraction and relaxation of blood vessels that augment the circulatory function of heart. The reduction of hormonal influx among women during postmenopausal phase can induce the risk of coronary heart disease. Osteoporosis and breast cancer are caused due to the agonist and antagonist mechanisms of estrogen (Fig. 11.2). The depletion of estrogenic level in postmenopausal women is elevated by HRT, which leads to the risk of breast cancer. Hence, there should be stability in the intensity of estrogen that can maintain the health of women during her postmenopausal phase. Similarly, women in their premenopausal period affected with breast and uterine cancer have the risk of developing osteoporosis due to the treatment in declining their estrogenic level. This responsive condition can be managed by the potentials of phytoestrogen, which are consumed as supplements in modern medicine. These naturally occurring compounds fetch an alternative remedy to artificial and potentially dangerous hormone replacement treatments. There is an imperative need to understand the mechanisms of phytoestrogens and comprehend the different uses of phytoestrogenic herbs. Phytoestrogen products act by adding the plants' own estrogen-like compounds to the body, helping to balance hormone levels.

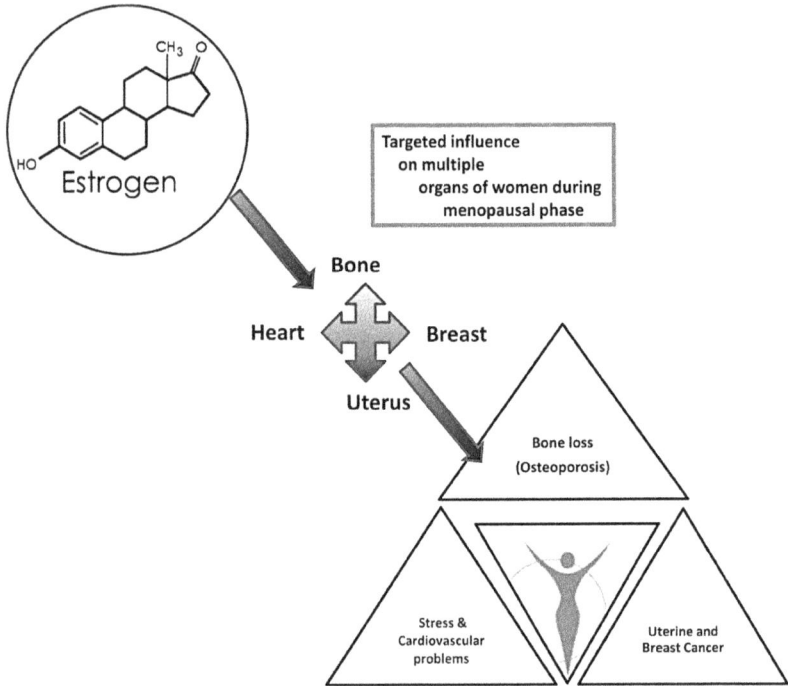

FIGURE 11.2 (**See color insert.**) Discrepancy of estrogen causing menaces among women in early stages of menopause.

Phytoestrogen supplements contain active chemical components such as isoflavones, coumestans, and lignans that work in a very similar way to the estrogen hormone in humans. Herbal phytoestrogens formulate to bind with the estrogenic receptor sites on human cells and compete with chemical estrogens that impair the system with adverse effects. The phytoestrogens employ both agonist and antagonist strategies in cells (Fig. 11.3) that react with the receptors based on the estrogen levels present in the body. If estrogen levels are low, then the phytoestrogen compounds bind to the site, effectively raising estrogen levels by mimicking the body's own estrogen compounds. Alternatively, if estrogen levels are high, phytoestrogens block ERs with their own weaker form of estrogen, effectively lowering estrogen levels.

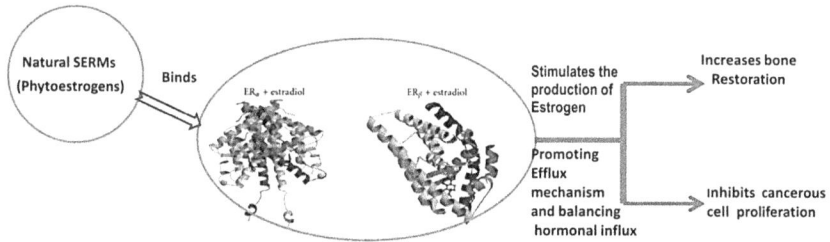

FIGURE 11.3 (See color insert.) An outline representing the mechanism of phytoestrogens as natural SERMs that compete with estrogen receptors.

11.9 COMPETENT PLANT-DERIVED COMPOUNDS TO COMBAT AILMENTS

Different classes of phytoestrogens and diverse compounds within each class affect the estrogen-mediated response in different ways. The flowchart in Figure 11.4 exemplifies the major classifications along with their structural illustration (Fig. 11.5).

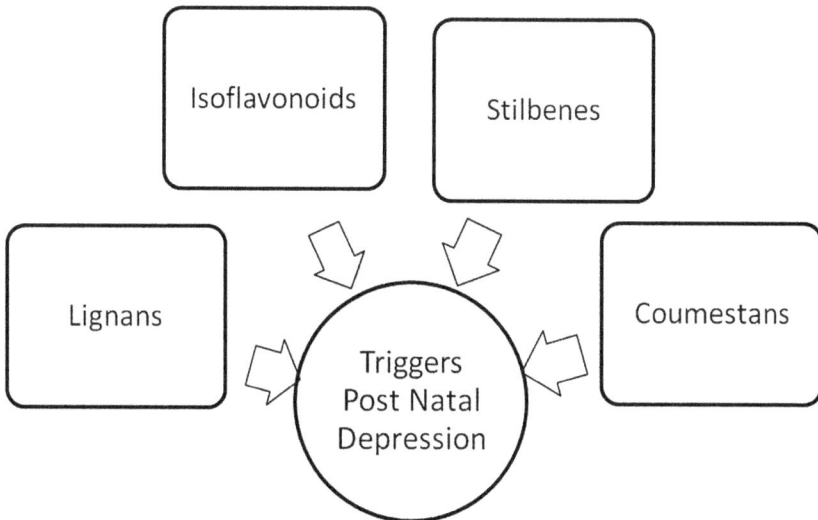

FIGURE 11.4 Categorization of phytoestrogens that induce estrogen-mediated response.

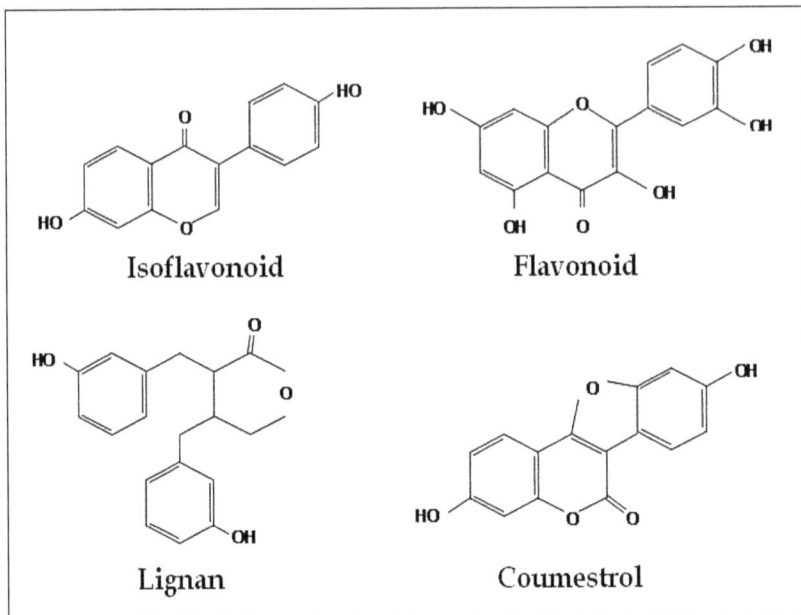

FIGURE 11.5   Chemical structures of the major phytoestrogens.

11.9.1   ISOFLAVONOIDS

Isoflavones are naturally occurring isoflavonoids, which are the most studied class of phytoestrogens and are found almost exclusively in the family of *Leguminosae*.[34] The flavonoids belong to large chemical class and are formed through the phenylpropanoid-acetate pathway by chalcone synthase and condensation reactions with malonyl-CoA. Isoflavones are produced through general phenylpropanoid pathway that produces flavonoid compounds in higher plants. The phenylpropanoid pathway begins from the amino acid phenylalanine, and an intermediate of the pathway, naringenin, is sequentially converted into the isoflavone genistein by two legume-specific enzymes: isoflavone synthase and a dehydratase. Similarly, another intermediate naringenin chalcone is converted to the isoflavone daidzein by sequential action of three legume-specific enzymes: chalcone reductase, type II chalcone isomerase, and isoflavone synthase. Plants use isoflavones and their derivatives as phytoalexin compounds to ward off disease-causing pathogenic fungi and other microbes. The isoflavonoids are subclass of flavonoids that differ in the position of one phenolic ring, which has shifted from C-3 to C-2. The isoflavonoids from legumes, that includes genistein-2 and daidzein, are most studied phytoestrogens. They can exist as glucosides or as glycones that are easily transported across intestinal epithelial cells and are hydrolyzed in the gut.[46]

11.9.2   LIGNANS

Lignans are estrogen-like chemicals and also act as antioxidants. They are compounds that form the building blocks of plant cell walls and phytoestrogens assist the regulation of body's estrogen production. When estrogen levels are too high, the lignans attach to the ERs, reducing the activity of body's natural estrogen hormones and blocking their effect in certain tissues. Research has shown that lignan phytoestrogens prevent some forms of cancer[17,23,73] by blocking estrogenic activity. The term lignan is used for a diverse class of phenylpropanoid dimers and oligomers. Secoisolariciresinol (SECO) and matairesinol-5 are two lignan dimers that are not estrogenic by themselves, but are readily converted to the mammalian lignans, enterodiol, and enterolactone, respectively, which are estrogenic.[31,80] The conversion occurs by gut microflora and the mammalian lignans are readily absorbed. The content of phytolignans is measurable in various body fluids such as urine, feces, and plasma.

11.9.3   STILBENES

Stilbenes are natural compounds such as flavonoids and belong to a subgroup polyphenols. They are produced through the phenylpropanoid-acetate pathway and are classified as resveratrol and pterostilbene. The main dietary source of phytoestrogenic stilbenes is resveratrol-4 and there are two isomers of resveratrol-4 (cis and trans), where only the trans form has been reported to be estrogenic.[86] It has a greater capacity to activate the ERβ than ERα.[28] Resveratrol-4 has shown agonistic and antagonistic activity in MCF-7 cells and the hamster ovarian cell line, CHOK1, transfected with human ERα and ERβ.[11,47,54] Resveratrol is a potential compound with profound health benefits. Observational studies reveal that it can compete against CVD and can act as an antioxidant and anti-inflammatory agent.[10,13]

11.9.4 COUMESTANS

Coumestrol and 4'-O-methoxycoumestrol are the predominant compounds that have exhibited estrogenic activity. They exhibit close structural resemblance with isoflavones. The main dietary source of coumestrol is legumes and low levels have been reported in Brussel sprouts and spinach.[27,32,48] Clover and soybean sprouts are reported to have the highest concentration, 28 and 7 mg/100 g dry wt., respectively; mature soybeans only have 0.12 mg/100 g dry wt. Although there are a large number of coumestans, only a small number has shown estrogenic activity, and their metabolism in human being is modest. Though, coumestans have not widely established its progeny from the group of isoflavonoids and anticipated as an efficient phytocompound to conjecture a wide source of well-being against ailments.[77]

11.10 THE DELIVERANCE OF MEDICINAL PLANT AS NATURAL SERMS

There are several potential medicinal plants that can enhance estrogen activity. The most prevalent phytoestrogenic contributors are soy products but are also found in other plant sources as shown below:

- Babchi (*Psoralea corylifolia*).
- Broccoli (*Brassica oleracea*).
- Flaxseed or linseed (*Linum usitatissimum L.*).
- Veldt grape (*Cissus quadrangularis*).

11.10.1 FLAXSEED OR LINSEED (LINUM USITATISSIMUM L.)

Flaxseed is the seed from *L. usitatissimum* commonly called flax of the family *Linnaceae*. It is a versatile blue flowered crop that is believed to be native to West Asia and the Mediterranean (Fig. 11.6). Flax originated in India and diffused over the globe for its usefulness and hardness.[21] Flax has been grown since the beginnings of civilization and people all over the world have celebrated its usefulness throughout the ages. Flax has varied uses from its oil seed to fibers, rope and textile industries to livestock feed, and health benefits to cosmetics and ornaments. In modern research, flax continues to surge forward in its recognition as a functional food.[66]

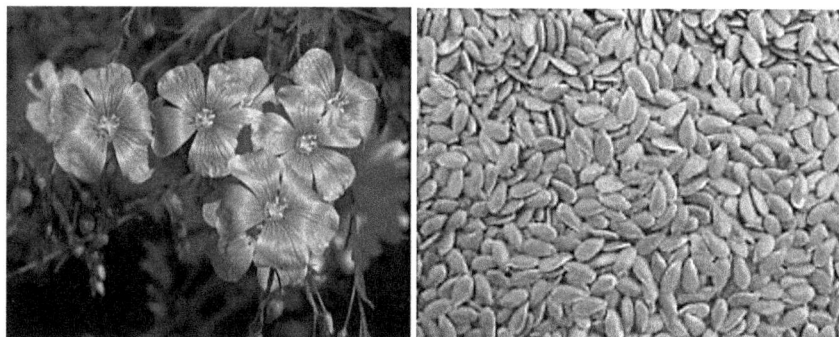

FIGURE 11.6 (See color insert.) Flax plant: flower and seed.

11.10.1.1 THE POTENTIAL PHYTOCONSTITUTIONS OF FLAXSEED

Flax varieties grown for human consumption are different from flax varieties grown to produce fiber for making linen.[95] Flaxseed varieties for human consumption are rich in omega-3 fat, dietary fiber, protein, both water- and fat-soluble vitamins, and one of the richest plant sources of lignans.[20]

Flaxseed oil is comprised healthy polyunsaturated fatty acids containing alpha-linoleic acid (LA), the essential omega-3 fatty acid, and LA, the essential omega-6 fatty acid. Our body is unable to synthesize them.[55] Whole flaxseeds are rich source of dietary fiber both soluble and insoluble.[30] Mucilage gum extracted from flaxseeds is a functional fiber that is used in cough syrups.[33] Amino acid composition of flaxseeds is viewed as the most important nutritional plant proteins with high content of all the essential amino acids. Flax is gluten free and hence can be tolerated by patients with celiac disease.[8]

Phenolics are group of compounds in plants that have varied functions. Flax is a rich source of three types of phenolics: phenolic acids, flavonoids, and lignans. Lignans are phytoestrogens with diphenolic ring structures resembling those of endogenous estrogens.[38,68] and have been shown to exert hormonal effects. The most prevailing lignan is secoisolariciresinol diglucoside (SDG). It does not exist in free form instead as a five-SDG molecule bound together with other molecules in the outer fiber layer of the seed.[58,81,99] The other lignans found are: matairesinol, pinoresinol, lariciresinol, isolariciresinol, and SECO. The lignans except isolariciresinol are converted by bacteria in the colon to the mammalian lignans, enterodiol, and enterolactone.[19]

11.10.1.2 *PHYTOESTROGENIC PROPERTIES OF FLAX: THE LIGNANS*

The mammalian lignans work by binding to ERs on cell membranes, same as the endogenous estrogens, and eventually affect the response of tissues. The mammalian lignans can act as either weak estrogens or they can oppose the actions of estrogen, depending on the presence of stronger estrogens such as estradiol.[36] In women, during her reproductive years when blood levels of endogenous estrogens are high, the lignans can bind to the ER and block the actions of endogenous estrogens as antagonists. After menopause, the levels of endogenous estrogens in the blood decrease naturally and it is when the lignans perform as weak estrogens, the agonists.[26]

Flax lignans and the mammalian lignans (enterodiol and enterolactone) are biologically active and help in increasing bone density. Metabolism of flax lignans has been illustrated by Clavel et al.[19] In a recent study of postmenopausal group of rodents with diabetes, the diet including flaxseed oil resulted in increased levels of bone-creating protein osteocalcin. They also found that levels of marker, deoxypyridinoline in urine, is associated with bone restoration, when the rats were given flaxseed oil.[101] In postmenopausal women, the mammalian lignans are responsible for maintaining bone mineral density and reducing bone loss.[65]

Diets high in lignans may help maintain good cognitive function in postmenopausal women who have anticancer and antiviral effects. They influence gene expression and may protect against estrogen-related diseases such as osteoporosis.[42,65,82] In postmenopausal women, estrogen deficiency is the major risk factor for osteoporosis. The studies suggest that the mammalian lignans of flax mimic the endogenous estrogens and bind to ERs of bone and hence improve osteocalcin;[3] reduce the risk of uterine fibroids in middle-aged women;[6] reduce breast cancer risk in women, and reduce the risk of acute fatal coronary events[100] and prostate cancer[103] in men. SDG is an antioxidant that scavenges hydroxyl ion free radicals.[9] Enterolactone activates the pregnane X receptor, which is involved in the metabolism of bile acids, steroid hormones, and has the ability to affect the metabolism of some drugs.[39] A study suggested that lignans, along with enterodiol and enterolactone, affect hormone receptors in breast tissue and reduce the risk of breast cancer. They inhibit the activity of aromatase, an enzyme involved in the production of estrogens, and protect against breast cancer.[12,105]

11.10.2 BABCHI (PSORALEA CORYLIFOLIA)

P. corylifolia is one of the notable medicinal plants in *Leguminosae* family. It is widely employed in Chinese medicinal system to treat various diseases. It is found in eastern Asia, China, Southern Africa, and throughout India. The parts of the whole plant have significant medicinal properties to cure various diseases.[84]

11.10.2.1 PHYTOCHEMICAL CONSTITUENTS OF P. CORYLIFOLIA

About 90 biocompounds were identified in *P. corylifolia* and categorized into five groups: coumarins, flavonoids, meroterpenes, benzofuran, and others.[44] Specific compounds (for instance: bakuchiol, corylifolin, psoralidin, and isobavachin) exhibited strong antioxidant activities. This plant has been used as an effective invigorant against impotence, menstruation disorder, coronary vasodilatory activity, vitiligo, psoriasis, leukoderma, leprosy, dermatitis, depression, hyperglycemia, and uterine hemorrhage. It also revealed antioxidant, antitumor, antibacterial, antiviral, diuretic, anthelmintic, laxative, stomachic, aphrodisiac, and diaphoretic properties.[44,52,111] Notably, the seed extract has been used as a treatment for bone fractures, osteomalacia, and osteoporosis.[84] The isoflavonoids content in *P. corylifolia* is much higher than the content in soybean. Isoflavones, particularly genistein, exhibits a vital effect in treating various cancer diseases. Health beneficial aspects of phytoconstituents present in *P. corylifolia* are mentioned by Lim et al.[51]

11.10.2.2 ESTROGENIC ACTIVITY OF P. CORYLIFOLIA

Preliminary use of estrogenic nutraceuticals is based on isoflavones, which are used to develop SERMs from plant-based HRT. Phytoestrogens from

P. corylifolia revealed ER-subtype selectivity in breast cancer cells. Moreover, psoralen and isopsoralen promoted the proliferation of MCF-7 cells, which confirmed the selectivity for ERs. Particularly, coumarin showed selectivity for ERα and flavonoids, and bakuchiol exhibit dual action on ERα and ERβ, and hence *P. corylifolia* may be used as a new source for SERMs.[110] In another study, ER agonist activity of bakuchiol and psoralidin was further confirmed by luciferase activity, which was contributed by estrogenic transactivation. The results suggested that *P. corylifolia* may be valuable as a novel ER modulator and has the potential to be used in alternative HRT formulations.[53] In addition to bakuchiol and psoralidin, the estrogenic activity of bavachin was characterized by ligand binding, reporter gene activation, and endogenous target gene regulation in CV-1 cells. Bavachin has the capability to increase the transcription level of estrogen-responsive genes and decrease the protein level, and hence effectively modulate cellular ER targets.[76] These kinds of estrogenic activities of *P. corylifolia* were accomplished by the presence of bakuchiol, bavachalcone, isobavachalcone, bavachromene, and isobavachromene since they have prynylated phenolic groups. Among these compounds, bakuchiol is the key compound with five times higher estrogenic activity than other compounds, thus suggesting their use in attenuating various symptoms of estrogen deficiency in postmenopausal women.[51] Corylin also has SERM-like behavior, which has not yet been studied in detail.[92]

11.10.3   VELDT GRAPE (CISSUS QUADRANGULARIS)

Veldt grape shrub (Fig. 11.7) belongs to the family *Vitaceae* and is traditionally called as "bone setter."[98] It had been widely used in India, Thailand, Sri Lanka, Java, and West Africa.[69,78] It is a climbing shrub and three variants are available in which two variants are consumed by human.[7] All parts of this plant play a pharmaceutical role in curing various diseases such as menstrual disorder, scurvy, gout, syphilis, venereal diseases, piles, bone fractures, asthma, dyspepsia, leucorrhoea, diarrhea, dysentery, and fistula.[7,69,78,98]

Although it is originated from West Africa, yet it is most frequently used medicinal plant in Ayurveda medicinal system of India. The most prominent beneficial role of this plant is healing of bone fracture. The alcoholic extraction of this plant showed bone healing properties in albino rats, dog,[60] and humans.[102] In addition, the chemical constituents of this plant influence the mineralization process of bone callus and provide rapid recovery.[88] Their healing speed is faster than ascorbic acid. They also showed potential antimicrobial activity against various pathogens.[93]

FIGURE 11.7 (See color insert.) Veldt grape plant.

11.10.3.1 CHEMICAL CONSTITUENTS OF VELDT GRAPE

Since this plant holds vast medicinal properties, its chemical elements have been studied in detail. Primarily, ascorbic acid, β-carotene, β-sitosterol, α-amyrin, δ-amyrone, quandragularins, anabolic steroids, flavonoids, indanes, calcium oxalate, triterpenoids, saponins, stilbenes, and their derivatives (such as resveratrol, piceatannol, pallidol, perthenocissin, and irridoids)[7,69,78,98] are present in considerable concentrations. Recently, more compounds (such as triterpene δ-amyrin acetate, hexadecanoic acid, stilbene glucoside trans-resveratrol-3-O-glucoside, 6-O-[2,3-dimethoxy]-trans-cinnamoyl catalpol, 6-O-meta-methoxy-benzoyl catalpol, quadra-gularin A, quercitin, quercitrin, and β-sitosterol glycoside)[93,98] have been identified.

Bone healing property is contributed by the presence of ascorbic acid and calcium, whereas sterols and tannins showed remarkable antioxidant activity. Luteolin and β-sitosterol inhibit the inflammatory effect and act as a potential inhibitors of cyclooxygenase and lipoxygenase pathways. Extract of this plant mimics the effect of aspirin in inhibiting writhing response that confirm its analgesic effects mediated by excitation of local nociceptors in the central nervous system in the hypothalamic region.[75] This analgesic effect was attributed by the presence of carotene, phytosterols, calcium, sitosterol, amyrin, and amyrone.[60] Its antioxidants activity brought significant reduction in weight, body fat, total cholesterol, low-density lipoprotein cholesterol, triglycerides, C-reactive protein and fasting blood glucose level, and serum lipids which resulted in improved cardiovascular health.[72] Extract of this plant enhanced the proliferation rate of marrow mesenchymal stem cells, which could stimulate osteoblastogenesis that prevents the bone disease osteoporosis.

11.10.3.2 ESTROGENIC ACTIVITY OF VELDT GRAPE

The estrogenic activity of *C. quadrangularis* has been studied in detail by Aswar et al. and others.[4,92] They studied the effect of *C. quadrangularis* formulation on estrogenic activity and sexual behavior of overiectomized female Wistar rats and concluded that the formulation had a moderate effect in the tested aspects by improving the reproductive behavior, uterine weight, and serum estrogen level along with vaginal cornification. Moreover, it also statistically resulted in the increase in bone thickness, bone mineral density, bone hardness, and serum estradiol level which ultimately prevented the bone loss. In addition, renal excretion of calcium and decreased calcium absorption during menopausal period were treated by improving the synthesis of calcium-binding protein by activating vitamin D_3 receptor through the phytoconstituents of *C. quadrangularis*. The presence of saponin in this formulation also affected the permeability of the small intestinal mucosal cells and had a positive effect on active nutrient support.[5]

11.10.4 BROCCOLI (BRASSICA OLERACEA)

Broccoli (Fig. 11.8) is considered as significant nutraceutical plant belonging to the family *Brassicaceae* (Cruciferae), which is widespread in

European region. Functional foods emphasize several health benefits that enforce anticarcinogenic properties.

FIGURE 11.8 **(See color insert.)** Broccoli.

11.10.4.1 HABITAT AND PHYTOCHEMICALS OF BROCCOLI

Brassica oleracea, variety Italica, belongs to the family *Brassicaceae* (Fig. 11.8). Broccoli is a rapidly growing annual plant that originated along the Atlantic seaboard of Western Europe and along the Mediterranean basin.[1] *B. oleracea* has been cultivated as a vegetable for more than 2500 years, and through selective breeding, particular characteristics of the plant have been developed. A number of types of vegetables have been derived from this wild stock through selection of favorable cultivars. Plants, in general, are known to be extremely rich in a variety of secondary metabolites with medicinal properties. The *Brassica* vegetables have many nutrients and bioactive substances, such as vitamins, minerals, fiber, carotenoids, bioflavonoids, sulfur, dithiolethiones, and glucosinolates.[85] Broccoli also provides many health-promoting properties, which attribute to its antioxidant and anticarcinogenic compounds. It is primarily composed of polyphenols, glucosinolates, sulforaphane, and selenium.[57]

Broccoli sprouts contain negligible quantities of indole glucosinolates that predominate in the mature vegetable and give rise to degradation products such as indole-3-carbinol with the capability to enhance tumorigenesis. Hence, small quantities of crucifer sprouts may protect against the risk of cancer as effectively as much larger quantities of mature vegetables of the same variety.[24] They are also known to contain a high content of flavonoids, vitamins, and mineral nutrients. Vitamin C is a good adjuvant in iron therapy but can interfere with the metabolism of some drugs and antineoplastic agents. The presence of these compounds has shown that broccoli provides immense benefits in protecting humans against cancer, and also assures to reduce the risk of specific cancers. One of the phytotherapeutic roles of broccoli is for skin diseases in which the juice of the leaves is used to treat warts.[67]

11.10.4.2   PHYTOESTROGENIC ASSETS OF B. OLERACEA

Brassica contains estrogen-like chemicals (phytoestrogens), which can bind to ERs on the reproductive tissue and exert anticancer effects. It is different from the common cruciferous vegetables that possess high levels of these constituents. The pharmacokinetics of broccoli explains that when hydrolysis takes place, glucoraphanin produces many products that include the bioactive isothiocyanate sulforaphane. Indole-3-carbinol also acts as a phytoestrogen (plant-based estrogens) and in this capacity, it can bind to ERs in the body, reducing the ability of stronger estrogens from overstimulating reproductive tissues such as breast, cervix, uterus, and in males, the prostate gland. Epidemiological studies consistently showed that a higher ingestion of indole-3-carbinol foods is highly associated with the prevention of reproductive organ cancers in women and men.[22,97,104] Indole-3-carbinol also promotes the metabolism of certain endogenous estrogens (estrone) into a safer, less cancer-promoting form (2-OH-estrone), further helping to reduce the risk of reproductive organ cancers. Some women naturally convert more of their estrone hormone to 16-hydroxyestrone, which has been shown to be a biomarker for increased risk of breast cancer, by some researchers. Supplementation with indole-3-carbinol has shown to alter genetic expression in such a way as to encourage greater activity of the enzyme that converts estrone into 2-hydroxyestrone. Thus, all women may benefit in this regard as the intake of indole-3-carbinol helps to improve the 2-hydroxy- to 16-hydroxyestrone ratio. Contemporary techniques such as nanotechnology can enhance the efficacy of the plant and it is inferred to undergo biological reduction with gold nanoparticles that act as an antimicrobial agent for both bacteria and fungi.[79] Hence, there is a need for profound exertion in initiating research to nurture the medicinal benefits and to persist the significance of phytoestrogens from broccoli.

11.11 SUMMARY

Phytoestrogen supplements have numerous benefits, which focus on increasing estrogen levels and balancing hormone levels within the body. Estrogen is commonly known as a female hormone. Estrogen, along with progesterone, helps to regulate the menstrual cycle and ensure fertility in a woman. HRT and estrogen replacement therapy can both relieve menopausal symptoms and are recommended for women in the postmenopausal phase. Since these treatments are considered artificial and potent to cause adverse effects, therefore phytoestrogen supplements are currently being used as a natural alternative to prevent CVD and osteoporosis by avoiding the risk of enduring breast cancer. Promotion and mobilization of research are required to prove the efficacy and potency of phytoestrogens that can ameliorate various ailments among women.

11.12 ACKNOWLEDGMENTS

The authors acknowledge the support offered by the management of Bishop Heber College, Tiruchirappalli and Stella Maris College, Chennai, Tamil Nadu, India. Facilities provided by the institutes for literature and data compilation are gratefully conceded.

KEYWORDS

- *Brassica oleracea*
- breast cancer
- *Cissus quadrangularis*
- coumestans
- estrogen replacement therapy
- functional foods
- hormone replacement therapy
- isoflavonoids
- lignans
- *Linum usitatissimum L.*
- menopause
- osteoporosis
- phytoestrogen
- *Psoralea corylifolia*
- selective estrogen receptor modulators
- stilbenes

REFERENCES

1. Anonymous. Broccoli. *Encyclopaedia Britannica;* 2015.
2. Ariana, P.; Arif, N. An Introduction to the Human Development Capability Approach. In *Freedom and Agency;* Deneulin, S., Shahani, L., Eds.; Earthscan/IDRC Publisher: Ottawa, 2009; Vol. 1, pp 228–245.
3. Arjmandi, B. H. The Role of Phytoestrogens in the Prevention and Treatment of Osteoporosis in Ovarian Hormone Deficiency. *J. Am. Coll. Nutr.* **2001,** *20,* 398S–402S.
4. Aswar, U. M.; Bhaskaran, S.; Mohan, V.; Bodhankar, S. L. Estrogenic Activity of Friedelin Rich Fraction (IND-HE) Separated from *Cissus quadrangularis* and Its Effect on Female Sexual Function. *Pharmacogn. Res.* **2010,** *2*(3), 138.
5. Aswar, U. M.; Mohan, V.; Bodhankar, S. L. Antiosteoporotic Activity of Phytoestrogen-Rich Fraction Separated from Ethanol Extract of Aerial Parts of *Cissus quadrangularis* in Ovariectomized Rats. *Indian J. Pharmacol.* **2012,** *44*(3), 345.
6. Atkinson, C.; Lampe, J. W.; Scholes, D. Lignan and Isoflavone Excretion in Relation to Uterine Fibroids: A Case-Control Study of Young to Middle-Aged Women in the United States. *Am. J. Clin. Nutr.* **2006,** *84,* 587–593.
7. Austin, A.; Kannan, R.; Jegadeesan, M. Pharmacognostical Studies on *Cissus quadrangularis* L. Variant I & II. *Anc. Sci. Life* **2004,** *23*(4), 33.
8. BeMiller, J. N.; Whistler, R. L.; Barkalow, D. G.; Chen, C. C. Aloe, Chia, Flaxseed, Okra, Psyllium Seed, Quince Seed, and Tamarind Gums. In *Industrial Gums: Polysaccharides and Their Derivatives,* BeMiller, J.; Whistler, R., Eds., 3rd ed.; Whistler Academic Press: New York, 1993; pp 227–256.
9. Bhathena, S. J.; Velasquez, M. T. Beneficial Role of Dietary Phytoestrogens in Obesity and Diabetes. *Am. J. Clin. Nutr.* **2002,** *76,* 1191–1201.
10. Borriello, A.; Cucciolla, V.; Della Ragione, F.; Galletti, P. Dietary Polyphenols: Focus on Resveratrol, a Promising Agent in the Prevention of Cardiovascular Diseases

and Control of Glucose Homeostasis. *Nutr. Metab. Cardiovasc. Dis.* **2010,** *20*(8), 618–625.

11. Bowers, J. L.; Tyulmenkov, V. V.; Jernigan, S. C.; Klinge, C. M. Resveratrol Acts as a Mixed Agonist/Antagonist for Estrogen Receptors Alpha and Beta. *Endocrinology* **2000,** 141, 3657–3667.

12. Brooks, J. D.; Thompson, L. U. Mammalian Lignans and Genistein Decrease the Activities of Aromatase and 17β-hydroxysteroid Dehydrogenase in MCF-7 Cells. *J. Steroid Biochem. Mol. Biol.* **2005,** *94,* 461–467.

13. Brownson, D. M.; Azios, N. G.; Fuqua, B. K.; Dharmawardhane, S. F.; Mabry, T. J. Flavonoid Effects Relevant to Cancer. *J. Nutr.* **2002,** *132,* 3482S–3489S.

14. Bryant, H. Selective Estrogen Receptor Modulators. *Rev. Endocr. Metab. Disord.* **2002,** *3,* 231–241.

15. Brzozowski, A. M.; Pike, A. C.; Dauter. Z.; Hubbard, R. E.; Bonn, T.; Engstrom, O.; et al. Molecular Basis of Agonism and Antagonism in the Oestrogen Receptor. *Nature* **1997,** *389,* 753–758.

16. Buzdar, A. U.; Hortobagyi, G. N. Tamoxifen and Toremifene in Breast Cancer: Comparison of Safety and Efficacy. *J. Clin. Oncol.* **1998,** *26,* 348–353.

17. Bylund, A.; Saarinen, N.; Zhang, J. X.; Bergh, A.; Widmark, A.; Johansson, A.; Lundin, E.; Adlercreutz, H.; Hallmans, G.; Stattin, P.; Mäkela, S. Anticancer Effects of a Plant Lignan 7-hydroxymatairesinol on a Prostate Cancer Model In Vivo. *Exp. Biol. Med.* **2005,** *230*(3), 217–223.

18. Choi, K. C.; Jeung. E. B. The Biomarker and Endocrine Disruptors in Mammals. *J. Reprod. Dev.* **2003,** *49*(5), 337–345.

19. Clavel, T.; Borrmann, D.; Braune, A.; et al. Occurrence and Activity of Human Intestinal Bacteria Involved in the Conversion of Dietary Lignans. *Anaerobe* **2006,** *12,* 140–147.

20. Dew, T. P.; Williamson, G. Controlled Flax Interventions for the Improvement of Menopausal Symptoms and Postmenopausal Bone Health: A Systematic Review. *Menopause* **2013,** *20*(11), 207–215.

21. Dharshini, Thirunalasundari, T.; Sumayaa, S. Evaluation of Biochemical Constituents of *Linum usitatissimum* by GC-MS. *Int. J. Sci. Res. Publ.* **2013,** *3*(4), 2250–3153.

22. Dhinmi, S. R.; Li, Y.; Upadhyay, S.; Koppolu, P. K.; Sarkar, F. H. Indole-3-carbinol (I3C)-Induced Cell Growth Inhibition, G1 Cell Cycle Arrest and Apoptosis in Prostate Cancer Cells. *Oncogene* **2001,** *20*(23), 2927–2936.

23. Dixion, R. A.; Ferreira, D. Genistein. *Phytochemistry* **2002,** *60*(3), 205–211.

24. Fahey; Zhang, Y.; Talalay. P. Broccoli Sprouts: An Exceptionally Rich Source of Inducers of Enzymes that Protect Against Chemical Carcinogens. *Proc. Natl. Acad. Sci. U. S. A.* **1997,** *94,* 10367–10372.

25. Fisher, B.; Constantino, J. P.; Wickerham, D. L.; Redmond, C. K.; Kavanah, M.; Cronin, M.; et al. Tamoxifen for Prevention of Breast Cancer: Report of the National Surgical Adjuvant Breast and Bowel Project P-1 Study. *J. Natl. Cancer Inst.* **1998,** *90,* 1371–1388.

26. Franco, O. H.; Burger, H.; Lebrun, C. E. I.; et al. Higher Dietary Intake of Lignans Is Associated with Better Cognitive Performance in Postmenopausal Women. *J. Nutr.* **2005,** *135,* 1190–1195.

27. Franke, A. A.; Custer, L. J.; Cerna, C. M.; Narala, K. K. Quantitation of Phytoestrogens in Legumes by HPLC. *J. Agric. Food Chem.* **1994,** *42,* 1905–1913.

28. Gehm, B. D.; McAndrews, J. M.; Chien, P. Y.; Jameson, J. L. Resveratrol, a Polyphenolic Compound Found in Grapes and Wine, Is an Agonist for the Estrogen Receptor. *Proc. Natl. Acad. Sci. U. S. A.* **1997,** *94,* 14138–14143.

29. Geserick, C.; Meyer. H. A.; Haendler, B. The Role of DNA Response Elements as Allosteric Modulators of Steroid Receptor Function. *Mol. Cell Endocrinol.* **2005,** *236*(1–2), 1–7.

30. Ghazanfarpour, M.; Sadegh, R.; Latifnejad, R.; Khadivzadeh, T.; Khorsand, I.; Afiat, M; Esmaeilizadeh, M. Effects of Flaxseed and *Hypericum perforatum* on Hot Flash, Vaginal Atrophy and Estrogen-Dependent Cancers in Menopausal Women: A Systematic Review and Meta-analysis. *Avicenna J Phytomed.* **2016,** *6*(3), 273–283.

31. Glitsø, L. V.; Mazur, W.; Adlercreutz, H.; Wähälä, K.; Mäkelä, T.; Sandström, B. Intestinal Metabolism of Rye Lignans in Pigs. *Br. J. Nutr.* **2000,** *84,* 429–437.

32. Gresele, P.; Cerletti, C.; Guglielmini, G.; Pignatelli, P.; de Gaetano, G.; Violi, F. Effects of Resveratrol and Other Wine Polyphenols on Vascular Function: An Update. *J. Nutr. Biochem.* **2011,** *22*(3), 201–211.

33. Gupta, C.; Prakash, D.; Gupta, S. Phytoestrogens as Pharma Foods. *Adv. Food Technol. Nutr. Sci. Open J.* **2016,** *2*(1), 19–31.

34. El-Shemy, H. Ed. *Soybean and Nutrition;* ISBN 978–953–307–536–5, Published: September 12, 2011 Under CC BY-NC-SA 3.0 License.

35. Hays, J.; Ockene, J. K.; Brunner, R. L.; Kotchen, J. M.; Manson, J. E.; Patterson, R. E.; Aragaki, A. K.; Shumaker, S. A.; Brzyski, R. G.; LaCroix, A. Z.; Granek, I. A.; Valanis, B.G. Women's Health Initiative Investigators Effects of Estrogen Plus Progestin on Health-Related Quality of Life. *N. Engl. J. Med.* **2003,** *348,* 1835–1837.

36. Hutchins, A. M.; Slavin, J. L. Effects of Flaxseed on Sex Hormone Metabolism. In *Flaxseed in Human Nutrition;* Thompson, L. U., Cunnane, S. C., Eds.; AOCS Press: Champaign, IL, 2003; pp 126–149.

37. Hwang, K. A.; Park, S. H.; Yi, B. R.; Choi, K. C. Gene Alterations of Ovarian Cancer Cells Expressing Estrogen Receptors by Estrogen and Bisphenol a Using Microarray Analysis. *Lab. Anim. Res.* **2011,** *27*(2), 99–107.

38. Institute of Medicine. Dietary Reference Intakes for Energy, Carbohydrate, Fiber, Fat, Fatty Acids, Cholesterol, Protein, and Amino Acids. National Academies Press: Washington, DC, 2002; pp 7-1–7-69 (Dietary Fiber), 8-1–8-97 (Fat and Fatty Acids).

39. Jacobs, M. N.; Nolan, G. T.; Hood, S. R. Lignans, Bacteriocides and Organochlorine Compounds Activate the Human Pregnane X Receptor (PXR). *Toxicol. Appl. Pharmacol.* **2005,** *209,* 123–133.

40. Jordan, V. C. Chemoprevention of Breast Cancer with Selective Oestrogen-Receptor Modulators. *Nat. Rev. Cancer* **2007,** *7,* 46–53.

41. Jordan, V. C.; Morrow, M. Tamoxifen, Raloxifene, and the Prevention of Breast Cancer. *Endocr. Rev.* **1999,** *20,* 253–278.

42. Khan, A. A.; Hodsman, A. B.; Papaioannou, A.; et al. Management of Osteoporosis in Men: An Update and Case Example. *CMAJ* **2007,** *176,* 345–348.

43. Khera, R.; Jain, S.; Lodha, R.; Ramakrishnan, S. Gender Bias in Child Care and Child Health: Global Patterns. *Arch. Dis. Child.* **2014,** *99*(4), 369–374.

44. Khushboo, P. S.; Jadhav, V. M.; Kadam, V. J.; Sathe, N. S. *Psoralea corylifolia* Linn.—"Kushtanashini". *Pharmacogn. Rev.* **2010,** *4*(7), 69.
45. Kian, T. M.; Rogatsky, I.; Tzagarakis-Foster, C.; Cvoro, A.; An, J.; Christy, R. J.; et al. Estradiol and Selective Estrogen Receptor Modulators Differentially Regulate Target Genes with Estrogen Receptors Alpha and Beta. *Mol. Biol. Cell* **2004,** *15*,1262–1272.
46. King, A.; Young, G. Characteristics and Occurrence of Phenolic Phytochemicals. *J. Am. Diet. Assoc.* **1999,** *99*, 213–218.
47. Klinge, C. M.; Risinger, K. E.; Watts, M. B.; Beck, V.; Eder, R.; Jungbauer, A. Estrogenic Activity in White and Red Wine Extracts. *J. Agric. Food Chem.* **2003,** *51*, 1850–1857.
48. Knuckles, B. E.; DeFremery, D.; Kohler, G. O. Coumestrol Content of Fractions Obtained During Wet Processing of Alfalfa. *J. Agric. Food Chem.* **1976,** *24*, 1177–1180.
49. Kong, E. H.; Pike, A. C.; Hubbard, R. E. Structure and Mechanism of the Oestrogen Receptor. *Biochem. Soc. Trans.* **2003,** *31*(Part 1), 56–59.
50. Lane, P. H. Estrogen Receptors in the Kidney: Lessons from Genetically Altered Mice. *Gend. Med.* **2008,** *5*(Suppl A), S11–18.
51. Lim, S. H.; Ha, T. Y.; Ahn, J.; Kim, S. Estrogenic Activities of *Psoralea corylifolia* L. Seed Extracts and Main Constituents. *Phytomedicine* **2011,** *18*(5), 425–430.
52. Liu, H.; Bai, Y. J.; Chen, Y. Y.; Zhao, Y. Y. Studies on Chemical Constituents from Seed of *Psoralea corylifolia*. *China J. Chin. Materia Medica* **2008,** *33*(5), 1410–1412.
53. Liu, X.; Nam, J. W.; Song, Y. S.; Viswanath, A. N. I.; Pae, A. N.; Kil, Y. S.; et al. Psoralidin, a Coumestan Analogue, as a Novel Potent Estrogen Receptor Signaling Molecule Isolated from *Psoralea corylifolia*. *Bioorg. Med. Chem. Lett.* **2014,** *24*(5), 1403–1406.
54. Lu, R.; Serrero, G. Resveratrol, a Natural Product Derived from Grape, Exhibits Antiestrogenic Activity and Inhibits the Growth of Human Breast Cancer Cells. *J. Cell Physiol.* **1999,** *179*, 297–304.
55. Lucas, E. A.; Wild, R. D.; Hammond, L. J.; Khalil, D. A.; Juma, S.; Daggy, B. P.; Stoecker, B. J.; Arjmandi, B. H. Flaxseed Improves Lipid Profile Without Altering Biomarkers of Bone Metabolism in Postmenopausal Women. *J. Clin. Endocrinol. Metab.* **2002,** *87*(4), 1527–1532.
56. MacEachron, A. Women's Health in the Post-2015 World: Ensuring No One Is Left Behind. BCUN News; Business Council for the United Nations; 2014. Retrieved July 13, 2016.
57. Mahn, A.; Reyes, A. An Overview of Health-Promoting C ompounds of Broccoli (*Brassica oleracea* var. Italica) and the Effect of Processing. *Food Sci. Technol. Int.* **2012,** *18*, 503–514.
58. Martin, J. H. J.; Crotty, S.; Warren, P.; Nelson, P. N. Does an Apple a Day Keep the Doctor Away Because a Phytoestrogen a Day Keeps the Virus at Bay? A Review of the Anti-viral Properties of Phytoestrogens. *Phytochemistry* **2007,** *68*, 266–274.
59. Marttunen, M. B.; Hietanen, P.; Titinen, A.; Roth, H. J.; Viinikka, L.; Ylikorkala, O. Effects of Tamoxifen and Toremifene on Urinary Excretion of Pyridinoline and Deoxypyridinoline and Bone Density in Postmenopausal Women with Breast Cancer. *Calcif. Tissue Int.* **1999,** *65*, 365–368.

60. Mate, G. S.; Naikwade, N. S.; Magdum, C. S.; Chowki, A. A.; Patil, S. B. Evaluation of Anti-nociceptive Activity of *Cissus quadrangularis* on Albino Mice. *Int. J. Green Pharm.* **2008,** *2*(2), 118–121.

61. McDonnell, D. P. The Molecular Pharmacology of Estrogen Receptor Modulators: Implications for the Treatment of Breast Cancer. *Clin. Cancer Res.* **2005,** *11,* 871s–877s.

62. McDonnell, D. P.; Wijayarate, A.; Chang, C.; Norris, J. D. Elucidation of the Molecular Mechanism of Action of Selective Estrogen Receptor Modulators. *Am. J. Cardiol.* **2002,** *90*(Suppl), 35F–43F.

63. McGuire, W. L.; Chamnes, G. C.; Fuqua, S. A. Estrogen Receptor Variants in Clinical Breast Cancer. *Mol. Endocrinol.* **1991,** *5,* 1571–1577.

64. Meegan, M. J.; Lloyd, D. G. Advances in the Science of Estrogen Receptor Modulation. *Curr. Med. Chem.* **2003,** *10,* 181–210.

65. Mer Harvi. Impact of Feeding Flaxseed Oil on Delaying the Development of Osteoporosis in Ovariectomized Diabetic Rats. *Int. J. Food Saf., Nutr. Publ. Health* **2009,** *2,* 189–201.

66. Mishra, N.; Mishra, V. N.; Devanshi. Natural Phytoestrogens in Health and Diseases. *JIACM* **2011,** *12*(3), 205–211.

67. Moreno, D. A. Chemical and Biological Characterization of Nutraceutical Compounds of Broccoli. *J. Pharm. Biomed. Anal.* **2006,** *4,* 1508–1522.

68. Muir, A. D. Flax Lignans—Analytical Methods and How They Influence Our Understanding of Biological Activity. *J. AOAC Int.* **2006,** *89,* 1147–1157.

69. Nagani, K. V.; Kevalia, J.; Chanda, S. V. Pharmacognostical and Phytochemical Evaluation of Stem of *Cissus quadrangularis* L. *Int. J. Pharm. Sci. Res.* **2011,** *2*(11), 2856.

70. Navarro, D.; Leon, L.; Chirino, R.; Fernandez, L.; Pestano, J.; Diaz-Chico, B. N. The Two Native Estrogen Receptor Forms of 8S and 4S Present in Cytosol from Human Uterine Tissues Display Opposite Reactivities with the Antiestrogen Tamoxifen Aziridine and the Estrogen Responsive Element. *J. Steroid Biochem. Mol. Biol.* **1998,** *64,* 49–58.

71. O'Regan, R. M.; Cisneros, G. M.; England, G. M.; MacGregor, J. L.; Muenzner, H. D.; Assikis, V. J.; et al. Effects of the Antiestrogens Tamoxifene, Toremifene and ICI 182,780 on Endometrial Cancer Growth. *J. Natl. Cancer Inst.* **1998,** *90,* 1552–1555.

72. Oben, J. E.; Enyegue, D. M.; Fomekong, G. I.; Soukontoua, Y. B.; Agbor, G. A. The Effect of *Cissus quadrangularis* (CQR-300) and a Cissus Formulation (CORE) on Obesity and Obesity-Induced Oxidative Stress. *Lipids Health Dis.* **2007,** *6*(1), 4.

73. Olsen, A.; Knudsen, K. E.; Thomsen, B. L.; Loft, S.; Stripp, C.; Overvad, K.; Møller, S.; Tjønneland, A. Plasma Enterolactone and Breast Cancer Incidence by Estrogen Receptor Status. *Cancer Epidemiol. Biomarkers Prev.* **2004,** *13*(12), 2084–2089.

74. Osz, J.; Brelivet, Y.; Peluso-Iltis, C.; Cura, V.; Eiler, S.; Ruff, M.; Bourguet, W.; Rochel N.; Moras, D. Structural Basis for a Molecular Allosteric Control Mechanism of Cofactor Binding to Nuclear Receptors. *Proc. Natl. Acad. Sci.* **2012,** *109*(10), E588–594.

75. Panthong, A.; Supraditaporn, W.; Kanjanapothi, D.; Taesotikul, T.; Reutrakul, V. Analgesic, Anti-inflammatory and Venotonic Effects of *Cissus quadrangularis* Linn. *J. Ethnopharmacol.* **2007,** *110*(2), 264–270.

76. Park, J.; Kim, D. H.; Ahn, H. N.; Song, Y. S.; Lee, Y. J.; Ryu, J. H. Activation of Estrogen Receptor by Bavachin from *Psoralea corylifolia*. *Biomol. Ther.* **2012**, *20*(2), 183–188.
77. Paul, C.; De Bruyne, T. Apers, S.; Vanden Berghe, D.; Pieters, L.; Vlietinck, A. J. Phytoestrogens: Recent Developments. *Planta Med.* **2003**, *69*, 589–599.
78. Pongboonrod, S. *Mai Thed Mung. Muang Thai, Bangkok;* Kaeseam-Bunnakit Printing: Thailand, 1995; pp 428–429.
79. Poornima, Krishnamurthy, P.; Prakash, P.; Margret, A. Gold Nanoparticles Synthesized by *Brassica oleracea* (Broccoli) Acting as Antimicrobial Agents Against Human Pathogenic Bacteria and Fungi. *Appl. Nanosci.* **2015**, *1*(Spl Issue), 1–7.
80. Qu, H.; Madl, R. L.; Takemoto, D. J.; Baybutt, R. C.; Wang, W. Lignans Are Involved in the Antitumor Activity of Wheat Bran in Colon Cancer SW480 Cells. *J. Nutr.* **2005**, *135*(3), 598–602.
81. Raffaelli, B.; Hoikkala, A.; Leppälä, E.; Wähälä, K. Enterolignans. *J. Chromatogr. B* **2002**, *777*, 29–43.
82. Raisz, L. G. Screening for Osteoporosis. *N. Engl. J. Med.* **2005**, *353*, 164–171.
83. Rowland, I.; Faughnan, M.; Hoey, L.; Wahala, K.; Williamson, G.; Cassidy, A. Bioavailability of Phyto-oestrogens. *Br. J. Nutr.* **2003**, *89*(Suppl 1), S45–S58.
84. Ruan, B.; Kong, L. Y.; Takaya, Y.; Niwa, M. Studies on the Chemical Constituents of *Psoralea corylifolia* L. *J. Asian Nat. Prod. Res.* **2007**, *9*(1), 41–44.
85. Samy, R. P.; Gopalakrishnakone, P. Therapeutic Potential of Plants as Anti-microbials for Drug Discovery. *Evid. Based Complement Alternat. Med.* **2010**, *7*, 283–294.
86. Setchell, K. D.; Lydeking-Olsen, E. Dietary Phytoestrogens and Their Effect on Bone: Evidence from In Vitro and In Vivo, Human Observational, and Dietary Intervention Studies. *Am. J. Clin. Nutr.* **2003**, *78*, 593S–609S.
87. Shapiro, C. L.; Recht, A. Side Effects of Adyuvant Treatment of Breast Cancer. *N. Engl. J. Med.* **2001**, *344*, 1997–2008.
88. Sharma, N.; Nathawat, R. S.; Gour, K.; Patni, V. Establishment of Callus Tissue and Effect of Growth Regulators on Enhanced Sterol Production in *Cissus quadrangularis* L. *Int. J. Pharmacol.* **2011**, *7*(5), 653–658.
89. Shiau, A. K.; Barstad. D.; Loria. P. M.; Cheng, L.; Kushner, P. J.; Agard, A. The Structural Basis of Estrogen Receptor/Coactivator Recognition and the Antagonism of this Action with Tamoxifen. *Cell* **1998**, *95*, 927–937.
90. Simoncini, T.; Genazzani, A. R.; Liao, J. K. Nongenomic Mechanisms of Endothelial Nitric Oxide Synthetase Activation by the Selective Estrogen Receptor Modulator Raloxifene. *Circulation* **2002**, *105*, 1368–1373.
91. Simoncini, T.; Varone, G.; Fornari, L.; Mannella, P.; Luisi, M.; Labrie. F.; et al. Genomic and Nongenomic Mechanisms of Nitric Oxide Synthesis Induction in Human Endothelial Cells by a Fourth-Generation Selective Estrogen Receptor Modulator. *Endocrinology* **2002**, *143*, 2052–2061.
92. Simons, R.; Gruppen, H.; Bovee, T. F.; Verbruggen, M. A.; Vincken, J. P. Prenylated Isoflavonoids from Plants as Selective Estrogen Receptor Modulators (phyto SERMs). *Food Funct.* **2012**, *3*(8), 810–827.
93. Singh, G.; Rawat, P.; Maurya, R. Constituents *of Cissus quadrangularis*. *Nat. Prod. Res.* **2007**, *21*(6), 522–528.
94. Skafar, D. F.; Zhao, C. The Multifunctional Estrogen Receptor-Alpha F Domain. *Endocrine* **2008**, *33*(1), 1–8.

95. Smeds, A. I.; Eklund, P. C.; Sjöholm, R. E.; et al. Quantification of a Broad Spectrum of Lignans in Cereals, Oilseeds, and Nuts. *J. Agric. Food Chem.* **2007,** *55*, 1337–1346.

96. Swedenborg, E.; Power, K. A.; Cai, W.; Pongratz, I.; Ruegg, J. Regulation of Estrogen Receptor Beta Activity and Implications in Health and Disease. *Cell Mol. Life Sci.* **2009,** *66*(24), 3873–3894.

97. Talaley, P.; Zhang, Y. Chemoprotection Against Cancer by Isothiocyanates and Glucosinolates. *Biochem. Soc. Trans.* **1996,** *24*, 806–810.

98. Thakur, A.; Jain, V.; Hingorani, L.; Laddha, K. S. Phytochemical Studies on *Cissus quadrangularis* Linn. *Pharmacogn. Res.* **2009,** *1*(4), 213.

99. Thompson, L. U., Ed. Analysis and Bioavailability of Lignans. In *Flaxseed in Human Nutrition*; AOCS Press: Champaign, IL, 2003; pp 92–116.

100. Touillaud, M. S.; Thiébaut, A. C. M.; Fournier, A.; et al. Dietary Lignan Intake and Postmenopausal Breast Cancer Risk by Estrogen and Progesterone Receptor Status. *J. Natl. Cancer Inst.* **2007,** *99*, 475–486.

101. Tristan, P.; Williamson, G. Controlled Flax Interventions for the Improvement of Menopausal Symptoms and Postmenopausal Bone Health. *Menopause* **2013,** *20*(11), 1207–1215.

102. Udupa, K. N.; Prasad, G. C. *Cissus quadrangularis* in Healing of Fractures: A Clinical Study. *J. Indian Med. Assoc.* **1962,** *38*(38), 590–593.

103. Vanharanta, M.; Voutilainen, S.; Lakka, T. A.; et al. Risk of Acute Coronary Events According to Serum Concentrations of Enterolactone: A Prospective Population-Based Case-Control Study. *Lancet* **1999,** *354*, 2112–2115.

104. Verhoeven, D. T.; Goldbohm, R. A.; Van Poppel, G.; et al. Epidemiological Studies on Brassica Vegetables and Cancer Risk. *Cancer Epidemiol. Biomarkers Prev.* **1996,** *5*, 733–748.

105. Wang, L. Q. Mammalian Phytoestrogens: Enterodiol and Enterolactone. *J. Chromatogr. B* **2002,** *777*, 289–309.

106. Wassmann, S.; Laufs, U.; Stamenkovic, D.; Linz, W.; Stasch, J. P.; Ahlbory, K.; et al. Raloxifene Improves Endothelial Dysfunction in Hypertension by Reduced Oxidative Stress and Enhanced Nitric Oxide Production. *Circulation* **2002,** *105*, 2083–2091.

107. Weiser, M. J.; Foradori. C. D.; Handa, R. J. Estrogen Receptor Beta in the Brain: From Form to Function. *Brain Res. Rev.* **2008,** *57*(2), 309–320.

108. Wickerham, L. Tamoxifen: An Update on Current Data and Where It Can Now be Used. *Breast Cancer Res. Treat.* **2002,** *75*(suppl 1), S7–12.

109. World Health Organization. Women's Health, 2016.

110. Xin, D.; Wang, H.; Yang, J.; Su, Y. F.; Fan, G. W.; Wang, Y. F.; et al. Phytoestrogens from *Psoralea corylifolia* Reveal Estrogen Receptor-Subtype Selectivity. *Phytomedicine* **2010,** *17*(2), 126–131.

111. Zhang, X.; Zhao, W.; Wang, Y.; Lu, J.; Chen, X. The Chemical Constituents and Bioactivities of *Psoralea corylifolia* Linn.: A Review. *Am. J. Chin. Med.* **2016,** *44*(01), 35–60.

112. Zhao, C.; Dahlman-Wright, K.; Gustafsson, J. A. Estrogen Receptor Beta: An Overview and Update. *Nucl. Recept. Signal* **2008,** *6*, e003.

113. Zhao, C.; Dahlman-Wright, K.; Gustafsson, J. A. Estrogen Signaling via Estrogen Receptor {Beta}. *J. Biol. Chem.* **2010,** *285*(51), 39575–39579.

INDEX

For Product Safety Concerns and Information please contact our EU
representative GPSR@taylorandfrancis.com
Taylor & Francis Verlag GmbH, Kaufingerstraße 24, 80331 München, Germany

www.ingramcontent.com/pod-product-compliance
Lightning Source LLC
Chambersburg PA
CBHW060752220326
41598CB00022B/2413

9 781774 631522